CROSSTALK AND INTEGRATION OF MEMBRANE TRAFFICKING PATHWAYS

Edited by **Roberto Weigert**

Crosstalk and Integration of Membrane Trafficking Pathways
Edited by Roberto Weigert

Published by InTech

Janeza Trdine 9, 51000 Rijeka, Croatia

Publishing Process Manager Romina Skomersic

Technical Editor Teodora Smiljanic

Cover Designer InTech Design Team

Additional hard copies can be obtained from orders@intechopen.com

Crosstalk and Integration of Membrane Trafficking Pathways, Edited by Roberto Weigert
p. cm.
ISBN 978-953-51-0515-2

For EU product safety concerns:
IN TECH d.o.o., Prolaz Marije Krucifikse Kozulić 3, 51000 Rijeka, Croatia,
info@intechopen.com or visit our website at intechopen.com.

Meet the editor

Dr Roberto Weigert is a principal investigator at the National Institutes of Health (NIH). He was born in Catania, where he got is degree in Physical Chemistry. He moved to the Consorzio Mario Negri Sud in Lanciano, where he started to work in Cell Biology and at the same time got his Ph.D. in Life Sciences with the Open University of London. In 2001, he moved to USA, where he was a Research Fellow at the National Heart Lung and Blood Institute (NIH). In 2006 he established the Intracellular Membrane Trafficking Unit at the National Institute of Dental and Craniofacial Research where he has developed novel microscopy-based technologies to study the patho-physiology of the oral cavity.

Contents

Preface

Membrane traffic is a broad field that studies the complex exchange of membranes that occurs inside the cell. Protein, lipids and other molecules traffic among intracellular organelles, and are delivered to, or transported from the cell surface by virtue of membranous carriers generally referred as "transport intermediates". These carriers have different shapes and sizes, and their biogenesis, modality of transport, and delivery to the final destination are regulated by a multitude of very complex molecular machineries, which include cytoskeletal elements, signaling complexes, and lipid modifying enzymes to name a few. A concept that has clearly emerged in the last decade is that each membrane pathway does not represent a close system, but is fully integrated with all the other trafficking pathways. The precise coordination and balance of membrane trafficking is an absolute requirement to ensure the correct functioning of the cell. The aim of this book is to provide the reader with a general overview of the extent of the crosstalk among the intracellular membrane pathways.

Although this book is far from being exhaustive, it conveys some of the novel ideas and directions in the membrane trafficking field through a combination of review, research, and technical articles. The book is organized in a series of chapters that will "transport" the reader from the endoplasmic reticulum, where proteins and lipids are synthesized, to the plasma membrane, where molecules are either delivered or internalized via endocytosis. Some of the basic steps of the secretory pathway are here reviewed, such as the transport of cargo molecules from the ER to the Golgi apparatus, the sorting of transport intermediates at the Trans-Golgi Network, and the delivery of secretory vesicles to the plasma membrane. Several new aspects have been highlighted: for example, a novel role for lipids as receptors which drive protein sorting, the diversity of the sorting events that occur at the level of the Golgi apparatus, and the cross talk between exocytosis and autophagy. Moreover, an interesting example of how the knowledge of these pathways is exploited to generate novel secretory routes to direct the synthesis of bio-molecules in "cell factories" is discussed. This journey continues with an introduction to the events originated from the plasma membrane and their relationship to exocytic and signaling pathways. This area is particularly important since the precise coordination of these processes is fundamental in regulating events that affect cell motility, proliferation, and cell division, whereas their imbalance leads to pathological outcomes, such as cancer or degenerative diseases.

I hope that the reader will find these chapters useful to get new perspectives on some of the current topics and issues in membrane trafficking, and to get a sense on the diversity of approaches and experimental systems that are used in this field.

Finally, I would like to thank all the people associated with the production of this book. First, the authors themselves for taking the time to prepare their chapters in a timely manner and in a format filled with technical details and novel and exciting concepts. It is not an easy task, and I sincerely thank them for their efforts. Second, I would like to thank Intech for the opportunity to edit this book, and its staff, that has made my job a real pleasure.

Roberto Weigert Ph.D.
Chief, Intracellular Membrane Trafficking Unit,
Oral and Pharyngeal Cancer Branch,
National Institute of Dental and Craniofacial Research,
National Institutes of Health, Bethesda
USA

1

Semi-Intact Cell Systems – Application to the Analysis of Membrane Trafficking Between the Endoplasmic Reticulum and the Golgi Apparatus and of Cell Cycle-Dependent Changes in the Morphology of These Organelles

Masayuki Murata and Fumi Kano
The University of Tokyo
Japan

1. Introduction

The endoplasmic reticulum (ER) and the Golgi apparatus both maintain their specific morphology, composition, and function in spite of the exchange of proteins and lipids between the two organelles through membrane trafficking. The morphology of the Golgi apparatus is closely linked to the balance between anterograde (ER-to-Golgi) and retrograde (Golgi-to-ER) transport. It has been reported that the inhibition of anterograde transport leads to the redistribution of Golgi components to the cytoplasm or the ER (Storrie et al., 1998; Ward et al., 2001; Miles et al., 2001). Inhibition of anterograde transport at the onset of mitosis also results in the relocation of Golgi enzymes to the ER (Zaal et al., 1999; Altran-Bonnet et al., 2006), although it is controversial as to whether the Golgi becomes integrated with the ER or whether they remain separate throughout mitosis (Lowe and Barr, 2007). Thus, in living cells, the rates of anterograde and retrograde transport between the two organelles appear to have a substantial effect on the morphology of the Golgi. The effect of the balance between anterograde and retrograde transport on the morphology of the ER remains to be explored. Recently, increasing evidence has suggested that alteration in Golgi morphology during mitosis is not a passive process but rather an active one, which is highly coordinated with entry into mitosis and its progression. On the basis of the concept that cell cycle-dependent membrane trafficking between the ER and Golgi during mitosis should be tightly coupled with the changes in their morphology, it will be important to elucidate the cell cycle-dependent regulation of membrane trafficking to understand the morphological changes in more detail.

However, in general, investigation of morphological changes in the Golgi and ER during mitosis is hampered by the fact that it is difficult to observe the precise morphology of organelles in mammalian cells during mitosis by light microscopy due to the round shape of the cells. The perturbation of the distinct characteristics of the two organelles during mitosis makes it difficult to analyze the efficiency of membrane trafficking between the organelles using quantitative microscopic methods, such as fluorescence recovery after photobleaching

(FRAP), etc. In addition, the fact that concerted morphological changes occur in the organelles simultaneously and transiently in a single cell during mitosis makes it extremely difficult to dissect these changes morphologically and biochemically. The analysis is complicated further by the asynchronous progression of the cell cycle in individual cells.

Herein, we describe a novel method that addresses the above-mentioned problems, namely, a semi-intact cell assay coupled with green fluorescence protein (GFP)-visualization techniques. By using the semi-intact cell system, we can observe the morphological changes that occur in "preexisting" organelles during mitosis more easily, and, at the same time, can investigate the effects of exogenously added antibodies, drugs, and recombinant proteins on the process. By reconstituting cell cycle-dependent morphological changes in organelles using interphase or mitotic phase cytosol, we can dissect processes that occur in an orchestrated manner in the cells, morphologically and biochemically, into elementary reactions, and investigate the biochemical requirements for each reaction.

2. Semi-intact cell assays

Semi-intact cells are cells whose plasma membrane has been permeabilized with detergent or toxins, and can be referred to as "cell-type test tubes" (Fig. 1). We use a bacterial pore-forming toxin, streptolysin O (SLO), to permeabilize the cells. At 4°C, SLO binds to cholesterol in plasma membranes. At warmer temperatures, SLO assembles to form amphiphilic oligomers, which results in the generation of small, stable transmembrane pores (Bhakdi et al., 1985). SLO-induced pores are approximately 30 nm in diameter, which is sufficiently large to allow immunoglobulin to enter into the cells (immunoglobulin G: 150 kDa). Protein complexes that are larger than immunoglobulin, such as homo- or hetero-oligomers, can enter the cells through the pores as individual subunits, and the complexes are reconstituted inside the semi-intact cells. After permeabilization, almost ~80% of the cytosol flows out through the pores into the medium. However, despite this loss of cytosol, the relative intracellular configuration of the cytoskeleton and organelles or between independent organelles can be maintained because damage to the membranes of the intracellular organelles caused by entry of SLO into the semi-intact cells can be minimized by washing away any excess SLO at 4°C before pore formation is initiated. In contrast, it is

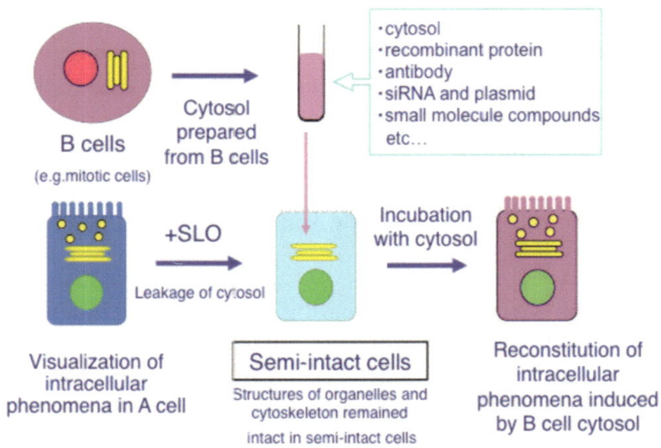

Fig. 1. A scheme of semi-intact cell assay

difficult to avoid damage to intracellular structures when cells are permeabilized with digitonin, a well-known pore-forming toxin, because digitonin-induced permeabilization is insensitive to temperature.

By exchanging cytoplasmic proteins with exogenously added proteins, antibodies or cytosol that has been prepared from cells at distinct stages of the cell cycle or differentiation, or from disease states, we can modulate the intracellular environment and reconstitute various physiological phenomena in semi-intact cells. The semi-intact cell method was originally established by Dr. Simons' group to study polarized vesicular trafficking in Madin-Darby canine kidney (MDCK) cells (Ikonen et al., 1995). We have refined the method by coupling it with GFP-visualization techniques and have established many types of assay for cell cycle-dependent changes in organelle morphology and membrane trafficking. Using our analytical system, we can manipulate intracellular conditions and then observe the resulting morphological changes in GFP-tagged organelles by fluorescence microscopy. In addition, we are able to dissect complex reaction processes in cells on a morphological basis and to investigate the biochemical requirements and kinetics of each process, for example the vesicular transport between the ER and the Golgi during mitosis. In particular, the maintenance of the integrity of the organelles and their configuration in the semi-intact cells enables us to analyze membrane trafficking in as intrinsic environment as possible.

3. Reconstitution of Golgi disassembly by mitotic cytosol in semi-intact cells

Disassembly of the Golgi during mitosis is a dynamic and highly regulated process, and is required for an equal partitioning of Golgi membranes into the two daughter cells. To investigate the biochemical requirements and kinetics of Golgi disassembly during mitosis, we reconstituted the process by adding mitotic cytosol prepared from *Xenopus* eggs to semi-intact cells and visualized it with GFP-tagged proteins (Kano et al., 2000). To this end, first, we produced the stable transfectant MDCK-GT, which continuously expresses mouse galactosyltransferase (GT) fused with GFP (GT-GFP). GT-GFP has been used to study Golgi membrane dynamics in living cells and has been characterized in detail (Cole et al., 1996). Next, we prepared semi-intact MDCK cells, which had been grown on polycarbonate membranes, incubated them with various types of cytosol, and observed the resulting changes in Golgi morphology. In semi-intact cells incubated with *Xenopus* interphase extracts, the Golgi apparatus forms perinuclear, tubular structures, which are typical of the Golgi apparatus in MDCK-GT cells. By contrast, incubation with *Xenopus* egg (M phase) extracts causes the Golgi to disassemble and the fluorescence of GT-GFP diffuses completely throughout the cytoplasm (see Fig. 2A, stage III). The diffuse staining pattern of GT-GFP that is observed, and which corresponds to small heterogeneous vesicular structures, is typical of Golgi membranes in living mitotic MDCK-GT cells.

Fluorescence microscopic observation of Golgi disassembly induced by mitotic cytosol in single MDCK-GT cells revealed that the disassembly process can be divided into three stages: stage I (intact), II (punctate), and III (dispersed) (Fig. 2). During stage I (intact), the tubular and stacked structures of a typical intact Golgi apparatus are observed in the perinuclear region through the apical to the middle part of the cells. During stage II (punctate), punctate Golgi structures are observed on mainly the apical side of the nucleus (Fig. 2A, stage II). These punctate structures are seldom seen in the basolateral cytoplasm. During stage III (dispersed), a diffuse staining pattern is observed throughout the cytoplasm

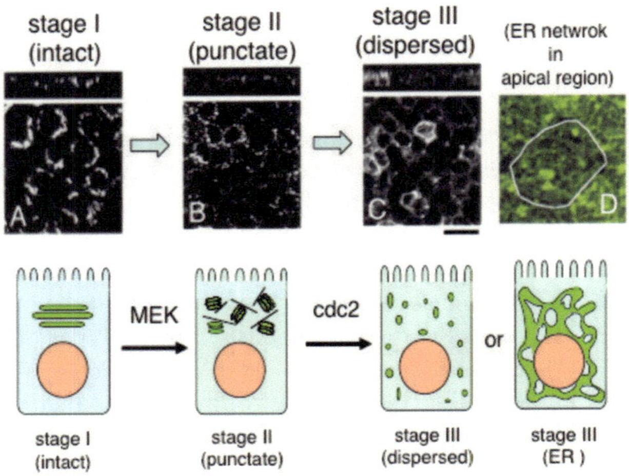

For the morphometric analysis, we divided the disassembly process into three stages based on the Golgi morphology: stage I (intact); intact perinuclear Golgi cisternae (A), stage II (punctate); punctate structures on the apical side of the nucleus (B), stage III (dispersed); highly dispersed Golgi membranes throughout the cytoplasm (C). Cells at each stage were observed by confocal microscope. The lower panel shows xy images in an apical region of the cells and the upper panel shows xz sectioning images. Bars:10 µm. In stage III cells, the relocation of Golgi component (GT-GFP) to the ER and nuclear envelopes was frequently observed in the apical region of the cells (D). Schematic model of Golgi disassembly was shown in cartoon. The morphological change in mitotic Golgi disassembly was dissected into two processes biochemically. The first process from stage I to stage II is mainly regulated by MEK1 and the second from stage II to stage III is mainly cdc2. In stage III Golgi, the vesiculated Golgi diffused throughout cytoplasm (dispersed) or the Golgi component (GT-GFP) was traslocated to the ER (ER).

Fig. 2. Morphological dissection of the Golgi disassembly process in semi-intact cells.

(Fig. 2A, stage III). Interestingly, in stage III cells, some GT-GFP appears to translocate to the ER/nuclear membranes (see the lower image in Fig. 2D). Electron microscopy confirmed the morphological differences in the Golgi apparatus between each stage. In particular, in stage II (punctate), small stacked cisternae are found in the apical region, but not in the basolateral region. The Golgi mini-stacks are 800 nm in diameter, and are associated with microtubules. Thus, the disassembly of the Golgi can be dissected into two processes morphologically (Fig. 2): the first process is the transition from stage I to stage II and the second is the transition from stage II to stage III.

Previously, two reports revealed that protein kinases, cdc2 and MEK1, are required for the Golgi disassembly process (Acharya et al., 1998; Lowe et al., 1998). However, the results of the two studies were not consistent with each other. We tested the effect of these two kinases on Golgi disassembly in semi-intact cells using mitotic cytosol that contained kinase inhibitors. Inactivation of cdc2 kinase in mitotic cytosol by the addition of butyrolactone (BL) arrests Golgi disassembly at stage II (punctate). In contrast, inactivation of MEK by PD98059 (PD) inhibits the initiation of Golgi disassembly (Fig. 3A). Next, we performed immunodepletion experiments using either cdc2- or MEK-depleted *Xenopus* egg mitotic extracts. Cdc2 was depleted from the extract by using Suc1-Sepharose beads, whereas MEK was depleted with rabbit anti-MEK polyclonal antibodies. In the presence of cdc2-depleted extract, punctate Golgi structures (stage II) are found in 90% of the cells. Furthermore, MEK-

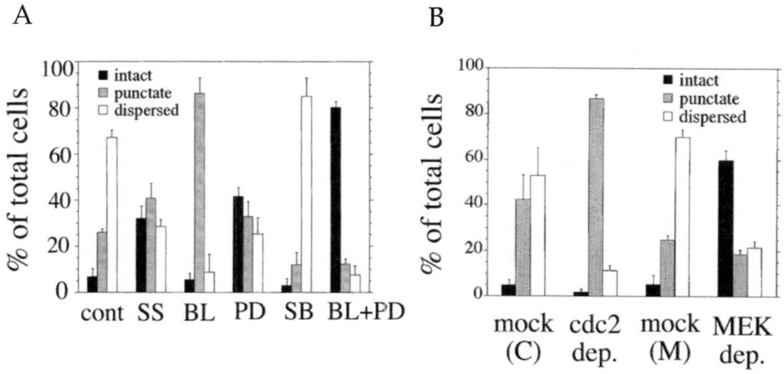

(A)Effect of protein kinase inhibitors on the Golgi disassembly. Semi-intact MDCK-GT cells were incubated with *Xenopus* egg extracts (mitotic cytosol) and ATP containing either no inhibitor (control), staurosporine (SS), butyrolactone l (BL), PD98059 (PD), SB203580 (SB), or BL+PD, respectively, at 33°C for 80 min. After incubation, the cells were fixed and morphometric analysis was performed. The protein concentration of the extract used was 5.0 mg/ml. 300 cells were counted in three randomly selected fields, and standard deviations are shown as vertical bars. (B) Inhibition of the Golgi disassembly by cdc2- or MEK-depleted *Xenopus* egg extracts (mitotic cytosol). Mock, cdc2- or MEK-depleted *Xenopus* egg extracts were applied to semi-intact cells and incubated at 33°C for 80 min. The cells were fixed and morphometric analysis was performed. Cdc2-depleted extracts arrested the disassembly process at stage II (punctate), and MEK-depleted extracts did so at stage I (intact).

Fig. 3. Sequential effect of MEK and cdc2 kinase on the mitotic Gogi disassembly.

depleted extract arrests the Golgi disassembly process at stage I (intact) in 60% of cells (Fig. 3B). We also confirmed the effect of each kinase on Golgi disassembly by using cdc2- or MEK-activated interphase *Xenopus* extract in semi-intact cells. The results were consistent with the effects of the kinase inhibitors and supported our model that cdc2 is responsible for the process from stage II to III, and MEK from stage I to II (Fig. 2).

However, more detailed studies using semi-intact cells revealed that each kinase does not correspond independently to the different steps. The two kinases might have overlapping functions, the first step is regulated mainly by MEK and the second step by cdc2. Interestingly, a delay in mitotic entry was observed upon inhibition of MEK1 activity. One of peripheral membrane proteins of the Golgi, GRASP55, was reported to be a substrate of MEK1 and can connect the Golgi stacks laterally into a ribbon as well as regulating mitotic progression (Feinstein and Linstedt, 2007, 2008). ERK1c and polo-like kinase-3, which are downstream of MEK1, have also been reported to contribute to Golgi disassembly during mitosis (Shaul and Seger, 2006; Xie et al., 2004). On the other hand, another peripheral Golgi protein GRASP65, which is known to function in cisternal stacking as well as in the lateral linking of stacks, was reported to be phosphorylated by cdc2 and the phosphorylation was required for entry into mitosis (Yoshimura et al., 2005; Preisinger, et al., 2005). Collectively, these results confirm our findings that the Golgi mini-stacks in stage II are generated from intact Golgi by the activation of MEK1, and the dispersed Golgi observed in stage III from the Golgi mini-stacks by the activation of cdc2. In addition, MEK1- or cdc2-dependent changes in Golgi morphology might be essential for mitotic entry.

Cdc2 is also likely to be involved in COPI-dependent disassembly of the Golgi during mitosis. At the onset of mitosis, the peripheral Golgi protein GM130 is known to be phosphorylated by cdc2; GM130 phosphorylation inhibits p115-dependent tethering and the subsequent fusion of Golgi-derived COPI-dependent vesicles to the Golgi cisternae (Lowe et al., 2000). Continuous budding without the fusion of new vesicles might reduce the cisternae rapidly and facilitate the mitotic disassembly of the Golgi. Thus, it is likely that the roles of MEK1 and cdc2 in mitotic Golgi disassembly involve both COPI-dependent and -independent processes.

4. Reconstitution of cell cycle-dependent morphological changes in the ER network in semi-intact cells

To investigate the cell cycle-dependent changes in the morphology of the ER network in mammalian cells, we created a clonal cell line derived from Chinese Hamster Ovary (CHO) cells that constitutively express GFP-HSP47 (CHO-HSP) and have a flat morphology when grown in culture, such that the cytoplasm is easy to visualize (Fig. 4). Using confocal microscopy, we found that, during interphase, GFP-HSP47 in CHO-HSP cells is associated with polygonal structures with three-way junctions that are located at the periphery of the cells and in the cisternae in the perinuclear region (Fig. 4, interphase). Interestingly, at the onset of mitosis, the ER appears to retain its network structure rather than being disrupted into vesicles, in contrast to the Golgi apparatus (Fig. 4, mitotic phase). Further observation with the fluorescence microscope revealed that the ER is partially severed at the onset of mitosis. In fact, recent advances in quantitative confocal and electron microscopy (EM) analyses by the application of electron tomography techniques have revealed that, in the mitotic ER, the tubules are shorter and more branched and cisternae are extended in comparison with the ER during interphase (Puhka et al., 2007; Lu et al., 2009). These results confirm our observations by fluorescence microscopy, and show that the cell cycle-dependent changes in ER morphology seem to be completely different from those in Golgi morphology.

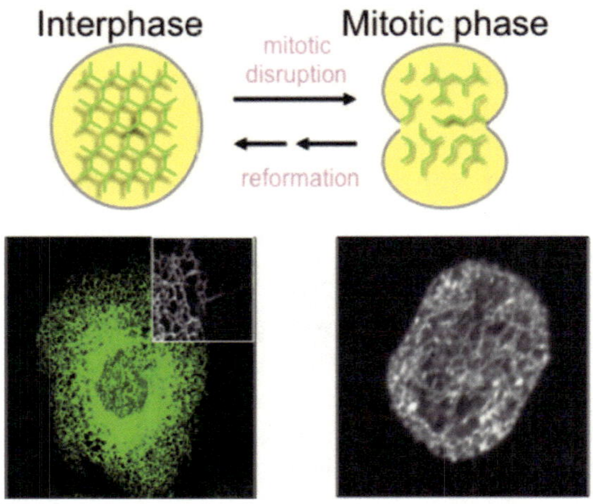

The ER network connected with three-way junctions is visible in interphase CHO-HSP cells (see inset of image for interphase ER network). The ER tubules fuse with each other to make new three-way junctions. At the onset of mitosis, the ER network appears to retain the network structure, and dose not seem to be disrupted into vesicles like mitotic Golgi vesicules. Further observation by fluorescence microscope enables us to discover the ER is partially severed during mitosis.

Fig. 4. Cell cycle-dependent morphological changes in the ER network.

We reconstituted the cell cycle-dependent changes in ER morphology in semi-intact cells, identified regulatory factors, and elucidated the mechanisms that underlie the morphological changes (Kano et al., 2005a and b). First, we preincubated CHO-HSP cells with nocodazole to disrupt the microtubules and then permeabilized them with SLO. Then, we incubated the semi-intact CHO-HSP cells with mitotic cytosol, which was prepared from synchronized mitotic L5178Y cells, and found that the continuous network of the ER was partially severed, as was seen in intact mitotic CHO-HSP cells. As shown in Fig. 5A, in the presence of mitotic cytosol, the ER network is partially disrupted and, thus, the connections between ER tubules are broken. In contrast, in the presence of interphase cytosol, the ER network remains intact. To quantify the partial disruption of the ER, we counted the number of three-way junctions per defined area (Fig. 5B). Mitotic cytosol induces a decrease in the number of three-way junctions, which signifies that the ER network is disrupted. In addition, we found that when nocodazole is not added the ER network remains intact even in the presence of mitotic cytosol. These results suggest that microtubules strengthen the integrity of the ER network, and that both depolymerization of microtubules and exposure to mitotic cytosol are necessary for the complete disruption of the ER network. Interestingly, we frequently observed dynamic tubulation/bifurcation of ER tubules in the presence of either mitotic or interphase cytosol, which suggests that partial disruption of the ER network by mitotic cytosol results from inhibition of the fusion process, rather than inhibition of tubulation/bifurcation (unpublished data).

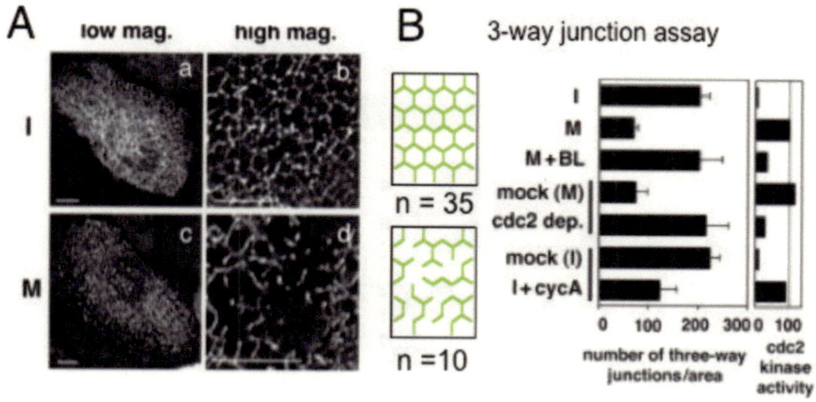

(A) CHO-HSP cells were pretreated with nocodazole and permeabilized with SLO. The cells were incubated with interphase (a, b) or mitotic (c, d) cytosol at 32°C for 40 min, and observed under a confocal microscope at low(a, c) or high (b, d) magnification. Bar = 10 µm (low mag.), 5 µm (high mag.). (B) After the incubation, to quantify the partial disruption of the ER network, we counted the number of three-way junctions per area and compared (three-way junction assay). Nocodazole-treated semi-intact CHO-HSP cells were incubated with interphase cytosol (I, cdc2 kinase activity was 0.27 units/µL), mitotic cytosol (M, 5.68 units/µL), mitotic cytosol containing 30 µm butyrolactone1 (M+BL, 1.86 units/µL), mock mitotic cytosol (mock (M), 6.55 units/µL), cdc2-depleted mitotic cytosol (cdc2 dep.1.66 units/µL), mock interphase cytosol (mock (I), 0.52 units/µL), or interphase cytosol treated with cyclin A (I+cycA, 5.18 units/µL). After the incubation, three-way junction assay was performed. Cdc2 kinase activity in each reaction mixture (means from two independent measurements) is shown in the right hand column, where 100% represents the value of cdc2 kinase activity in mitotic cytosol. Cdc2 kinase activity correlated with the disruption of ER network.

Fig. 5. Disruption of the ER network by mitotic cytosol in semi-intact CHO-HSP cells.

With regard to the fusion process, some cytosolic proteins or their regulators that are downstream of cdc2 kinase are thought to be inactive in mitotic cytosol (Lowe et al. 1998; Kano et al. 2000a). Extrapolating from these findings, the disruption of the ER network by mitotic cytosol *in vitro* could also result from the blocking of fusion events by cdc2 kinase-mediated phosphorylation. One of the candidates for this inhibition is p47, a cofactor of p97, which mediates the fusion of Golgi membranes (Kondo et al. 1997). More recently, Uchiyama et al. (2003) found that Ser140 of p47 was selectively phosphorylated by cdc2 kinase and that this phosphorylation was involved in Golgi disassembly during mitosis. They also found that a non-phosphorylated form of p47, p47 (S140A), which is referred to as p47NP, inhibited mitotic Golgi disassembly. Phosphorylation of p47 by cdc2 dissociated the p97/p47 fusion complex from membranes, which has the potential to inhibit membrane fusion between ER tubules. To test this, we investigated the effect of p47NP on the partial disruption of the ER network that is induced by mitotic cytosol in semi-intact cells. In the presence of p97/p47NP, the partial disruption of the ER network was inhibited. Therefore, we concluded that the disruption of the ER during mitosis depends on phosphorylation of p47 by cdc2 (Fig. 6).

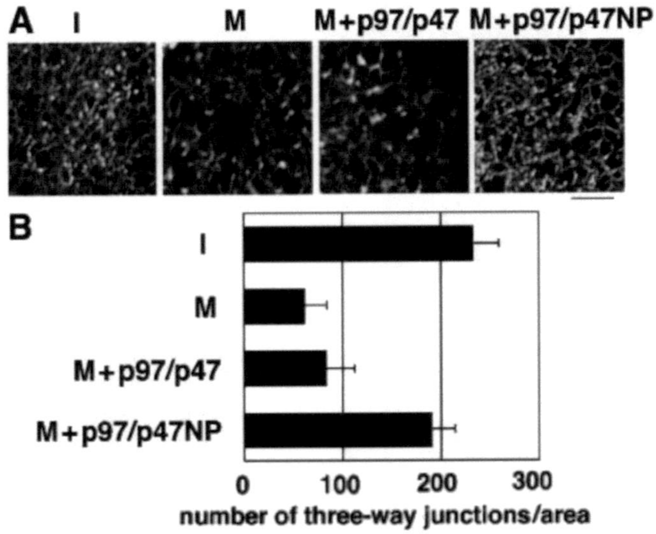

(A) Nocodazole-treated semi-intact CHO-HSP cells were incubated with interphase cytosol (I), mitotic cytosol (M), mitotic cytosol +recombinant p97 and p47 (M+p97/p47), mitotic cytosol +p97 and mutated p47S140 A (M+p97/p47NP), at 32°C for 40 min. The cells were observed by confocal microscopy (A) and were subjected to a three-way junction assay (B). Bar = 10 µm.

Fig. 6. Phosphorylation of p47 by cdc2 results in the disruption of the ER network.

Next, we reconstituted the reformation of the ER network after cell division (Kano et al., 2005b). First, semi-intact CHO-HSP cells were incubated with mitotic cytosol to induce the partial disruption of the ER network. After the mitotic cytosol had been removed, the semi-intact cells were incubated with interphase cytosol, which led to the reformation of the ER network (Fig. 7). This reformation is induced by the fusion protein complexes in the interphase cytosol. There are two well-characterized protein complexes involved in intracellular membrane fusion: the NSF/SNAP complex and the p97/p47 complex.

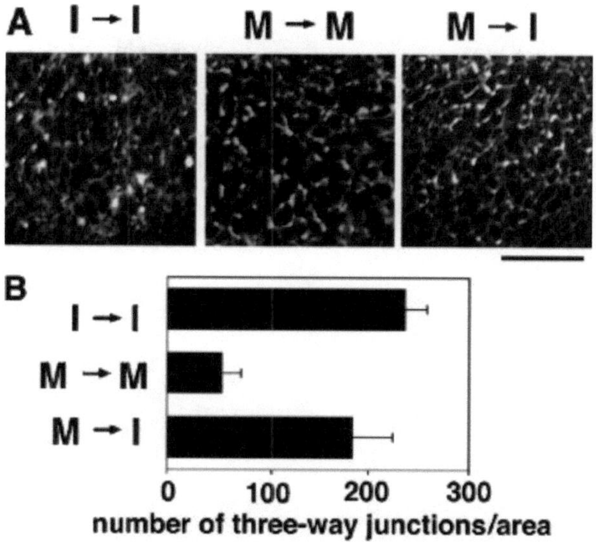

Nocodazole-treated, semi-intact CHO-HSP cells were incubated with interphase (I→) or mitotic (M→) cytosol at 32°C for 40 min. After washing out the cytosol with cold TB, the cells were further incubated with interphase (I→I, M→I), or mitotic cytosol (M→M). The cells were fixed, and images were acquired by confocal microscopy (A) or were subjected to a three-way junction assay (B). Bar = 10 μm.

Fig. 7. Reformation of the ER network from the disrupted ER tubules by interphase cytosol.

Antibodies against NSF or p47 that inhibit NSF/SNAP- or p97/p47-mediated intracellular membrane fusion also inhibit the reformation of the ER that is induced by interphase cytosol, which indicates that the reformation process is regulated by both the NSF/SNAP complex and the p97/p47 complex (Fig. 8A). These results were confirmed by the following experiment. Both NSF and p97 are sensitive to the alkylating reagent N-ethylmaleimide (NEM), and NEM treatment inactivates the fusion of Golgi vesicles. Interphase cytosol was treated with NEM and the effect of the cytosol on ER reformation was examined. As expected, the NEM-treated cytosol did not induce reformation of the ER network. However, the addition of recombinant NSF/SNAP and p97/p47 complexes to the NEM-treated interphase cytosol restored its ability to induce reformation of the network (Fig. 8B). Next, we incubated semi-intact CHO-HSP cells in which the ER network had been disrupted by the addition of mitotic cytosol with a mixture of NSF/SNAP and p97/p47 complexes only (without NEM-treated cytosol), and found that the ER network was not reformed fully. This suggested that other factor(s) in NEM-treated cytosol are required for the ER reformation. We found that VCIP135, a deubiquitinating enzyme that transiently associates with the p97/p47 complex, is necessary for the reformation but p115, which plays a role in tethering Golgi-derived vesicles to Golgi cisternae, is not. Consequently, this result signified that the NSF/SNAP and p97/p47/VCIP135 complexes are the minimal factors required for the ER reformation. In addition, we found that the order of action of the NSF/SNAP and p97/p47/VCIP135 complexes is crucial for the reformation (Fig. 8B). When we incubated the disrupted ER network with the NSF/SNAP complex first, and then the p97/p47/VCIP135 complex, the ER network reformed. However, when the disrupted network was incubated with the p97/p47/VCIP135 complex first, and then the NSF/SNAP complex, the network remained disrupted.

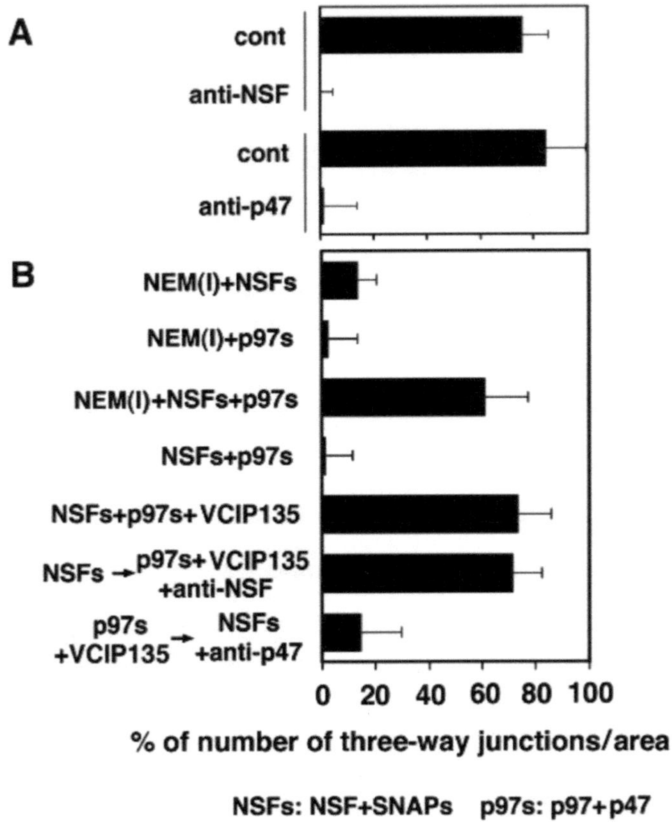

(A) The ER reformation assay was performed using interphase cytosol with pre-immune serum (cont), interphase cytosol with anti-NSF antibody (anti-NSF), interphase cytosol with anti-p47 antibody (anti-p47). (B) The ER reformation assay was performed using NEM-treated interphase cytosol (NEM(I)), NEM-treated interphase cytosol with NSF+SNAPs (NEM(I)+NSFs), NEM(I) cytosol with p97+p47 (NEM(I) +p97s), NEM(I) cytosol with both (NEM(I)+NSFs+p97s), NSFs+p97s, or NSFs+ p97s+VCIP135 at 32°C for 80 min. When cells were incubated in a sequential manner, semi-intact cells, treated with mitotic cytosol were incubated with NSFs or p97s+VCIP135 at 32°C for 40 min, washed with 2M KCl in TB for 25 min, then further incubated with p97s+VCIP135+anti-NSF antibodies or NSFs+anti-p47 antibodies at 32°C for 40 min, respectively (NSFs→p97s+VCIP135+anti-NSF, p97s+VCIP135→NSFs+anti-p47). The efficiency of ER network reformation under each condition were estimated by three-way junction assay.

Fig. 8. NSF/SNAP complex and p97/p47 complex are required for ER network reformation.

Fig. 9 shows a schematic model of the cell cycle-dependent morphological changes in the ER network. As mentioned above, we found that the cdc2-dependent phosphorylation of p47 induces the partial disruption of the ER network during mitosis. Interestingly, the reformation of the ER network is not accomplished by a single fusion reaction, but rather requires two sequential fusion reactions. The first fusion event is mediated by the NSF/SNAP complex, which creates intermediate membranous structures between the disrupted ER tubules. These intermediate structures can be observed only by EM. The second fusion event is mediated by the p97/p47/VCIP135 complex, which induces the fusion of connected ER tubules to form three-way junctions.

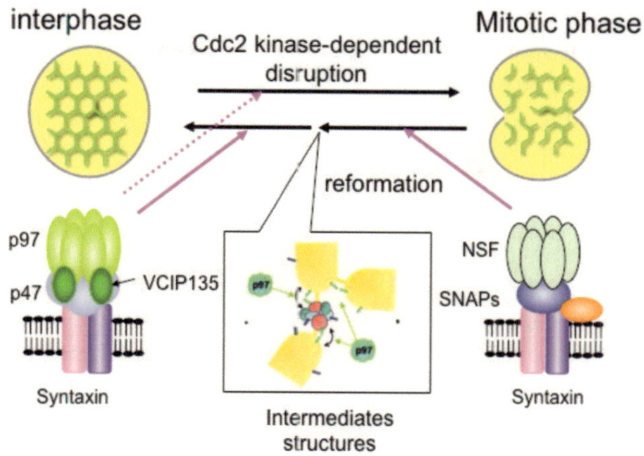

The process of ER network reformation was dissected into two elementary process. Firstly, disrupted ER tubules, induced by mitotic cytosol, are connected by two types of intermediate structures; fine junctions or vesicle aggregates. Both of these intermediate structures are created by NSFcomplex and may function as a "connecting" system. Secondly, the "connected" ER tubules completely fuse with each other directly or through the intermediate structures indirectly to form three-way junctions. This process is dependent on p97/p47 and VCIP135. Syntaxin family is involved in both NSF- and p97-mediated fusion processes with unidentified vesicle-localizing receptors.

Fig. 9. Schematic model for the ER network reformation process.

Given that the NSF/SNAP and p97/p47/VCIP135 complexes are required for the reassembly of the Golgi apparatus after mitosis, it is not surprising that both fusion complexes play a crucial role in the reformation of disrupted ER networks. During Golgi reassembly, the p97/p47 complex is reported to generate single, long cisternae, whereas the NSF/SNAP complex fuses membranes into much shorter but stacked cisternae (Tang et al., 2008). As described above, in our EM study of ER reformation, fine junctions that connected two individual tubules or vesicle aggregates were frequently observed as intermediate structures between the disrupted ER tubules. It is likely that these intermediate structures are produced by the NSF/SNAP complex from Golgi vesicles. The intermediates disappeared within 5 min of incubation with the p97/p47/VCIP135 complex, and formed long tubular structures with bifurcations (three-way junctions), which were also seen in intact cells. Thus, the two fusion complexes seem to generate similar membrane products during both Golgi reassembly and ER reformation.

It might be important to note that p47 contains a UBA (ubiquitin-associated) domain and recruits p97 to monoubiquitinated substrates (Meyer et al., 2002). Taking into consideration the fact that VCIP135 is a deubiquitinating enzyme, regulation of the balance between ubiquitination-deubiquitination might be involved in the disassembly and reformation of the ER network during mitosis, as is the case for Golgi reassembly.

5. Reconstitution of cell cycle-dependent anterograde or retrograde transport between the ER and the Golgi in semi-intact cells

Using our semi-intact cell system, we reconstituted cell cycle-dependent vesicular transport between the ER and the Golgi (Kano et al., 2009). To measure the vesicular transport between the ER and Golgi, we used FRAP. Firstly, we established CHO-GT cells, in which GT-GFP was stably expressed. GT-GFP is trafficked between the ER and Golgi by vesicular transport, but in the steady state, GT-GFP is mainly localized to the Golgi apparatus, with a small proportion in the ER. For FRAP, the fluorescence of GT-GFP in the Golgi region is bleached by repetitive laser illumination. After bleaching, fluorescence in the Golgi area is recovered due to anterograde transport of GT-GFP from the ER to the Golgi. By measuring the fluorescence recovery in the Golgi area, we can estimate the extent of transport of GT-GFP from the ER to the Golgi (Fig.10). To examine retrograde transport from the Golgi to the ER, the fluorescence in the ER region is bleached and the fluorescence recovery of the ER (whole area of the cell except for the nucleus) is determined (Fig. 11). We confirmed that fluorescence recovery in the ER region is not due to the appearance of newly synthesized GT-GFP by pretreating the cells with cycloheximide to inhibit protein synthesis. In fact, when both the Golgi and ER regions were photobleached simultaneously, no fluorescence recovery was observed in the ER region (F. K., unpublished results). Figure 12 shows representative kinetic curves for the anterograde and retrograde transport of GT-GFP that were obtained from the assay. In the presence of mitotic cytosol, anterograde transport is

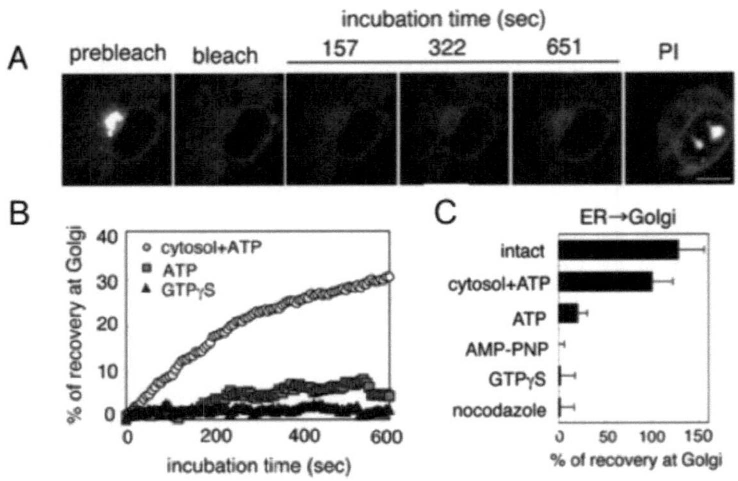

(A) CHO-GT cells were pretreated with cycloheximide (CHX). GT-GFP in the Golgi region of a single semi-intact CHO-GT cell was photobleached by laser illumination (bleach). The cell was incubated in the presence of cytosol/ATP for the indicated times (sec). PI, propidium iodide. Scale bar: 10 μm. (B) Kinetics of fluorescence recovery after photobleaching in the Golgi region in the presence of cytosol/ATP (cytosol+ATP), an ATP regenerating system only (ATP), or cytosol/ATP plus 1 mM GTPγS (GTPγS). After CHX treatment, GT-GFP in the Golgi region was photobleached, and the semi-intact cells were incubated at 32°C for 10 minutes in the presence of cytosol/ATP. Cells were then treated with 10 μg/ml brefeldin A (BFA) for 30 minutes to relocate the Golgi-localized GT-GFP to the ER. This indicated that the recovered fluorescent structure was the Golgi apparatus. (C) Semi-intact CHO-GT cells that had been treated with CHX were incubated with cytosol/ATP (cytosol+ATP), an ATP regenerating system only (ATP), cytosol plus 1 mM AMP-PNP (AMP-PNP), or cytosol/ATP plus 1 mM GTPγS (GTPγS), and then subjected to the anterograde transport assay. In addition, intact CHO-GT cells were treated with nocodazole (nocodazole), and then subjected to the anterograde transport assay. In our anterograde transport assay, we confirmed that the fluorescence recovery in the Golgi was attributable to the anterograde transport of GT-GFP alone, and was not affected by retrograde transport from the Golgi.

Fig. 10. Reconstitution of anterograde transport of GT-GFP in semi-intact cells.

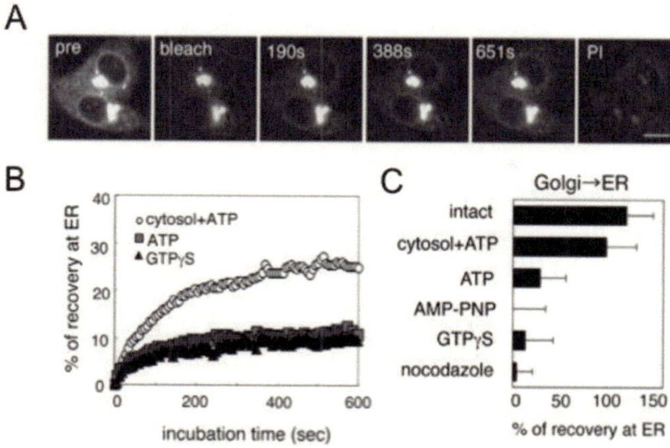

(A) After cycloheximide (CHX) treatment to inhibit protein synthesis, GT-GFP within the ER region was photobleached by laser illumination (bleach). Semi-intact cells were then incubated with cytosol/ATP at 32°C for the indicated times (sec). PI represents propidium iodide. Scale bar: 10 µm. (B) Semi-intact CHO-GT cells that had been treated with CHX were incubated with cytosol/ATP (cytosol+ATP), an ATP regenerating system only (ATP), cytosol plus 1 mM AMP-PNP (AMP-PNP), or cytosol/ATP plus 1 mM GTPγS (GTPγS), and then subjected to the retrograde transport assay. In addition, intact CHO-GT cells were treated with nocodazole (nocodazole), and then subjected to the retrograde transport assay.

Fig. 11. Reconstitution of the retrograde transport of GT-GFP in semi-intact cells.

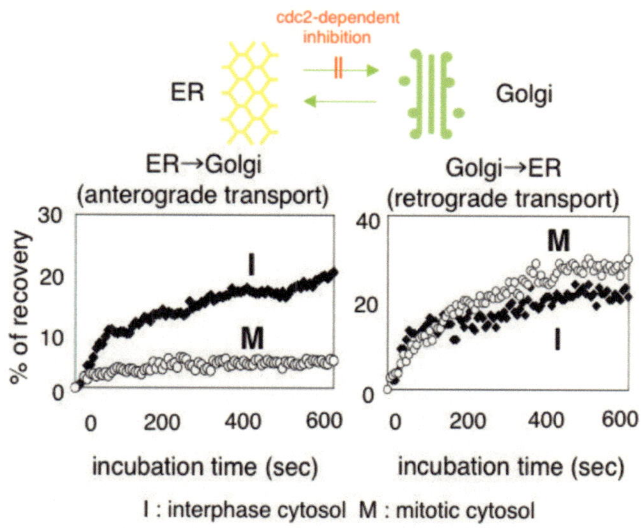

The antrograde or retrograde transport assay described in Fig.10 and 11 were performed using interphase or mitotic cytosol. In the transport kinetics graph, representative kinetics of fluorescence recovery after photobleaching was shown in the presence of interphase (I) or mitotic (M) cytosol and ATP-generating system. In the presence of mitotic cytosol, the anterograde transport was selectively inhibited, but the retrograde one remained intact. By using cdc2-depleted mitotic cytosol, we also found that the mitotic inhibition of the anterograde transport was dependent on cdc2 kinase.

Fig. 12. Reconstitution of anterograde or retrograde transport of GT-GFP in the presence of interphase or mitotic cytosol.

selectively inhibited, whereas retrograde transport remains intact. In addition, we found that cdc2-depleted mitotic cytosol induces anterograde transport normally, which indicates that the mitotic inhibition of anterograde transport is also dependent on cdc2 kinase.

Next, we examined which process in the anterograde transport is inhibited by mitotic cytosol. For this purpose, we focused on the cell cycle-dependent changes in the morphology of ER exit sites. ER exit sites (ERES) are specialized membrane domains in the ER from which vesicles that contain cargo proteins destined for the Golgi bud. To visualize ERES, we established CHO-YIP cells, which stably express the ERES resident protein Yip1A as a fusion with GFP (Fig. 13A). Yip1A belongs to the Yip family of proteins, which contain the Yip domain, and is thought to have a role in membrane trafficking. Although we have confirmed recently that endogenous Yip1A is localized to the ERGIC (ER-Golgi intermediate compartment) by immunofluorescence using an anti-Yip1A antibody, we also observed that in CHO cells Yip1A fused with GFP accumulates at ERES as bright punctate structures throughout the ER network (Fig. 13A). By observing the morphological changes in ERES that occur in CHO-YIP cells during the cell cycle, we found that ERES are disrupted and Yip1A-GFP is dispersed throughout the ER network at the onset of mitosis. After cell division, the diffuse Yip1A-GFP fluorescence accumulates at the ERES in daughter cells (Fig. 13B). The disruption of ERES during mitosis suggests that vesicle budding from ERES does not occur during mitosis. To test this, we reconstituted the mitotic disassembly of ERES by adding mitotic cytosol to semi-intact CHO-YIP cells and investigated the biochemical requirements for the disassembly process. As shown in Fig. 14, the addition of mitotic

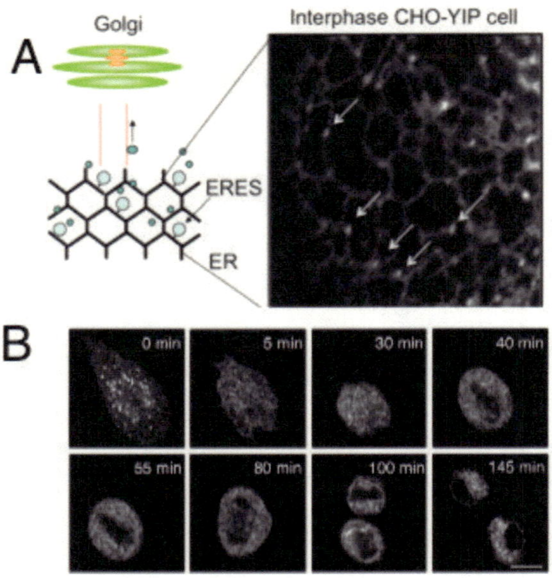

(A) ERES, which is the specialized membrane domain in the endoplasmic reticulum where vesicles that contained cargo proteins budded, were visualized using Yip1A-GFP in CHO-YIP cells (arrows). In CHO-YIP cells the majority of the Yip1A-GFP fluorescence accumulated as bright punctate structures throughout the ER network. (B) Observing the morphological changes of the ERES during cell cycle. The ERES was disrupted and Yip1A-GFP was dispersed throughout the ER network at the onset of mitosis. This proves that the vesicle budding at ERES failed to occur during mitosis.

Fig. 13. ER exit sites (ERES) visualized by Yip1A-GFP in CHO-YIP cells.

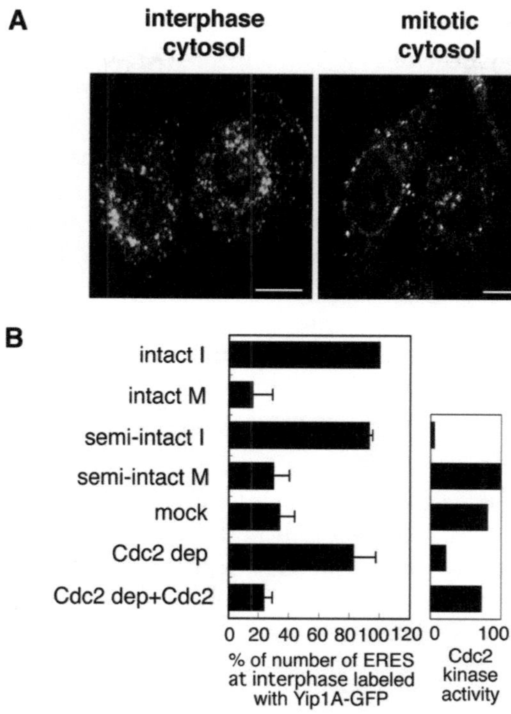

(A) Semi-intact CHO-YIP cells were incubated with interphase or mitotic cytosol with an ATP-regenerating system at 32°C for 20 min and then viewed by a confocal microscope. Bar, 10μm. (B) Number of ERES in interphase cells (intact I) or mitotic cells (intact M) was calculated using the ERES disassembly assay (see in details in Kano et al., 2004). Semi-intact CHO-YIP cells were incubated with interphase cytosol (semi-intact I), mitotic cytosol (semi-intact M), mock mitotic cytosol (mock), Cdc2-depleted mitotic cytosol (Cdc2 dep), or Cdc2-depleted mitotic cytosol plus 72 U of Cdc2/cyclin B (Cdc2 dep +Cdc2). After the incubation, the cells were subjected to the ERES disassembly assay. Cdc2 kinase activity in each reaction mixture (means from two independent measurements) is shown in the right hand column, in which 100% represents the value of Cdc2 kinase activity in mitotic cytosol. Three independent assays were performed and the means and standard deviations are plotted in the graph.

Fig. 14. Biochemical requirements for ERES disassembly by mitotic cytosol.

cytosol causes ERES to disassemble and reduces their number significantly, perhaps by as much as 80%, as compared with interphase cytosol. In addition, cdc2-depleted mitotic cytosol does not induce the disassembly of ERES. Furthermore, addition of recombinant cdc2 kinase to mitotic cytosol from which cdc2 has been depleted restores the ability of the cytosol to induce disassembly. Taken together, the results show that the mitotic disruption of ERES also depends on the activation of cdc2 kinase.

One of the target proteins of cdc2 kinase that might be involved in the mitotic disassembly of ERES is p47. p47 has been reported to play a crucial role in the *de novo* formation of ERES *in vitro* (Roy et al., 2000) and in the formation of ER network from microsomal membrane vesicles (Hetzer et al., 2001) or the reformation of disrupted ER networks after mitosis (Kano et al., 2005b). On the basis of these results, we hypothesized that p47 is also required for the maintenance of ERES and that the disassembly of preexisting ERES is controlled by phosphorylation of p47, in a cdc2-dependent manner. To test this, we assayed ERES disassembly in semi-intact cells incubated with mitotic cytosol and ATP in the presence of p97 and p47 or p97 and p47NP. The mitotic disassembly of ERES is partially blocked in the presence of p97 and p47. However, in the presence of p97 and p47NP, disassembly is completely inhibited. We also examined the effect of p97/p47NP on the dissociation of Sec13, one of the components of COPII vesicles, from ERES, and confirmed that the dissociation of Sec13 induced by mitotic cytosol is substantially inhibited by p97/p47NP and partially inhibited by p97/p47. These results suggest that the cdc2-dependent phosphorylation of p47 plays a crucial role in the disassembly of preexisting ERES. We assumed that the complete inhibition of the disassembly by p97/p47NP would preserve the anterograde transport of VSVGts045-GFP. However, anterograde transport is only partially restored under these conditions. Other factors, such as microtubule-dependent motor proteins, probably contribute to the specific block of anterograde transport during mitosis.

These results suggest that the phosphorylation of p47 triggers the disassembly of ERES and corresponding inhibition of anterograde transport. Recently, Hughes and Stephens (2010) found that Sec16A, which interacts with Sec23 and Sec13 and serves to optimize COPII assembly at ERES, remains associated with ERES throughout mitosis. They suggested that the Sec16A at ERES facilitates full assembly of COPII vesicles during anaphase, which precedes the reassembly of the Golgi apparatus during telophase.

The mechanism that regulates the partitioning of Golgi vesicles during mitosis remains controversial. Two mechanisms have been proposed: 1) the Golgi fragmentation model by Warren (1993) and 2) the Golgi absorption model by Zaal et al. (1999). In the first model, the distinction between the Golgi and the ER persists during mitosis and the components of the two organelles are inherited independently. In the second model, some components of the Golgi translocate to and merge with the ER, and thus are inherited as components of the ER/nuclear envelope. Although both models are supported extensively by evidence from several elegant experimental systems, the issue remains to be resolved (Atlan-Bonnet et al., 2006; Bartz et al., 2008). Our retrograde transport assay using semi-intact cells revealed that mitotic cytosol is competent to induce mitotic processes, including vesicle budding, vesicle transport, and vesicle fusion, and results in the translocation of Golgi components to the ER. In addition, as described in the experiments on mitotic Golgi disassembly in semi-intact cells, during the late steps of Golgi disassembly (most likely at stage III), substantial amounts of GT-GFP can be observed in the nuclear envelope and ER-like networks, having translocated from the Golgi. Furthermore, staining of the nuclear envelope with GT-GFP is more extensive at stage III than at stage II. These findings suggest that the relocation of Golgi components (in this case, GT-GFP) to the ER does occur during Golgi disassembly, at least in mitotic MDCK cells, and most likely takes place during the later steps of disassembly. However, it is difficult for us to address whether the Golgi membranes are converted into vesicles during disassembly or fuse with the ER. Given that GT-GFP is known to be a cargo for COPI-independent retrograde transport from the Golgi to the ER, the translocation of GT-GFP to the ER during mitosis might be attributed to COPI-independent Golgi disassembly, which is frequently observed as the transformation of cisternae into an extensive tubular network in a cell-free system or semi-intact cells (Misteli et al., 1995).

6. Schematic model for the coupling of ER and Golgi biogenesis to vesicular transport during the cell cycle

On the basis of our results and those of others, we have developed a hypothesis about the relationship between the cell cycle-dependent morphological changes in the ER and Golgi and the regulation of vesicular transport between the ER and Golgi in mammalian cells (Fig. 15). During the earlier steps of mitosis, activation of cdc2 kinase induces the phosphorylation of p47 and other proteins. The phosphorylated p47 causes the disassembly or vesiculation of the Golgi apparatus, and concurrently causes the partial severance of the ER network and disassembly of ERES. The disassembly of ERES inhibits anterograde transport from the ER to the Golgi, but retrograde transport remains intact. As a result, some components of the Golgi apparatus translocate to the ER network during the early phase of mitosis. The severed ER network and vesiculated Golgi membranes are easily distributed into two daughter cells, on a stochastic basis. After mitosis, the vesiculated Golgi membranes fuse with one another and reform intact Golgi stacks. In addition, the severed ER tubules are quickly reformed into a continuous ER network. These processes of reformation depend on the membrane fusion reactions that are mediated by the NSF/SNAP and p97/p47/VCIP135 systems. It is very interesting that, during Golgi reformation after mitosis, the two membrane fusion complexes act concurrently, whereas, in the reformation of the ER network, the two complexes act sequentially. It is likely that the reformation of ERES might involve the dephosphorylation of p47 given that the activity of cdc2 kinase diminishes in late mitosis and anterograde transport resumes in conjunction with Golgi reassembly and ER reformation after cell division. Probably, Sec16A, which remains associated with ERES throughout mitosis, facilitates the resumption of anterograde transport.

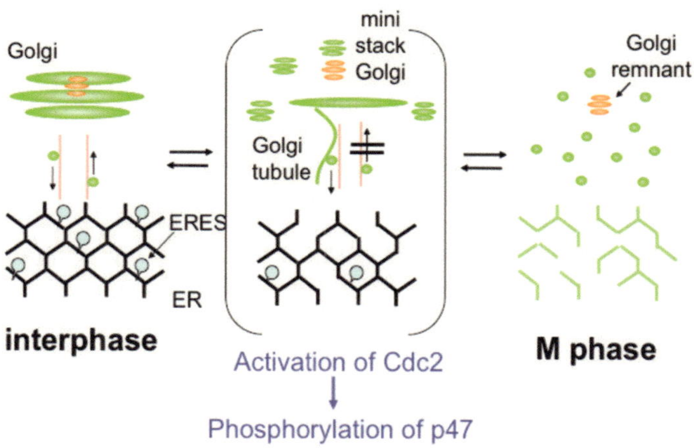

Fig. 15. Schematic model of cell cycle-dependent organelle morphology coupling and membrane traffics.

7. Complementary usage of *in vitro* reconstitution assays and semi-intact cell assay

In mitotic cells, orchestrated changes in Golgi and ER morphology occur simultaneously and are coupled with the re-arrangement of the cytoskeleton. Thus, it is likely that the roles of certain proteins in mitotic cells might be masked. Our semi-intact cell assay is suitable for investigating the biochemical requirements of specific processes, which might be masked by the orchestrated physiological reactions. For example, the ER disassembly assay revealed that a p97/p47-mediated fusion process plays a crucial role in the maintenance of the ER network when microtubules are disrupted by nocodazole. In another case, the ER reformation assay revealed that reformation of the ER is accomplished even in nocodazole-treated semi-intact cells. When microtubules are intact, the contribution of the fusion process to the maintenance or reformation of the ER network appears to be masked. Our transport assay also revealed that, even in the presence of mitotic cytosol, the retrograde transport of GT-GFP occurs normally when microtubules remain intact, but ERES are disassembled easily under these conditions (Fig.14). The findings suggest that mitotic cytosol can facilitate retrograde transport as long as microtubule integrity is maintained, but anterograde transport ceases rapidly in the presence of mitotic cytosol. Thus, the ability to manipulate the cytoskeleton easily in semi-intact cell systems will be useful in elucidating the role of the cytoskeleton in the process of morphological change in organelles or membrane trafficking during mitosis.

Many *in vitro* reconstitution assays have been developed to investigate the biochemical requirements for the maintenance or mitotic alteration of Golgi or ER morphology, and a variety of key molecules have been identified using these methods (Acharya et al.,1998; Lowe et al., 1998,2000; Hetzer et al., 2001). Our semi-intact cell assays will be useful for confirming the precise role of these molecules in the maintenance or alteration of morphology under conditions in which the configuration between organelles and the cytoskeleton is almost the same as in living cells. Thus, our assays will provide additional spatial information about where the molecules function or where the biological reactions occur in cells. By applying the Golgi disassembly assay in semi-intact cells, we found that punctate Golgi structures (Fig. 3, stage II Golgi), which are produced mainly by MEK1 from cisternal Golgi and are referred as to Golgi mini-stacks, are found mainly on the apical side of the nucleus and are associated with apical microtubules. Given that the spatial configuration of the cell is virtually unaffected in semi-intact cells, the semi-intact cell assays are superior to *in vitro* reconstitution assays for investigating the anterograde or retrograde transport between the Golgi and the ER.

There are some differences between *in vitro* reconstitution systems and our semi-intact cell system. For example, an *in vitro* ER formation assay developed by Dreier and Rapoport (2000) revealed that the characteristic polygonal structure of the ER was formed from microsomal membranes. However, the *in vitro* network produced in their assay appeared to be slightly different from the ER network formed in CHO-HSP cells. The length of one side of the three-way junctions was approximately 5 μm in their reconstituted network, compared to 1–1.5 μm in our intact or semi-intact CHO-HSP cells. We have frequently observed that this length varies with the cellular conditions. For example, following serum starvation, the length appears to be greater than 5 μm (F. K., unpublished data).

Collectively, our semi-intact cell assays are superior to *in vitro* reconstitution assays in terms of obtaining morphological or spatial information, but *in vitro* assays are more appropriate for determining biochemical requirements than semi-intact cell assays. Using both assays together will enable us to identify the key molecules involved in morphological changes, which might be masked by the orchestrated processes that occur during mitosis, and to elucidate the underlying mechanisms more precisely on the basis of morphological data.

8. Conclusions

The mechanisms that regulate the cell cycle-dependent changes in Golgi morphology in mammalian cells have been studied extensively (see reviews, Wei & Seemann, 2009). In terms of the relationship between Golgi morphology and membrane trafficking, the size and morphology of the Golgi are thought to be determined mainly by the membrane influx/efflux ratio. Thus, the characteristic features of Golgi morphology could depend on the stage of the cell cycle, cell type or intracellular conditions (Sengupta & Linstedt, 2011). In contrast, many aspects of the regulation of the morphology of the ER network remain poorly understood. The ratio of membrane influx/efflux at the ER seems to affect ER morphology less than the ratio at the Golgi affects Golgi morphology because a large amount of membrane is retained in ER structures and this could have a buffering effect on ER morphology. Unlike the case of the Golgi, a variety of ER stress responses might be induced by the aberrant accumulation of secreted proteins in the ER by the inhibition of anterograde transport, and these responses might cause not only ER dysfunction but also the change in its morphology. Furthermore, accumulating evidence suggests that the communication between the early secretory organelles and plasma membrane exists. For example, signaling by growth factors (e.g. MAPK/ERK) at plasma membranes affects the early secretory pathway (anterograde transport) via the ERES (Farhan et al., 2010). Thus, it is important to investigate the overall balance of membrane trafficking between the relevant organelles, as well as the plasma membrane, to elucidate the changes in Golgi and ER morphology that occur during mitosis more fully. The quantitative analysis of membrane trafficking while the spatial configuration of cells is maintained will be of increasing significance. Therefore, our semi-intact cell assays will provide one suitable tool for studying the regulatory mechanisms of membrane trafficking, not only during mitosis, but also under other cellular conditions, for example, disease conditions.

9. References

Acharya, U.; Mallabiabarrena, A.; Acharya, J.K. & Malhotra. V. (1998). Signaling via mitogen-activated protein kinase kinase (MEK1) is required for Golgi fragmentation during mitosis. *Cell*, 92, pp.183–192.

Altan-Bonnet, N.; Sougrat, R.; Liu, W.;, Snapp, E.L.; Ward, T. & Lippincott-Schwartz, J. (2006). Golgi inheritance in mammalian cells is mediated through endoplasmic reticulum export activities. *Mol Biol Cell.*, 17(2), pp. 990-1005.

Bartz, R.; Sun, L.P.; Bisel, B.; Wei, J.H. & Seemann, J. (2008). Spatial separation of Golgi and ER during mitosis protects SREBP from unregulated activation *EMBO J.*, 27(7), pp. 948-55.

Bhakdi, S.; Tranum-Jensen, J. & Sziegoleit, A. (1985). Mechanism of membrane damage by streptolysin-O. *Infect Immun.*, 47(1), pp. 52-60.

Cole, N.B.; Smith, C.L.; Sciaky, N.; Terasaki, M.; Edidin, M. & Lippincott-Schwartz, J. (1996). Diffusional mobility of Golgi proteins in membranes of living cells. *Science*, 273, pp. 797–801.

Dreier, L. & Rapoport, T.A. (2000). In vitro formation of the endoplasmic reticulum occurs independently of microtubules by a controlled fusion reaction. *J Cell Biol.*,148(5), pp. 883-898.

Farhan, H.; Wendeler, M.W.; Mitrovic, S.; Fava, E.; Silberberg, Y.; Sharan, R.; Zerial, M. & Hauri, HP. (2010). MAPK signaling to the early secretory pathway revealed by kinase/phosphatase functional screening *J Cell Biol.*, 189(6), pp. 997-1011.

Feinstein, T.N. & Linstedt, A.D. (2007). Mitogen-activated protein kinase kinase 1-dependent Golgi unlinking occurs in G2 phase and promotes the G2/M cell cycle transition. *Mol Biol Cell.* 18(2), pp. 594-604.

Feinstein, T.N. & Linstedt, A.D. (2008). GRASP55 regulates Golgi ribbon formation. *Mol Biol Cell.*, 19(7), pp. 2696-2707.

Hetzer, M.; Meyer, H.H.; Walther, T.C.; Bilbao-Cortes, D.; Warren, G. & Mattaj, I.W. (2001). Distinct AAA-ATPase p97 complexes function in discrete steps of nuclear assembly. *Nat. Cell Biol.*, *3*, pp. 1086–1091.

Hughes, H. & Stephens, D.J. (2010). Sec16A defines the site for vesicle budding from the endoplasmic reticulum on exit from mitosis. *J Cell Sci.*, 123(Pt 23), pp. 4032-8.

Ikonen, E.; Tagaya, M.; Ullrich, O.; Montecucco, C. & Simons, K. (1995). Different requirements for NSF, SNAP, and Rab proteins in apical and basolateral transport in MDCK cells. *Cell.* 81, pp. 571–580.

Kano, F.; Takenaka, K.; Yamamoto, A.;Nagayama, K.; Nishida, E. & Murata, M. (2000a). MEK and Cdc2 kinase are sequentially required for Golgi disassembly in MDCK cells by the mitotic *Xenopus* extracts. *J. Cell Biol.*, pp. 357–368.

Kano, F.; Sako, Y.; Tagaya, M.; Yanagida, T. & Murata, M. (2000b). Reconstitution of brefeldin A-induced Golgi tubulation and fusion with the endoplasmic reticulum in semi-intact Chinese hamster ovary cells. *Mol. Biol. Cell*, 11, pp. 3073–3087.

Kano, F.; Tanaka, A.R.; Yamauchi, S.; Kondo, H. & Murata, M. (2004). Cdc2 kinase-dependent disassembly of endoplasmic reticulum (ER) exit sites inhibits ER-to-Golgi vesicular transport during mitosis. *Mol. Biol. Cell.* 15, pp. 4289-4298.

Kano, F.; Kondo, H.; Yamamoto, A.; Tanaka, A.R.; Hosokawa, N.; Nagata, K. & Murata, M. (2005). The maintenance of the ER network is regulated by p47, a cofactor of p97, through phosphorylation by cdc2 kinase. *Genes Cells*, 10, pp. 333–344.

Kano, F.; Kondo, H.; Yamamoto, A.; Kaneko, Y.; Uchiyama, K.; Hosokawa, N.; Nagata, K. & Murata, M. (2005). NSF/SNAPs and p97/p47/VCIP135 are sequentially required for cell cycle-dependent reformation of nthe ER network. *Genes Cells*, 10, pp. 989-999.

Kano,F.; Yamauchi,S.; Yoshida,Y.; Watanabe-Takahashi,M.; Nishikawa, K.; Nakamura, N. & Murata, M. (2009). Yip1A regulates the COPI-independent retrograde transport from the Golgi apparatus to the endoplasmic reticulum. *J. Cell Sci.*, 122, pp. 2218-2227.

Kondo, H.; Rabouille, C.; Newman, R.; Levine, T.P.; Pappin, D.; Freemont, P. & Warren, G. (1997). p47 is a cofactor for p97-mediated membrane fusion. *Nature.* 388(6637), pp. 75-8.

Lowe, M.; Rabouille, C.; Nakamura, N.; Watson, R.; Jackman, M.; Jamsa, E.; Rahman, D.; Pappin, D.J. & Warren, G. (1998). Cdc2 kinase directly phosphorylates the cis-Golgi matrix protein GM130 and is required for Golgi fragmentation in mitosis. *Cell*. 94, pp. 783–793.

Lowe, M.; Gonatas, N.K. & Warren, G. (2000). The mitotic phosphorylation cycle of the cis-Golgi matrix protein GM130. *J Cell Biol.*, 149(2), pp. 341-356.

Lu, L.; Ladinsky, M.S. & Kirchhausen, T. (2009). Cisternal organization of the endoplasmic reticulum during mitosis. *Mol Biol Cell*, 20(15), pp. 3471-3480.

Meyer, H.H.; Wang, Y. & Warren, G. (2002). Direct binding of ubiquitin conjugates by the mammalian p97 adaptor complexes, p47 and Ufd1-Npl4. *EMBO J.*, 21(21), pp. 5645-5652.

Misteli, T. & Warren. G. (1995). A role for tubular networks and a COP I-independent pathway in the mitotic fragmentation of Golgi stacks in a cell-free system. *J Cell Biol.*, pp. 1027-1039.

Preisinger, C.; Körner, R.; Wind, M.; Lehmann, W.D.; Kopajtich, R. & Barr, F.A. (2005). Plk1 docking to GRASP65 phosphorylated by Cdk1 suggests a mechanism for Golgi checkpoint signalling. *EMBO J.*, pp. 753-65.

Puhka, M.; Vihinen, H.; Joensuu, M. & Jokitalo, E. (2007). Endoplasmic reticulum remains continuous and undergoes sheet-to-tubule transformation during cell division in mammalian cells. *J Cell Biol.*, 179(5), pp. 895-909.

Roy, L.; Bergeron, J.J.; Lavoie, C.; Hendriks, R.; Gushue, J.; Fazel, A.; Pelletier. A.; Morré, D.J.; Subramaniam, V.N.; Hong, W. & Paiement, J. (2000). Role of p97 and syntaxin 5 in the assembly of transitional endoplasmic reticulum. *Mol Biol Cell*, 11(8), pp. 2529-42.

Sengupta,D. & Linstedt, A.D. (2011). Control of Organelle Size: The Golgi Complex. *Annu. Rev. Cell Dev. Biol.*, 27, pp. 5.1-5.21.

Shaul, Y.D. & Seger, R. (2006). ERK1c regulates Golgi fragmentation during mitosis. *J Cell Biol.*, 172(6), pp. 885-897.

Tang, D.; Mar, K.; Warren, G. & Wang, Y. (2008). Molecular mechanism of mitotic Golgi disassembly and reassembly revealed by a defined reconstitution assay. *J Biol Chem.*, 283(10), pp. 6085-6094.

Uchiyama, K.; Jokitalo, E.; Lindman, M.; Jackman, M.; Kano, F.; Murata, M.; Zhang, X. & Kondo, H. (2003). The localization and phosphorylation of p47 are important for Golgi disassembly-assembly during the cell cycle. *J Cell Biol.*, 161, pp. 1067-1079.

Warren, G. (1993). Membrane partitioning during cell division. *Annu. Rev. Biochem.* 62, pp. 323–348.

Wei, J.H. & Seemann, J. (2009). Mitotic division of the mammalian Golgi apparatus. *Semin. Cell Dev. Biol.*, 20(7), pp. 810-816.

Xie, S.; Wang, Q.; Ruan, Q.; Liu, T.; Jhanwar-Uniyal, M.; Guan, K. & Dai, W. (2004). MEK1-induced Golgi dynamics during cell cycle progression is partly mediated by Polo-like kinase-3. *Oncogene*, 23(21), pp. 3822-3829.

Yoshimura, S,; Yoshioka, K.; Barr, F.A.; Lowe, M.; Nakayama, K.; Ohkuma, S. & Nakamura, N. (2005). Convergence of cell cycle regulation and growth factor signals on GRASP65. *J Biol Chem.*, 280(24), pp. 23048-23056.

Zaal, K.J.M.; Smith, C.L.; Polishchuk, R.S.; Altan, N.; Cole, N.B.; Ellenberg, J.; Hirschberg, K.; Presley, J.F.; Roberts, T.H.; Siggia, E.;Phair, A,D. & Lippincott-Schwartz, J. (1999). Golgi membranes are absorbed into and reemerge from the ER during mitosis. *Cell.* 99, pp. 589–601.

2

Morphogenesis and Dynamics of Post-Golgi Transport Carriers

Roman S. Polishchuk[1,*] and Elena V. Polishchuk[1,2]
[1]*Telethon Institute of Genetics and Medicine, Naples*
[2]*Institute of Protein Biochemistry, Naples*
Italy

1. Introduction

The identities of many intracellular organelles and of specific domains of the cell surface rely on the delivery of proteins and lipids through biosynthetic or/and endocytic pathways to the sites of their specific activities. The Golgi complex serves as a central station in the biosynthetic pathway, from where proteins are sorted towards their different destinations, such as various domains of the cell surface or the endosomal-lysosomal system. To be delivered from the Golgi complex to their target compartments, cargo proteins are incorporated into dynamic membrane-bound organelles that are generally known as 'post-Golgi carriers'. Given that these post-Golgi carriers have such an important role in the process of intracellular transport their morphology, living dynamics and molecular composition became the subjects of significant interest over the last decade.

Post-Golgi carriers (PGCs) were originally discovered and described as a result of the development of green fluorescent protein (GFP) technology and live-cell imaging (Lippincott-Schwartz et al., 2000). The first few fluorescently tagged cargo proteins observed in living cells revealed a new world of highly dynamic structures traveling from the Golgi complex to the plasma membrane (Hirschberg et al., 1998; Nakata et al., 1998). With time, the list of molecules that could be visualized *in vivo* expanded greatly, to expose the unexpected complexity of the post-Golgi transport pathways. However, in mammalian cells, most of PGSs have several common features that are independent of the pathway(s) to which they belong.

PGCs form from membrane domains of the Golgi complex that lack resident Golgi enzymes, and there are known as 'PGC precursors' (Hirschberg et al., 1998; Keller et al., 2001; Polishchuk et al., 2003; Puertollano et al., 2003). The shapes and sizes of PGCs that can even carry the same cargo vary across a relatively wide range. Most that were seen under light microscopy were clearly larger that plasma membrane (PM)-associated clathrin vesicles and 100-nm-diameter fluorescent beads (Lippincott-Schwartz et al., 2000). Indeed, while the smaller PGSs can usually have an extension of 300 nm to 400 nm, some large carriers can reach dozens of microns in length. Video microscopy has revealed that many of these

* Corresponding Author

carriers are globular in appearance, although they are frequently stretched into tubular shapes during their translocation through the cytosol. Thus, PGCs have been frequently referred to as 'pleiomorphic' structures. PGCs use microtubules to move towards particular locations within the cell. Although carriers can form and support post-Golgi transport without association with microtubules, the correct targeting of cargo proteins is usually compromised under such conditions (Kreitzer et al., 2003; Rindler et al., 1987)

The life cycle of a PGC consists of three stages: (i) formation; (ii) transition through the cytosol; and (iii) docking and fusion with the target membrane (Fig. 1) (Polishchuk et al., 2000). In this chapter we take you on a journey with newly born PGCs, to follow them through all of the stages of their life cycle.

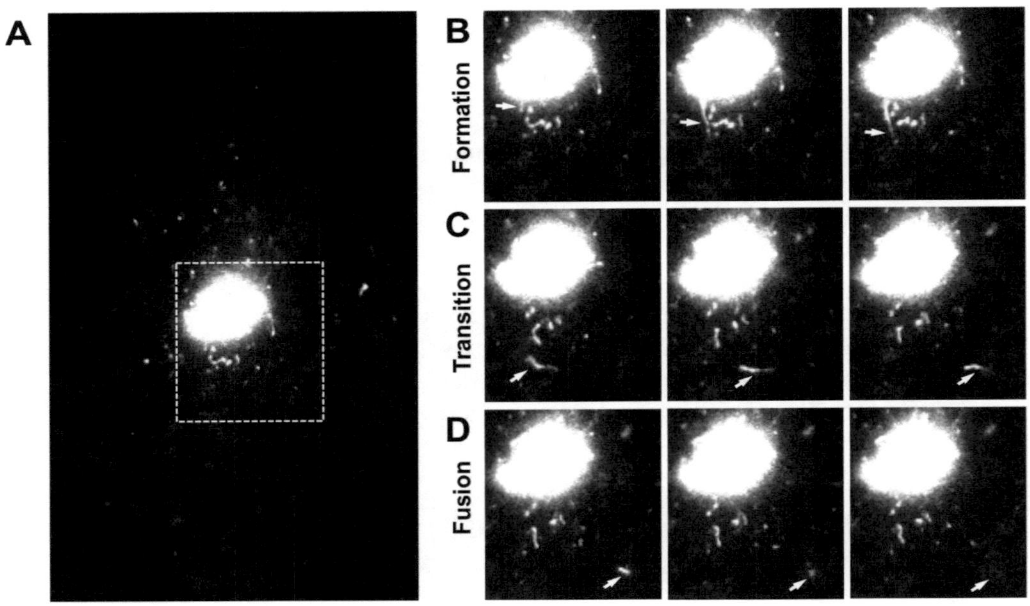

A. An example of PGCs dynamics, as taken in the area (dashed box) of a living cell expressing VSVG-GFP upon release of this chimeric protein from the Golgi complex. **B.** Representative images extracted from a time-lapse sequence (corresponding to the area outlined by the dashed box in panel A) show the dynamics of PGC formation from the Golgi complex (arrows). **C.** Transition of the same newly formed PGC through the cytoplasm (arrows). **D.** Fusion of the same PGC (panels B, C) with the target membrane, as shown in these three subsequent time-lapse images by arrows.

Fig. 1. The life cycle of a PGC.

2. PGC formation

The process PGC morphogenesis occurs at the level of the most distal Golgi compartment, known as the trans-Golgi network (TGN), and this was characterized in detail using a combination of time-lapse and electron microscopy (EM). This process comprises three main steps: (i) formation of specialized TGN export domain, known as the PGC precursor; (ii) extrusion of this domain along microtubules; and (iii) fission of the export domain to generate a free carrier (Polishchuk et al., 2003).

2.1 Morphogenesis of PGC precursors at the TGN

The first step of PGC formation coincides with the segregation of the cargo proteins from the Golgi-resident enzymes. This appears to be a common process for proteins that exit the Golgi complex to head towards different post-Golgi compartments, such as the basolateral PM (Hirschberg et al., 1998; Polishchuk et al., 2003), the apical PM (Keller et al., 2001), and the endosomal-lysosomal system (Puertollano et al., 2003). These enzyme-free Golgi domains usually exhibit a tubular structure and contain *bona-fide* TGN markers (Polishchuk et al., 2003; Puertollano et al., 2003). At the EM level, this segregation of the cargo proteins from the Golgi-resident enzymes corresponds to the transition of the cargo-containing compartments from cisterna-like into a tubular network morphology (Polishchuk et al., 2003).

This thus raises the question of the mechanism by which the originally flat Golgi membranes are converted into highly bent, tubular-reticular TGN structures. A long time ago, Rambourg and Clermont (Rambourg and Clermont, 1990) noted that the cisternae in the middle of a Golgi stack appear quite 'solid' and contain just a few small fenestrae (Fig. 2A). During progression toward the *trans* face of the Golgi complex, both the number and size of these fenestrae increase (Rambourg and Clermont, 1990). Thus the *trans*-most Golgi cisternae generally look like flat tubular webs (Fig. 2A, B), which frequently 'peel off' from the Golgi stack (Rambourg and Clermont, 1990) (Fig. 2C, D), and these can then be easily transformed into TGN membranes by a few fission events (see Fig. 2B). A similar conversion of the Golgi compartments was seen to occur along the Golgi stack in the yeast *Pichia pastoris* (Mogelsvang et al., 2003).

The mechanisms behind this transformation are not yet completely clear. The cisterna-like morphology of the Golgi compartments can be stabilized through the formation of large polymers formed by the Golgi enzymes (Nilsson et al., 1996). Indeed, truncation of the protein domains responsible for enzyme oligomerization results in a loss of the regular Golgi morphology (Nilsson et al., 1996). Thus, a gradual reduction in Golgi enzyme polymers in the *trans*-Golgi compartments would favor transformation of cargo-containing cisternae into networks of tubular membranes. This process can be further assisted by changes in the lipid composition of the *trans*-Golgi membranes, which include the input of material from the endocytic system (Pavelka et al., 1998) and the local activities of TGN-specific lipid-modifying enzymes or lipid-transfer proteins (De Matteis and Luini, 2008). Therefore, the membranes become thicker in the TGN and they thus fail to provide a favorable environment for the short transmembrane domains of the Golgi enzymes (Munro, 1995). As a result, the Golgi enzymes get pushed out of the tubular TGN back towards the core Golgi regions that are composed of the stacked cisternae.

The bending of the flat cisternae membranes into TGN tubules might be also facilitated by the action of specific proteins. Various roles of structural proteins in membrane deformation/ tubulation have been widely recognized (McMahon and Gallop, 2005). These proteins act either by forcing membrane curvature or by sensing and stabilizing it. As an example of the former, coat proteins (such as clathrin) polymerize into curved structures that can bend membrane domains (Antonny, 2006). Alternatively, such TGN proteins as the FAPPs can insert their amphipathic helices into the outer leaflet of a lipid bilayer, thereby increasing the positive membrane curvature (Lenoir et al., 2010). Other rigid curved proteins, or protein modules (such as the BAR domain), can bind to curved membranes and stabilize them by electrostatic interactions (Cullen, 2008; McMahon and Gallop, 2005).

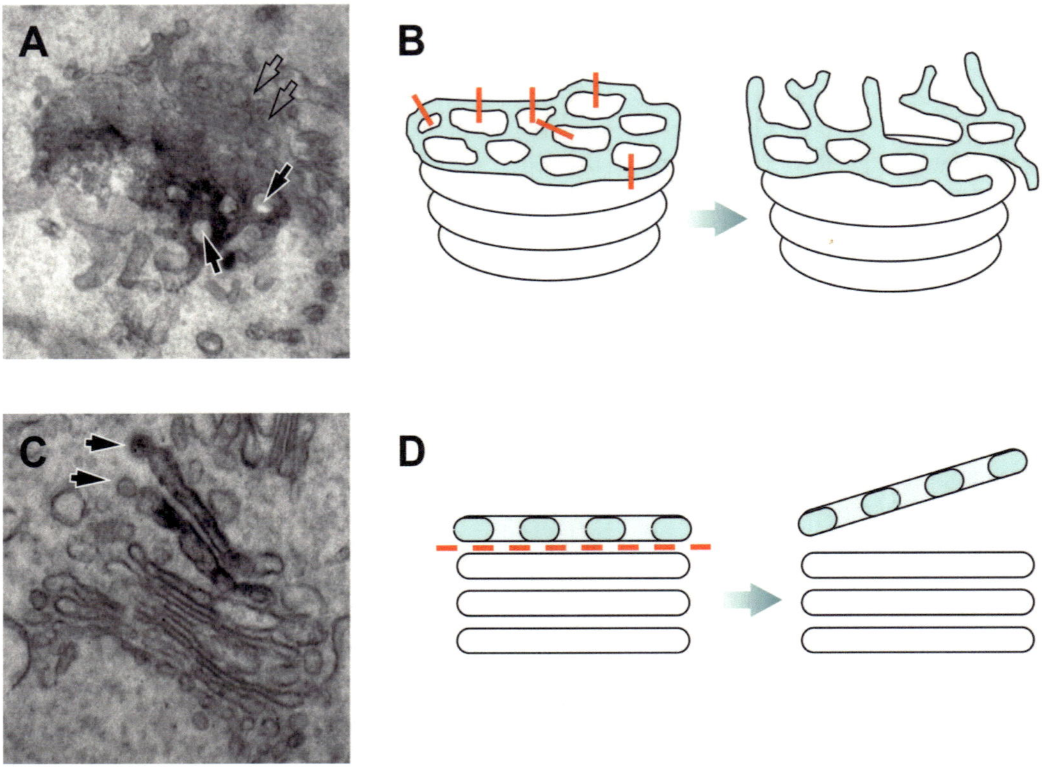

A. Thin *en-face* section of the Golgi stack in cells transfected with a TGN38-HRP chimera. The solid arrows show the increase in size of the fenestrae of a TGN38-positive cisterna (dark staining) over the fenestrae of an unstained cisterna (empty arrows). **B.** Scheme showing how highly fenestrated *trans*-most Golgi cisternae can be transformed into tubular network through a few fission events (red lines). **C.** Thin section across the Golgi complex in cells transfected with a TGN38-HRP chimera to show two of the *trans*-most cisternae (arrows) detaching from the rest of the stack. **D.** The process of cisternae peeling off from the rest of the Golgi stack might be determined by a loss of the stacking proteins/ mechanisms (red dashed line) in the *trans*-Golgi compartment.

Fig. 2. Transformation of Golgi cisternae into TGN.

On the other hand, changes in lipid content, and hence membrane curvature, can be modulated via lipid-metabolizing enzymes that reside at the Golgi complex (De Matteis and Luini, 2008). The transmembrane or inter-organelle transfer of lipids can contribute to the generation of particular lipid environments in the membranes of the TGN. In this respect, it is important to note that numerous contact sites between the ER and the trans-cisternae of the Golgi complex have been detected by EM tomography (Ladinsky et al., 1999). Such contact sites can be used for lipid transfer between the ER and the *trans*-Golgi, which can be mediated by specific lipid-transfer proteins, such as CERT and OSBP (De Matteis et al., 2007).

Finally, conversion of the Golgi cisternae into tubular PGC precursors at the TGN can be accompanied by the loss of the stacking mechanisms at the *trans* side of the Golgi complex. The two main proteins that are involved in the maintenance of cisterna stacking, GRASP65 and GRASP55, have been detected mainly in the *cis* and medial Golgi, rather than in the *trans*-Golgi (Barr et al., 1997; Shorter et al., 1999). Therefore, this intercisternal 'glue' migh be gradually lost as a cisterna progresses towards the *trans* pole of the Golgi complex. This has been confirmed both in mammals and yeast by observations that the *trans*-most cisterna frequently peels off from the main Golgi stack (Clermont et al., 1995; Mogelsvang et al., 2003) (Fig. 2C, D). It is likely that two or more mechanisms act in synergy here, to allow the conversion of flat Golgi cisternae into the tubulo-reticular PGC precursors at the TGN.

In addition to the formation of tubular domains at the exit face of the Golgi complex, this is also the level at which the cargo proteins that are directed to different post-Golgi destinations need to be sorted. The classical view in the membrane transport field implies that this sorting at the TGN (as well as throughout the whole secretory pathway) is driven mainly by the coat-adaptor-protein machinery, which interacts specifically with amino-acid signals of certain transmembrane cargo proteins; this then provides the mechanical force for the budding and fission of transport vesicles (Mellman and Warren, 2000).

This holds true for endo-lysosome-directed carriers, which have been scrupulously characterized. In contrast, PGCs carrying a cargo like the G-protein of vesicular stomatitis virus (VSVG) can be formed in a coat-protein-independent and AP-independent manner. Both VSVG-positive PGCs and their precursors do not exhibit β-COP or γ-, δ- and ε-adaptins at their membranes (Polishchuk et al., 2003). Other clathrin adaptors, such as the GGAs, are excluded from VSVG carriers as well (Polishchuk et al., 2003; Puertollano et al., 2003). Similarly, coats and adaptors have never been detected on PGCs that are carrying proteins to the apical PM surface in polarized cells (Kreitzer et al., 2003). Thus, these carriers should form either by virtue of some still-unknown adaptors that have not yet been visualized by EM, or by their association with specific lipid microdomains that are involved in sorting (Simons and Gerl, 2010). This might be the case for proteins directed to the apical surface of the PM in polarized epithelial cells, the concentration of which at the TGN appears to be through their partition into cholesterol-rich and sphingolipid-rich membrane domains, which are known as 'membrane rafts' (Simons and Gerl, 2010).

Thus, the TGN represents a mosaic of different export domains, which strongly resemble free PGCs in their molecular composition. EM has also revealed structural similarities between carriers and their precursors at the TGN. For example, while the PGCs that carry VSVG mainly have a tubular morphology, several can have a complex structure and even contain clearly visible fenestrae (Polishchuk et al., 2003; Polishchuk et al., 2000), as would be expected of membranes that derive from a protrusion of the TGN. Indeed, PGC precursors visualized using correlative light-electron microscopy (CLEM) comprise tubular segments that consist of complex branching and fenestrated membranes remaining continuous with the parent membranes of the Golgi stack (Polishchuk et al., 2003). Similarly, carriers containing the apical cargo protein HA frequently exhibit tubular morphology, as do the HA-positive domains at the TGN (Puertollano et al., 2001). Thus structural similarities with their precursor at the TGN appear to be a common feature of the different types of PGCs. This strongly suggests that PGCs form by the fission of these precursor domains (or large parts of them) from the rest of the TGN membranes.

2.2 Elongation of PGC precursors

After the initial budding from the donor compartment, PGC precursors frequently undergo further extension, which in some cases, can reach to lengths of over a dozen microns. This process is usually coupled to the loading of the cargo, and it allows the adapting of bulky proteins into budding PGCs; e.g. rigid 300-nm-long procollagen rods (Polishchuk et al., 2003), and long tubular multimers of von Willebrand blood coagulation factor (Zenner et al., 2007).

The process of tubule elongation can be supported via the recruitment of various scaffold proteins. Clathrin has been shown to polymerize into tubular shapes (Zhang et al., 2007). In addition, local activities of lipid-modifying enzymes might be required to maintain the production of specific lipids that favor a tubule-like conformation of membranes, and therefore, to support PGC growth (Brown et al., 2003). Finally, elongation of PGC precursors is frequently assisted by the cytoskeleton. A number of actin-associated and microtubule-associated motor proteins appear to be implicated in the pulling out of tubular structures from donor membranes (Kreitzer et al., 2000; Sahlender et al., 2005). Remarkably, microtubules and kinesin alone appear to be sufficient to trigger the formation of long membrane tubules from liposomes *in vitro* (Roux et al., 2002) and this process has been shown to operate for PGC elongation (Polishchuk et al., 2003).

2.3 Fission of PGCs from the TGN

The dynamics of the PGC fission process appears to be fairly complex. Live-cell imaging has revealed PGC fission frequently to coincide with the mechanical pulling out of carrier precursor from the TGN along microtubules (Polishchuk et al., 2003). Apparently, this pulling force that the molecular motors such as kinesin (see below) can apply to the TGN membranes is important in the extension of PGC precursors from the Golgi body and for the later fission of the PGC (Kreitzer et al., 2000; Polishchuk et al., 2003). In cell-free systems, addition of kinesin to Golgi membranes (and even to liposomes) together with microtubules induces the formation of tubule-like membranes that are similar to PGC precursors (Roux et al., 2002), while a block in kinesin function in cells by microinjection of an inhibitory antibody (Kreitzer et al., 2000) or expression of a 'headless' kinesin mutant (Nakata and Hirokawa, 2003) prevents PGC formation from the Golgi complex. Kinesin has been seen to be associated with the tip of PGC precursors, although it can also attach to other points along the membrane of a PGC precursor (Polishchuk et al., 2003). The movement of kinesins along microtubules can then create tension within a PGC precursor that will facilitate the fission process (Shemesh et al., 2003). Indeed, based on *in-vitro* data, membranes under tension have recently been proposed to have an important role in fission (Roux et al., 2006). The extension of tubular PGC precursors might also result in looser lipid packing in its membrane. This exposes the membranes of a budding PGC to easier access by proteins that promote membrane fission at the TGN, such as dynamin and CtBP1-S/BARS (Corda et al., 2006; McNiven et al., 2000). Nonetheless, PGCs can also form when microtubules have been destroyed by nocodazole treatment; in this case, the pulling force to create membrane tension in fission-prone regions might be applied by the actin motors (Miserey-Lenkei et al., 2010; Sahlender et al., 2005).

Live-cell imaging and CLEM have also shown that fission does not take place randomly along the membranes of PGC precursors. Fission frequently takes place at the thinnest and geometrically simplest regions of the elongating PGC membranes (Fig. 3A), which at the EM level correspond to thin tubular segments of membranes (Polishchuk et al., 2003). In contrast, fission does not take place at the TGN regions with a complex morphology (i.e., in

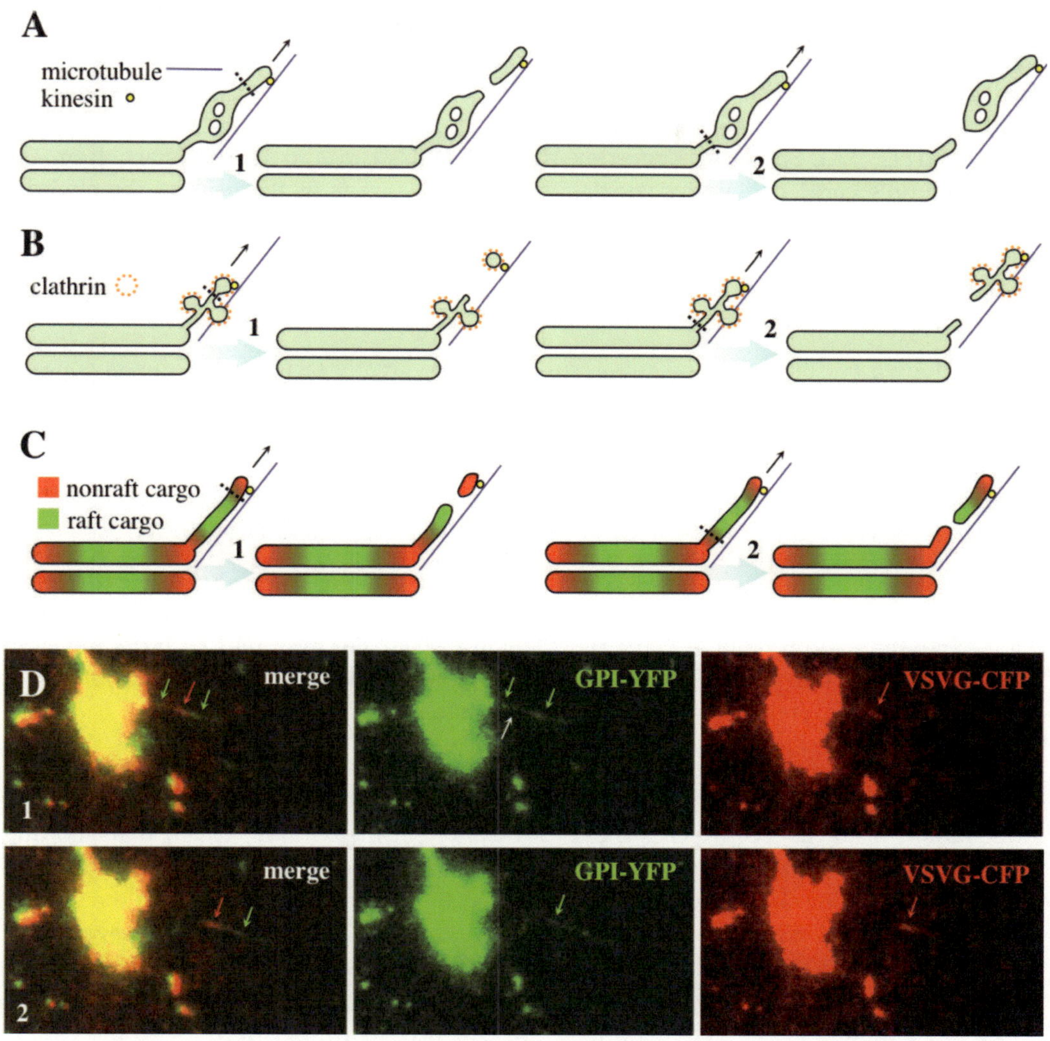

A. TGN precursors of post-Golgi carriers are pulled along microtubules by kinesin. The fission of the carriers (dashed line) occurs at the thinnest parts of the PGC precursor, which correspond to thin tubular segments of the TGN membrane at the EM level. In contrast, fission does not take place at TGN regions with a complex morphology (i.e., containing tubular networks and fenestrae, or in thick vacuolar regions). If fission occurs close to the tip of a PGC precursor, the carrier will be smaller in size (panel 1). In contrast, larger PGCs can be formed by cleavage at the base of a PGC precursor (panel 2). **B**. PGCs directed to endosomes can detach from the TGN as simple clathrin-coated vesicles when the fission (dashed line) occurs at the neck of the clathrin-coated bud (1). Alternatively, entire chunks of the TGN membrane containing 2-3 clathrin-coated buds can be cleaved from the Golgi complex (2). **C**. In budding PGCs that contain raft and non-raft proteins and lipids, fission (dashed line) can occur at the fission-prone border between raft and nonraft areas, which leads to the formation of PGCs with either single (1) or several (2) membrane microdomains. **D.** The subsequent frames extracted from the time-lapse sequence to demonstrate budding (1) and fission (2) of PGC that contains different membrane microdomains, which are labeled with either the raft cargo GPI-YFP (green arrow) or with the nonraft cargo VSVG-CFP (red arrows).

Fig. 3. Fission of post-Golgi transport carriers.

those containing tubular networks and branching tubules, or in thick vacuolar regions). Obviously, the precise points of fission will define not only the composition of a PGC carrier, but also its morphology. If fission occurs close to the tip of the TGN tubule, a carrier will be smaller in size. In contrast, larger PGCs can form by cleavage at the base of a PGC precursor (Fig. 3A). Similarly, endosome-directed PGCs can apparently detach from the TGN as simple cargo-containing vesicles if the fission occurs at the neck of the clathrin-coated buds (Fig. 3B). However, many clathrin-positive PGCs have a grape-like morphology (a tubule with several 'buds'), which suggests that entire chunks of TGN membranes that contain 2-3 clathrin-coated buds can be cleaved from the Golgi complex (Polishchuk et al., 2006).

Fission along a budding PGC precursor might also be greatly facilitated by heterogeneous lipid microdomains (Roux et al., 2005). Indeed, liposome tubules have been shown to break into small pieces at the border between phosphatidylcholine-enriched and cholesterol-enriched regions (Roux et al., 2005). Given that multiple lipid microdomains can be seen along a forming PGC, it appears that large post-Golgi carriers might contain several regions that are filled with raft and non-raft proteins (Polishchuk et al., 2004) (Fig. 3C, D).

Finally, fission of PGCs might be regulated by the cargo proteins themselves. According to a recent hypothesis, in some cases fission appears not to occur until the cargo is completely loaded into the budding tubule (Bard and Malhotra, 2006). For example, if a large and rigid procollagen rod is still present in the neck of a budding tubule, this neck can apparently not be constricted sufficiently (by dynamin or coat proteins) to trigger fission.

3. PGC transition to the target membrane

After fission from the TGN, PGCs move to their target membranes. Given that budding PGCs are associated with motor proteins, these can dock onto microtubules and use them as the 'highways' to reach their destination. In this context, different members of the kinesin superfamily (Kamal et al., 2000; Nakata and Hirokawa, 2003; Teng et al., 2005), or even other microtubule motors, such as dynein (Tai et al., 1999), have been shown to drive post-Golgi transport of specific cargo towards their acceptor compartment. Such high fidelity of cargo selection by motors at the TGN, and as a consequence, its further delivery to the correct surface or intracellular domain, might be regulated by interactions of motor proteins directly with the cargo (Kamal et al., 2000; Teng et al., 2005) or with components of the TGN sorting machinery (Nakagawa et al., 2000). For instance, transport of HA and annexin 13b to the apical PM surface in epithelial cells relies on the raft-associated motor protein KIFC3 (Noda et al., 2001). KIF13A has been shown to operate in the other post-Golgi route that directs the mannose-6-phosphate receptor from the Golgi complex to the endosomes (Nakagawa et al., 2000). A number of neuronal proteins, such as bAPP, GAP43 and SNAP25, require KIF5 for their correct targeting (Nakata and Hirokawa, 2003). The microtubule minus-end-directed motor dynein has been described as supporting rhodopsin transport in rod photoreceptors (Tai et al., 1999; Yeh et al., 2006).

Of note, the sorting of specific cargo to either axons or dendrites by different kinesins has been demonstrated in a single individual neuron (Nakata and Hirokawa, 2003). KIF5 has been shown to carry VSVG-containing PGCs to the axon, while KIF17A provides the delivery of the Kv2.1 ion channel to dendrites (Nakata and Hirokawa, 2003). In such cases, how do the different motors know where to deliver these specific cargoes? Apparently motor heads of KIF5 and KIF17A can bind with higher affinity to different subsets of microtubules, which will provide directional cues for polarized axonal transport (Nakata and Hirokawa, 2003). Different populations of microtubules have been found also in other cell types, and therefore, these might serve as highways for polarized cargo delivery (Nakata and Hirokawa, 2003; Spiliotis et al., 2008).

In this context, microtubule architecture appears to play a significant role in targeting of TGN-derived PGCs to their correct acceptor membrane(s). Several molecular players, such as Par-1 (Cohen et al., 2004) and septins (Spiliotis et al., 2008), have been shown to regulate microtubule organization in vertical or horizontal arrays during the polarization of columnar (MDCK cells) or planar (hepatocytes) epithelia, respectively. As a consequence of the divers microtubule architecture, same apical proteins can be delivered from the Golgi complex directly to the correct surface domain of MDCK cells, while in hepatocytes, these apical markers first appear at the basolateral surface, and then transcytosis is used for their apical delivery (Cohen et al., 2004).

The other important issue that needs to be addressed is whether any sorting processes take place in the PGCs *en route* to their acceptor compartment. This occurs, for example, when the mannose-6-phosphate receptor is sorted from the maturing secretory granules by clathrin-coated vesicles (Klumperman et al., 1998). So several approaches have been used to determine whether similar sorting events occur with PGCs. Mature TGN-derived carriers can be arrested before their fusion with the PM, either by microinjection of an anti-NSF antibody or by treatment with tannic acid (Polishchuk et al., 2003; Polishchuk et al., 2004). In contrast to secretory granules, a comparison of mature with newly formed PGCs did not reveal significant variations in either their ultrastructure or their molecular composition (Polishchuk et al., 2003; Polishchuk et al., 2004). Similarly, mature carriers directed from the Golgi complex to endosomes accumulated in cells upon endosome ablation. However, they did not show any significant transformation, except for a slight reduction in the area covered by clathrin (Polishchuk et al., 2006).

Live-cell imaging of MDCK cells has revealed that PCGs that contain both the basolateral marker VSVG-CFP and the apical marker GPI-YFP do not sort out each of these proteins into separate structures, but deliver both of these proteins to the PM (Polishchuk and Mironov, 2004). GPI-YFP is then sorted from the basolateral surface to the apical surface through transcytosis. On the other hand, segregation of proteins from their common PGC into two separate carriers has also been documented (Jacob and Naim, 2001). This suggests that sorting from the PGC does happen, but that it is likely to depend on the nature of the transported cargo proteins.

The complexity of sorting events in the post-Golgi space became more evident with discovery that certain cargoes can pass through an endosomal intermediate before their arrival at the PM. Such indirect 'through-endosome' trafficking of cargo to the cell surface might be significantly facilitated by close association of the TGN membranes with a number of the endocytic compartments in the perinuclear area of a cell (Pavelka et al., 1998). The list of the proteins using this pathway has been updated recently, and it has now been shown that in MDCK epithelial cells, VSVG, the LDL receptor, and E-cadherin can be detected in endosomes before their exit to the PM (Ang et al., 2004; Lock and Stow, 2005). These findings, however, raise a number of further questions. The first is whether this indirect transport route exists in different cells. The second is whether different cargoes move through the same endosomal compartment on their way to the cell surface in epithelial cells. This appears not to be a case, as a number of proteins (such as VSVG and the LDL receptor) have been reported to use a Rab8-positive sub-population of endosomes as an intermediate station on their way to the basolateral PM in epithelial cells (Ang et al., 2004), while others cargoes (such as E-cadherin, for example) move to the PM through a Rab11-containing endocytic compartment (Lock and Stow, 2005). Furthermore, ablation of the different endocytic compartments through horseradish peroxidase (HRP)-catalyzed crosslinking has revealed a number of apical proteins to take various through-endosome routes to reach the cell surface (Weisz and Rodriguez-Boulan, 2009). It remains to be understood, however, whether any cross-talk between these routes exists. Unfortunately, detailed characterization of PGCs that operate either to or from these intermediate endocytic stations is still missing. Moreover, directionality and selectivity of the post-Golgi routes, and therefore the PGC properties, might change upon cell differentiation. For example, at the early stages in MDCK cells, GPI-anchored proteins are transported to the basolateral surface of the PM, from where that tend to be transcytosed to the apical surface (Polishchuk et al., 2004). Later, however, the cells tend to divert GPI-anchored proteins into a direct Golgi-to-apical surface route (Paladino et al., 2006). Further efforts need to be made to understand this interplay of the different mechanisms that define the PGC path from the TGN to the target compartment.

4. Docking and fusion of PGCs with acceptor membranes

To complete their movement across a cell, PGCs need to fuse with their acceptor membrane to deliver their contents. Docking and fusion of PGCs with the cell surface has been studied by total internal reflection (TIRF) microscopy, which allows the visualization of intracellular events within very narrow (50-150 nm) distance from the PM. This TIRF analysis demonstrated that after the docking heterogeneously sized PGCs, these usually fuse completely with the PM (Schmoranzer et al., 2000; Toomre et al., 2000). However, occasionally, larger tubular PGCs can fuse with the PM only partially, at their tips (Schmoranzer et al., 2000; Toomre et al., 2000), using a kind of 'kiss-and-run' mechanism. Interestingly, some PGCs also remain attached to microtubules even as their fusion with the PM initiates (Schmoranzer and Simon, 2003).

PGC fusion with the PM might not be randomly distributed, and might instead be concentrated at several 'hot spots' along the PM (Toomre et al., 2000). In general, the delivery of the PGCs is frequently directed to rapidly growing membrane surfaces. In motile cells, this process is restricted to the leading edge of the cell (Polishchuk et al., 2004; Schmoranzer et al., 2003). Visualization of exocytosis in polarized cells has revealed that the vertical growth of epithelia relies on the directed delivery of PGCs to the lateral surface of the columnar cells, where the tethering factors, such as mammalian exocyst or *Drosophila* DLG, reside (Kreitzer et al., 2003; Lee et al., 2003; Polishchuk et al., 2004). Similarly, the exocyst can define PGC docking sites in neurite growth cones of differentiating neurons (Vega and Hsu, 2001). Interestingly, during tissue biogenesis, a number of cells (e.g. epithelial cells) show an incredible flexibility in terms of their transport routes (Mostov et al., 2003; Rodriguez-Boulan et al., 2005). In epithelia growing on a filter support for 2-3 days, GPI-anchored proteins (which are normally apically targeted) have been found together with basolateral markers within the same carriers docked onto the lateral membranes of the MDCK cells (Polishchuk et al., 2004). Only after their arrival at the basolateral surface of the PM were these proteins sorted to the apical domain of the PM by transcytosis (Polishchuk et al., 2004). However, after 4 days in culture, MDCK cells apparently start to switch delivery of GPI-anchored proteins from a transcytotic to direct route (Paladino et al., 2006). Similarly, in thyroid epithelial cells, the delivery of dipeptidylpeptidase-IV to the apical PM surface changes from transcytosis to a direct route as the epithelial monolayer matures (Zurzolo et al., 1992). This might happen because the target patch for the PGC fusion forms at the lateral surface of epithelial cells earlier then at the apical targeting patch (Mostov et al., 2003; Nelson, 2003; Rodriguez-Boulan et al., 2005). Indeed, Sec6 undergoes recruitment to the sites of cell-to-cell junctions as soon as subconfluent cells start to contact each other (Yeaman et al., 2004), while expression of the apical sorting machinery components occurs later during the process of cell polarization (Halbleib et al., 2007). Thus, during the early stages of polarization, most of PGCs fuse near junctional complexes at the lateral domain of the PM (Kreitzer et al., 2003; Polishchuk et al., 2004), which contributes to fast vertical elongation of a cell within the epithelial sheet (Mostov et al., 2003; Nelson, 2003; Rodriguez-Boulan et al., 2005).

The precise spatial targeting of PGCs to the correct PM area might be important for processes that contribute to correct tissue development, such as, for example, the parallel alignment of collagen fibers in a tendon (Canty et al., 2004). The cellular mechanism of this alignment is thought to involve the assembly of intracellular collagen fibrils within PGCs. The PGCs carrying procollagen subsequently connect to the extracellular matrix via finger-like projections of the PM, which are known as fibripositors, and which are oriented along the axis of the tendon (Canty et al., 2004). Interestingly, actin filament disassembly results in the rapid loss of fibripositors and in the subsequent disappearance of intracellular fibrils. In this case, a significant proportion of collagen fibrils are found to no longer be orientated with the long axis of the tendon. This suggests an important role for the actin cytoskeleton in the alignment of PGC delivery and in the further organization of the collagenous extracellular matrix in the embryonic tendon (Canty et al., 2006).

In brain tissue, transformation of the contact between an axon and a dendrite into a synapse is accompanied by the accumulation of the synaptic machinery at the site, with these delivered in TGN-derived carriers. In cultured hippocampal neurons, PGCs are linked via spectrin to clusters of the neural cell adhesion molecule (NCAM) in the PM. These complexes are trapped at sites of initial neurite-to-neurite contact within several minutes of the formation of the initial contact. The accumulation of PGCs at contacts with NCAM-deficient neurons is reduced when compared with wild-type cells, which suggests that NCAM mediates the anchoring of intracellular organelles in nascent synapses (Sytnyk et al., 2002).

5. Concluding remarks

The extensive characterization of PGC morphology by the combination of live-cell imaging and EM has provided significant advances in our understanding of the mechanisms that operate during PGC morphogenesis and the other steps of the PGC lifecycle. It appears that the type and size of a cargo can strongly impact on the architecture of a PGC, its path to the target compartment, and way of its fusion with the acceptor membrane. This is achieved through the interplay of the cargo molecules with the components of the sorting and transport machineries at the TGN and in the more distal compartments of the post-Golgi transport routes. Therefore, the integrity of the mechanisms involved in cargo sorting into PGCs, and its further delivery to the correct target destination plays a fundamental role in the maintenance of cell homeostasis, as well as in the organization of specific tissue architecture and function.

6. Acknowledgments

The authors would like to thank C.P. Berrie for critical reading of the manuscript, and A. Luini and A. De Matteis for helpful discussions. We acknowledge financial support from the AIRC Italy (grant IG10233) and Telethon Italy (grants TGM11CB4 and GTF08001).

7. References

Ang, A.L., Taguchi, T., Francis, S., Folsch, H., Murrells, L.J., Pypaert, M., Warren, G., and Mellman, I. (2004). Recycling endosomes can serve as intermediates during transport from the Golgi to the plasma membrane of MDCK cells. J Cell Biol *167*, 531-543.

Antonny, B. (2006). Membrane deformation by protein coats. Curr Opin Cell Biol *18*, 386-394.

Bard, F., and Malhotra, V. (2006). The formation of TGN-to-plasma-membrane transport carriers. Annu Rev Cell Dev Biol *22*, 439-455.

Barr, F.A., Puype, M., Vandekerckhove, J., and Warren, G. (1997). GRASP65, a protein involved in the stacking of Golgi cisternae. Cell *91*, 253-262.

Brown, W.J., Chambers, K., and Doody, A. (2003). Phospholipase A2 (PLA2) enzymes in membrane trafficking: mediators of membrane shape and function. Traffic *4*, 214-221.

Canty, E.G., Lu, Y., Meadows, R.S., Shaw, M.K., Holmes, D.F., and Kadler, K.E. (2004). Coalignment of plasma membrane channels and protrusions (fibripositors) specifies the parallelism of tendon. J Cell Biol *165*, 553-563.

Canty, E.G., Starborg, T., Lu, Y., Humphries, S.M., Holmes, D.F., Meadows, R.S., Huffman, A., O'Toole, E.T., and Kadler, K.E. (2006). Actin filaments are required for fibripositor-mediated collagen fibril alignment in tendon. The Journal of biological chemistry *281*, 38592-38598.

Clermont, Y., Rambourg, A., and Hermo, L. (1995). Trans-Golgi network (TGN) of different cell types: three-dimensional structural characteristics and variability. Anat Rec *242*, 289-301.

Cohen, D., Brennwald, P.J., Rodriguez-Boulan, E., and Musch, A. (2004). Mammalian PAR-1 determines epithelial lumen polarity by organizing the microtubule cytoskeleton. The Journal of cell biology *164*, 717-727.

Corda, D., Colanzi, A., and Luini, A. (2006). The multiple activities of CtBP/BARS proteins: the Golgi view. Trends Cell Biol *16*, 167-173.

Cullen, P.J. (2008). Endosomal sorting and signalling: an emerging role for sorting nexins. Nat Rev Mol Cell Biol *9*, 574-582.

De Matteis, M.A., and Luini, A. (2008). Exiting the Golgi complex. Nat Rev Mol Cell Biol *9*, 273-284.

De Matteis, M.A., Di Campli, A., and D'Angelo, G. (2007). Lipid-transfer proteins in membrane trafficking at the Golgi complex. Biochim Biophys Acta *1771*, 761-768.

Halbleib, J.M., Saaf, A.M., Brown, P.O., and Nelson, W.J. (2007). Transcriptional modulation of genes encoding structural characteristics of differentiating enterocytes during development of a polarized epithelium in vitro. Molecular biology of the cell *18*, 4261-4278.

Hirschberg, K., Miller, C.M., Ellenberg, J., Presley, J.F., Siggia, E.D., Phair, R.D., and Lippincott-Schwartz, J. (1998). Kinetic analysis of secretory protein traffic and characterization of golgi to plasma membrane transport intermediates in living cells. J Cell Biol *143*, 1485-1503.

Jacob, R., and Naim, H.Y. (2001). Apical membrane proteins are transported in distinct vesicular carriers. Current biology : CB *11*, 1444-1450.

Kamal, A., Stokin, G.B., Yang, Z., Xia, C.H., and Goldstein, L.S. (2000). Axonal transport of amyloid precursor protein is mediated by direct binding to the kinesin light chain subunit of kinesin-I. Neuron *28*, 449-459.

Keller, P., Toomre, D., Diaz, E., White, J., and Simons, K. (2001). Multicolour imaging of post-Golgi sorting and trafficking in live cells. Nature cell biology *3*, 140-149.

Klumperman, J., Kuliawat, R., Griffith, J.M., Geuze, H.J., and Arvan, P. (1998). Mannose 6-phosphate receptors are sorted from immature secretory granules via adaptor protein AP-1, clathrin, and syntaxin 6-positive vesicles. The Journal of cell biology *141*, 359-371.

Kreitzer, G., Marmorstein, A., Okamoto, P., Vallee, R., and Rodriguez-Boulan, E. (2000). Kinesin and dynamin are required for post-Golgi transport of a plasma-membrane protein. Nat Cell Biol *2*, 125-127.

Kreitzer, G., Schmoranzer, J., Low, S.H., Li, X., Gan, Y., Weimbs, T., Simon, S.M., and Rodriguez-Boulan, E. (2003). Three-dimensional analysis of post-Golgi carrier exocytosis in epithelial cells. Nat Cell Biol *5*, 126-136.

Ladinsky, M.S., Mastronarde, D.N., McIntosh, J.R., Howell, K.E., and Staehelin, L.A. (1999). Golgi structure in three dimensions: functional insights from the normal rat kidney cell. The Journal of cell biology *144*, 1135-1149.

Lee, O.K., Frese, K.K., James, J.S., Chadda, D., Chen, Z.H., Javier, R.T., and Cho, K.O. (2003). Discs-Large and Strabismus are functionally linked to plasma membrane formation. Nat Cell Biol *5*, 987-993.

Lenoir, M., Coskun, U., Grzybek, M., Cao, X., Buschhorn, S.B., James, J., Simons, K., and Overduin, M. (2010). Structural basis of wedging the Golgi membrane by FAPP pleckstrin homology domains. EMBO Rep *11*, 279-284.

Lippincott-Schwartz, J., Roberts, T.H., and Hirschberg, K. (2000). Secretory protein trafficking and organelle dynamics in living cells. Annu Rev Cell Dev Biol *16*, 557-589.

Lock, J.G., and Stow, J.L. (2005). Rab11 in recycling endosomes regulates the sorting and basolateral transport of E-cadherin. Mol Biol Cell *16*, 1744-1755.

McMahon, H.T., and Gallop, J.L. (2005). Membrane curvature and mechanisms of dynamic cell membrane remodelling. Nature *438*, 590-596.

McNiven, M.A., Cao, H., Pitts, K.R., and Yoon, Y. (2000). The dynamin family of mechanoenzymes: pinching in new places. Trends Biochem Sci *25*, 115-120.

Mellman, I., and Warren, G. (2000). The road taken: past and future foundations of membrane traffic. Cell *100*, 99-112.

Miserey-Lenkei, S., Chalancon, G., Bardin, S., Formstecher, E., Goud, B., and Echard, A. (2010). Rab and actomyosin-dependent fission of transport vesicles at the Golgi complex. Nature cell biology *12*, 645-654.

Mogelsvang, S., Gomez-Ospina, N., Soderholm, J., Glick, B.S., and Staehelin, L.A. (2003). Tomographic evidence for continuous turnover of Golgi cisternae in Pichia pastoris. Molecular biology of the cell *14*, 2277-2291.

Mostov, K., Su, T., and ter Beest, M. (2003). Polarized epithelial membrane traffic: conservation and plasticity. Nat Cell Biol *5*, 287-293.

Munro, S. (1995). An investigation of the role of transmembrane domains in Golgi protein retention. The EMBO journal *14*, 4695-4704.

Nakagawa, T., Setou, M., Seog, D., Ogasawara, K., Dohmae, N., Takio, K., and Hirokawa, N. (2000). A novel motor, KIF13A, transports mannose-6-phosphate receptor to plasma membrane through direct interaction with AP-1 complex. Cell *103*, 569-581.

Nakata, T., and Hirokawa, N. (2003). Microtubules provide directional cues for polarized axonal transport through interaction with kinesin motor head. The Journal of cell biology *162*, 1045-1055.

Nakata, T., Terada, S., and Hirokawa, N. (1998). Visualization of the dynamics of synaptic vesicle and plasma membrane proteins in living axons. The Journal of cell biology *140*, 659-674.

Nelson, W.J. (2003). Adaptation of core mechanisms to generate cell polarity. Nature *422*, 766-774.

Nilsson, T., Rabouille, C., Hui, N., Watson, R., and Warren, G. (1996). The role of the membrane-spanning domain and stalk region of N-acetylglucosaminyltransferase I in retention, kin recognition and structural maintenance of the Golgi apparatus in HeLa cells. J Cell Sci *109 (Pt 7)*, 1975-1989.

Noda, Y., Okada, Y., Saito, N., Setou, M., Xu, Y., Zhang, Z., and Hirokawa, N. (2001). KIFC3, a microtubule minus end-directed motor for the apical transport of annexin XIIIb-associated Triton-insoluble membranes. The Journal of cell biology *155*, 77-88.

Paladino, S., Pocard, T., Catino, M.A., and Zurzolo, C. (2006). GPI-anchored proteins are directly targeted to the apical surface in fully polarized MDCK cells. The Journal of cell biology *172*, 1023-1034.

Pavelka, M., Ellinger, A., Debbage, P., Loewe, C., Vetterlein, M., and Roth, J. (1998). Endocytic routes to the Golgi apparatus. Histochemistry and cell biology *109*, 555-570.

Polishchuk, E.V., Di Pentima, A., Luini, A., and Polishchuk, R.S. (2003). Mechanism of constitutive export from the golgi: bulk flow via the formation, protrusion, and en bloc cleavage of large trans-golgi network tubular domains. Mol Biol Cell *14*, 4470-4485.

Polishchuk, R., Di Pentima, A., and Lippincott-Schwartz, J. (2004). Delivery of raft-associated, GPI-anchored proteins to the apical surface of polarized MDCK cells by a transcytotic pathway. Nat Cell Biol *6*, 297-307.

Polishchuk, R.S., and Mironov, A.A. (2004). Structural aspects of Golgi function. Cell Mol Life Sci *61*, 146-158.

Polishchuk, R.S., Polishchuk, E.V., Marra, P., Alberti, S., Buccione, R., Luini, A., and Mironov, A.A. (2000). Correlative light-electron microscopy reveals the tubular-saccular ultrastructure of carriers operating between Golgi apparatus and plasma membrane. J Cell Biol *148*, 45-58.

Polishchuk, R.S., San Pietro, E., Di Pentima, A., Tete, S., and Bonifacino, J.S. (2006). Ultrastructure of long-range transport carriers moving from the trans Golgi network to peripheral endosomes. Traffic *7*, 1092-1103.

Puertollano, R., Martinez-Menarguez, J.A., Batista, A., Ballesta, J., and Alonso, M.A. (2001). An intact dilysine-like motif in the carboxyl terminus of MAL is required for normal apical transport of the influenza virus hemagglutinin cargo protein in epithelial Madin-Darby canine kidney cells. Molecular biology of the cell *12*, 1869-1883.

Puertollano, R., van der Wel, N.N., Greene, L.E., Eisenberg, E., Peters, P.J., and Bonifacino, J.S. (2003). Morphology and dynamics of clathrin/GGA1-coated carriers budding from the trans-Golgi network. Mol Biol Cell *14*, 1545-1557.

Rambourg, A., and Clermont, Y. (1990). Three-dimensional electron microscopy: structure of the Golgi apparatus. Eur J Cell Biol *51*, 189-200.

Rindler, M.J., Ivanov, I.E., and Sabatini, D.D. (1987). Microtubule-acting drugs lead to the nonpolarized delivery of the influenza hemagglutinin to the cell surface of polarized Madin-Darby canine kidney cells. The Journal of cell biology *104*, 231-241.

Rodriguez-Boulan, E., Kreitzer, G., and Musch, A. (2005). Organization of vesicular trafficking in epithelia. Nat Rev Mol Cell Biol *6*, 233-247.

Roux, A., Cappello, G., Cartaud, J., Prost, J., Goud, B., and Bassereau, P. (2002). A minimal system allowing tubulation with molecular motors pulling on giant liposomes. Proc Natl Acad Sci U S A *99*, 5394-5399.

Roux, A., Cuvelier, D., Nassoy, P., Prost, J., Bassereau, P., and Goud, B. (2005). Role of curvature and phase transition in lipid sorting and fission of membrane tubules. Embo J *24*, 1537-1545.

Roux, A., Uyhazi, K., Frost, A., and De Camilli, P. (2006). GTP-dependent twisting of dynamin implicates constriction and tension in membrane fission. Nature *441*, 528-531.

Sahlender, D.A., Roberts, R.C., Arden, S.D., Spudich, G., Taylor, M.J., Luzio, J.P., Kendrick-Jones, J., and Buss, F. (2005). Optineurin links myosin VI to the Golgi complex and is involved in Golgi organization and exocytosis. J Cell Biol *169*, 285-295.

Schmoranzer, J., and Simon, S.M. (2003). Role of microtubules in fusion of post-Golgi vesicles to the plasma membrane. Mol Biol Cell *14*, 1558-1569.

Schmoranzer, J., Goulian, M., Axelrod, D., and Simon, S.M. (2000). Imaging constitutive exocytosis with total internal reflection fluorescence microscopy. J Cell Biol *149*, 23-32.

Schmoranzer, J., Kreitzer, G., and Simon, S.M. (2003). Migrating fibroblasts perform polarized, microtubule-dependent exocytosis towards the leading edge. J Cell Sci *116*, 4513-4519.

Shemesh, T., Luini, A., Malhotra, V., Burger, K.N., and Kozlov, M.M. (2003). Prefission constriction of Golgi tubular carriers driven by local lipid metabolism: a theoretical model. Biophys J *85*, 3813-3827.

Shorter, J., Watson, R., Giannakou, M.E., Clarke, M., Warren, G., and Barr, F.A. (1999). GRASP55, a second mammalian GRASP protein involved in the stacking of Golgi cisternae in a cell-free system. The EMBO journal *18*, 4949-4960.

Simons, K., and Gerl, M.J. (2010). Revitalizing membrane rafts: new tools and insights. Nature reviews Molecular cell biology *11*, 688-699.

Spiliotis, E.T., Hunt, S.J., Hu, Q., Kinoshita, M., and Nelson, W.J. (2008). Epithelial polarity requires septin coupling of vesicle transport to polyglutamylated microtubules. The Journal of cell biology *180*, 295-303.

Sytnyk, V., Leshchyns'ka, I., Delling, M., Dityateva, G., Dityatev, A., and Schachner, M. (2002). Neural cell adhesion molecule promotes accumulation of TGN organelles at sites of neuron-to-neuron contacts. J Cell Biol *159*, 649-661.

Tai, A.W., Chuang, J.Z., Bode, C., Wolfrum, U., and Sung, C.H. (1999). Rhodopsin's carboxy-terminal cytoplasmic tail acts as a membrane receptor for cytoplasmic dynein by binding to the dynein light chain Tctex-1. Cell *97*, 877-887.

Teng, J., Rai, T., Tanaka, Y., Takei, Y., Nakata, T., Hirasawa, M., Kulkarni, A.B., and Hirokawa, N. (2005). The KIF3 motor transports N-cadherin and organizes the developing neuroepithelium. Nature cell biology *7*, 474-482.

Toomre, D., Steyer, J.A., Keller, P., Almers, W., and Simons, K. (2000). Fusion of constitutive membrane traffic with the cell surface observed by evanescent wave microscopy. J Cell Biol *149*, 33-40.

Vega, I.E., and Hsu, S.C. (2001). The exocyst complex associates with microtubules to mediate vesicle targeting and neurite outgrowth. J Neurosci *21*, 3839-3848.

Weisz, O.A., and Rodriguez-Boulan, E. (2009). Apical trafficking in epithelial cells: signals, clusters and motors. Journal of cell science *122*, 4253-4266.

Yeaman, C., Grindstaff, K.K., and Nelson, W.J. (2004). Mechanism of recruiting Sec6/8 (exocyst) complex to the apical junctional complex during polarization of epithelial cells. Journal of cell science *117*, 559-570.

Yeh, T.Y., Peretti, D., Chuang, J.Z., Rodriguez-Boulan, E., and Sung, C.H. (2006). Regulatory dissociation of Tctex-1 light chain from dynein complex is essential for the apical delivery of rhodopsin. Traffic *7*, 1495-1502.

Zenner, H.L., Collinson, L.M., Michaux, G., and Cutler, D.F. (2007). High-pressure freezing provides insights into Weibel-Palade body biogenesis. J Cell Sci *120*, 2117-2125.

Zhang, F., Yim, Y.I., Scarselletta, S., Norton, M., Eisenberg, E., and Greene, L.E. (2007). Clathrin adaptor GGA1 polymerizes clathrin into tubules. J Biol Chem *282*, 13282-13289.

Zurzolo, C., Le Bivic, A., Quaroni, A., Nitsch, L., and Rodriguez-Boulan, E. (1992). Modulation of transcytotic and direct targeting pathways in a polarized thyroid cell line. The EMBO journal *11*, 2337-2344.

3

Phosphatidylinositol Bisphosphate Mediated Sorting of Secretory Granule Cargo

Douglas S. Darling, Srirangapatnam G. Venkatesh,
Dipti Goyal and Anne L. Carenbauer
University of Louisville, Louisville, Kentucky
USA

1. Introduction

Every cell must sort and transport proteins. This is true for soluble proteins as well as proteins that are in membranes, each of which need to be directed to appropriate subcellular or extracellular destinations in order to perform their essential functions. In eukaryotes, selective trafficking contributes to maintaining the different compositions of different membranes such as apical and basolateral plasma membranes, as well as directing appropriate proteins to lysosomes, endosomes, multivesicular bodies, or other intracellular compartments. The normal physiology of the cell is critically dependent on selective trafficking of proteins and membranes between different transport pathways within the cell. Other chapters in this book focus on the mechanics of transporting cargo membranes, including the molecular aspects of vesicle fusion to specific target membranes. This chapter will focus on the importance and mechanisms of sorting luminal cargo into different pathways, i.e., the "selective" aspect of selective trafficking, particularly with respect to exocrine secretion.

Selective trafficking of new proteins is largely achieved by budding of vesicles from the trans-Golgi network (TGN) for transport to specific organelles or to specific regions of the plasmalemma. Different terminology is used for these vesicles depending on their size, histological appearance, contents, or cell type. Granules (including dense-core secretory granules, DCSG) are secretory vesicles present in endocrine, exocrine, immune, and neuroendocrine cells, responsible for both storage and secretion of proteins. Lymphocytes, dendritic cells, and natural killer cells also contain secretory lysosomes for the release of lysosomal enzymes (Stanley and Lacy 2010), and neurons contain peptidergic synaptic vesicles (Park and Loh 2008; Park et al. 2011). However, all of these types of vesicles serve the same broad purpose of transporting specific cargo to specific destinations by an appropriately regulated pathway.

The lipid membranes of these vesicles carry tightly associated cytoplasmic proteins (termed coat proteins) which not only help form the vesicle, but also direct the vesicle to the correct destination (De Matteis and Luini 2008; Santiago-Tirado and Bretscher 2011; Wilson et al. 2011). The matrix of coat proteins on the cytosolic face of the membrane contributes to the bending of the TGN membrane during budding of the vesicle. This matrix is formed by multiple interactions, including binding of coat proteins to phosphatidylinositol phosphates

(PtdInsPs) in the TGN membrane, interactions between the coat proteins, and binding to integral proteins of the TGN membrane. Importantly, populations of vesicles are distinguished by the presence of specific combinations of coat proteins, such as clathrin, Adaptor Proteins (AP1-4), FAPP1/2, GGAs, ARF, v-snares, and synaptotagmin. In the parotid gland, VAMP2, VAMP8, syntaxin4/6, and synaptotagmin decorate the cytoplasmic side of secretory granules (Fujita-Yoshigaki et al. 2006; Wang et al. 2007). These different coat proteins on different vesicles direct the vesicles to the correct target membranes. For example, FAPP2 is critical for constitutive apical trafficking, whereas FAPP1 directs basolateral trafficking (Vieira et al. 2005). The coat proteins also mediate interactions with other proteins, including actins, to mediate transportation of that vesicle.

Having vesicles destined for different targets raises the central question of how does the correct cargo get put into just the correct type of vesicle? These post-TGN vesicles carry integral membrane proteins, which are one type of cargo delivered by this process. Transmembrane cargo proteins (such as MPR300) are localized by direct interactions with coat proteins (such as GGAs) on the outside of the forming vesicle as it buds from the TGN (Ghosh et al. 2003). Sorting sequences which mediate these interactions have been identified on the cytosolic tails of many transmembrane cargo proteins (Folsch et al. 2009). Hence, the problem of sorting membrane cargo proteins to the correct vesicle has an elegant solution based on direct interactions of transmembrane cargo with the coat protein complex which identifies that vesicle and targets it to the correct destination (recently reviewed in (De Matteis and Luini 2011)).

Importantly, the lumen of the vesicle contains a different type of cargo composed of specific soluble proteins. Luminal cargo proteins include lysosomal enzymes, hormones, cytokines, neurotransmitters, digestive enzymes, and salivary proteins. As can be seen from this list, soluble cargo proteins are present in a variety of different secretory cell types. These luminal cargo proteins cannot interact directly with the coat proteins on the cytoplasmic (outer) surface of the vesicle membrane; therefore, other mechanisms must be involved to localize the correct soluble cargo protein into the vesicle destined for the correct target, and not into the incorrect vesicles. This is an information transfer problem, i.e., how to get the information encoded by the cytoplasmic coat proteins (which determine the destination of the vesicle) to select the appropriate luminal cargo proteins.

Information transfer for sorting is a typical problem in any distribution network and must be solved by companies involved with distribution, such as UPS or FedEx. The need for solutions to such problems is reflected by the growing number of Logistics and Distribution programs at universities. Notably, eukaryotic cells developed solutions to these logistics problems many eons ago. For an exocrine or endocrine cell, the problem is how to get the lysosomal enzymes (soluble cargo proteins) into a forming TGN vesicle destined for the lysosome, and secretory cargo proteins into a different TGN vesicle destined for the plasmalemma. This requires the transfer of information from the cytosolic side of the forming vesicle membrane to the luminal side of the membrane.

While there are some good model systems, we do not have a clear understanding of the molecular mechanisms that direct the sorting of soluble cargo proteins between different vesicles. Nonetheless, this is an important issue since all eukaryotic cells produce several different types of vesicles at the TGN (Folsch et al. 2009), and many cell types secrete proteins by specific pathways such as apical versus basolateral pathways. Changes in trafficking not only affect the physiology of the cell, but also embryonic development (Shilo and Schejter 2011) and disease. This chapter will focus on sorting (selective trafficking) for secretion of luminal cargo proteins. We will review some of the general aspects of selective trafficking, and then build on that background by focusing on our recent work suggesting a novel mechanism for sorting in the parotid salivary gland.

2. Biogenesis during trans-Golgi network vesicle trafficking

Secreted proteins are translated at the rough endoplasmic reticulum and transit from the ER, through the ER-Golgi Intermediate Compartment (ERGIC) to the Golgi and subsequently the trans-Golgi network (TGN) (Shilo and Schejter 2011). Post-translational modifications such as glycosylation occur in the ER and Golgi. Membranes on the trans side of the Golgi apparatus form dynamic tubular reticular structures having a large surface area (De Matteis and Luini 2008). This network of saccules and tubes is continuously remodeled such that both the structure and size of the TGN varies depending on the secretory activity and the cell type (Trucco et al. 2004). Selective trafficking at the TGN will sort cargo into vesicles or carrier tubules (De Matteis and Luini 2008), and this sorting requires the genesis of carrier vesicles targeted to specific membranes within the cell. As noted above, these vesicles are distinguished by the combination of coat proteins on the cytosolic face of the membrane which determine the target membrane for that vesicle. In some pathways additional sorting occurs at the recycling endosome (reviewed in (De Matteis and Luini 2008; Santiago-Tirado and Bretscher 2011)).

The initiating event in vesicle biogenesis may be driven by the local membrane lipid composition where asymmetry in the types of lipid in the two faces of the membrane bilayer can induce bending (van Meer and Sprong 2004). Initiation may also involve membrane rafts, which are reported to be present in the TGN of all cells (Park and Loh 2008), and on membranes of secretory granules (Hosaka et al. 2004; Lang 2007; Guerriero et al. 2008). While membrane rafts are well characterized to play important roles in endocytosis at the plasmalemma, their role in vesicle formation at the TGN is not as well understood. Many vesicle coat proteins have been localized to lipid rafts on vesicles (Puri and Roche 2006). This has been suggested to be important for the formation of the coat protein complexes on the TGN for the initial creation of vesicle buds and selective trafficking (Simons and Sampaio 2011). For example, SNARE proteins are enriched in cholesterol-dependent rafts in beta-cells and PC-12 cells (Lang 2007). However, technical issues have called into question the validity of some methods for isolation of 'lipid rafts', leading to a more stringent definition and the term 'membrane rafts' (Lang 2007). Nonetheless, even with the more stringent approach, SNAREs such as VAMP2 and VAMP3 are enriched in membrane rafts of vesicles. Membrane rafts can contain different complements of proteins due to specific protein interactions. One model for the formation of vesicle buds is that membrane rafts on the TGN (with associated transmembrane cargo proteins) coalesce creating a lipid domain that is favorable to bending the membrane, and containing transmembrane proteins which can facilitate the decoration of the cytosolic face of the membrane with adaptor and other coat proteins (De Matteis and Luini 2008; Simons and Sampaio 2011). Testing the relevance of this model to selective sorting for secretion in living cells is important, and requires determining whether trafficking vesicles in different pathways (e.g., regulated secretion versus constitutive secretion pathways) contain different types of membrane rafts, or that some pathways lack membrane rafts. Importantly, Guerriero et al. recently found that raft-independent and raft-associated proteins collect in distinct sites at the Golgi, and likely enter different vesicles (Guerriero et al. 2008).

2.1 Phosphatidylinositol phosphates in biogenesis of secretory vesicles

In all cell types, the earliest events that are strongly linked to the biogenesis of secretory vesicles is the binding of coat proteins to phosphatidylinositol phosphate lipids (PtdInsP) and PI-kinases in the TGN membrane. The phosphoinositides and small GTPases of the Arf and Rab families define the identity of the membrane and recruit additional coat proteins (Di Paolo and De Camilli 2006; De Matteis and Luini 2008; Santiago-Tirado and Bretscher 2011). PtdInsPs are recognized as being critical for selective trafficking of vesicles within cells (Di Paolo and De Camilli 2006; D'Angelo et al. 2008; Vicinanza et al. 2008; Graham and Burd 2011). Phosphatidylinositol comprises less than 10% of membrane phospholipids, and the individual phosphorylated forms total less than 1.5% of lipids (Di Paolo and De Camilli 2006; Roth 2004). Phosphatidylinositol can be phosphorylated on any combination of carbons 3, 4, or 5 of the inositol ring (Fig. 1). The most abundant, PtdIns4P, occurs at approximately 0.05% of membrane lipids, whereas the low abundance forms such as PtdIns(3,4)P_2 or PtdIns(3,4,5)P_3 are approximately 0.0001% each (Cullen 2011). Subcellular pools of the 7 different PtdInsPs have diverse regulatory roles in cytoskeleton remodeling, second messenger signaling, endosomal trafficking, membrane trafficking, osmotic stress, nuclear signaling, and other aspects of cell physiology (Godi et al. 2004; Balla and Balla 2006; Di Paolo and De Camilli 2006). Cellular effects can be mediated by signaling through production of second messengers (diacylglycerol, and inositol trisphosphate). However, many cellular effects are mediated by the localized anchoring of cytosolic proteins having specific PtdInsP-binding domains (e.g., PH, FYVE, PX, ENTH-domains) (De Matteis et al. 2005; Balla and Balla 2006). For example, Arf1 directly binds and recruits PI4-Kinases to the TGN. This produces PtdIns4P which aids in recruitment of coat proteins such as AP1, GGAs, VAMPs, and FAPP1/2 most of which interact with PtdIns4P as well as other proteins in the coat matrix (Balla and Balla 2006). This interaction provides identity to the membrane (reviewed in (Santiago-Tirado and Bretscher 2011)). PI3K-C2α is also present on the TGN. Each type of PtdInsP is localized to specific membranes within the cell. The most abundant phosphoinositides, PtdIns4P, and PtdIns(4,5)P_2, are predominantly localized to the Golgi and plasma membrane, respectively, whereas PtdIns3P and PtdIns(3,5)P_2 are predominantly localized to early and late endosomes (Di Paolo and De Camilli 2006; Santiago-Tirado and Bretscher 2011). PtdIns(3,4)P_2 is rare in resting cells, but is present in the plasmalemma and multivesicular and early endosomes (Roth 2004; Di Paolo and De Camilli 2006). PtdIns(3,4)P_2 is not noted as being present in the Golgi or TGN (De Matteis et al. 2005), however, this

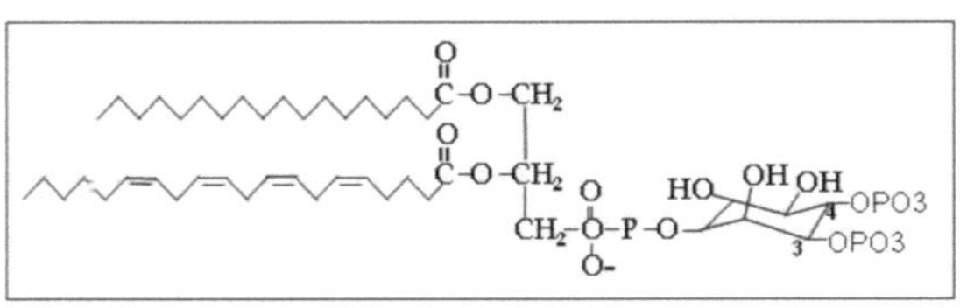

Fig. 1. Structure of Phosphatidylinositol (3,4)bisphosphate. PtdIns(3,4)P_2 is an exceedingly rare membrane lipid that is highly localized to specific subcellular membranes within the cell.

has not been well characterized in any cell type. Recently, PtdIns(3,4)P_2 was found to be transiently synthesized at the plasmalemma as a second messenger of platelet-derived growth factor (PDGF) (Hogan et al. 2004). Importantly, our understanding of the roles of PtdInsPs derives primarily from studies of yeast and a few mammalian cell lines. It is unclear whether these generalizations carry over to well differentiated cells in tissues. For example, we find high levels of expression of PtdIns(3,4)P_2 in parotid secretory granule membranes.

Discrete and dynamic localization of PtdIns-kinases (PI-kinases) and PI-phosphatases regulate the production of PtdInsPs. Of these enzyme families, PI4-kinases have the dominant roles in Golgi function and secretion, although PI3-kinases and PI5-kinases also have roles in secretion (Wang et al. 2003; Roth 2004; Balla and Balla 2006). Several PI4-kinases are localized to the Golgi, TGN, or endosomes, and form complexes with several coat proteins including GGA and FAPP1/2 (D'Angelo et al. 2008). Careful immunofluorescence co-localization studies with MDCK cells have found PI4KIIα in the TGN and PI4KIIIβ in the cis/medial Golgi (Weixel et al. 2005). This is consistent with PtdIns4P being the most abundant PtdInsP in the Golgi. PI4KIIIβ is recruited by the coat protein Arf1 to the Golgi, and in turn PI4KIIIβ recruits Rab11 and the PtdIns4P-binding protein FAPP1 (Godi et al. 2004). Knock-down of PI4KIIα by RNAi has little effect on intra-Golgi transit, but inhibits TGN export of vesicles (Wang et al. 2003). These studies are interpreted as showing that PI4-kinases produce PtdIns4P on the cytosolic leaflet of the TGN and vesicles, which anchors essential adaptor proteins (Santiago-Tirado and Bretscher 2011). Importantly, class I PI3-kinases are also present on Golgi membranes, and are essential for tumor necrosis factor (TNF) secretion by macrophages (Low et al. 2010). Similarly, PI3-kinase was localized to secretory granules of PC12 cells by both immunofluorescence and cell fractionation (Meunier et al. 2005). Transfection of a PtdIns3P-binding domain (FYVE), or a catalytically-inactive PI3-kinase, blocked regulated secretion, possibly by interfering with fusion of the granule at the plasmalemma (Meunier et al. 2005). In summary, certain PI-kinases decorate the Golgi and TGN, interact with coat proteins, and produce PtdInsPs. These PtdInsPs are important for vesicle trafficking, however, the identities and roles of PI-kinases have not been investigated in cells that are highly specialized for bulk exocrine secretion, such as the parotid.

3. Selective trafficking for secretion

The presence of multiple different trafficking pathways for secretion has been described in many cell types (Dikeakos et al. 2007; De Matteis and Luini 2008; Park et al. 2008; Folsch et al. 2009; Perez et al. 2010; Stanley and Lacy 2010; Lacy and Stow 2011; Santiago-Tirado and Bretscher 2011). The best characterized of these pathways have a specific cargo protein that is a unique marker of that route, which is essential for molecular characterization (Lara-Lemus et al. 2006). For example, the cytolytic protein perforin of natural killer cells undergoes polarized secretion into the immune synapse, whereas the same cells secrete TNF in a non-polarized pattern (Reefman et al. 2010) thereby marking a different pathway. Also, some pituitary gonadotropes segregate luteinizing hormone into separate granules than follicle-stimulating hormone for different regulated secretion dynamics (Nicol et al. 2004). Alternatively, pathways can be marked by the use of different coat proteins. For example, the coat protein FAPP1 is essential for secretion by the basolateral pathway, whereas FAPP2 directs vesicles to the apical pathway (Godi et al. 2004; Vieira et al. 2005). Importantly, mutation of proteins involved with sorting between trafficking pathways can cause disease. This is also seen with non-genetic diseases, such as pancreatitis during which inappropriate basolateral (endocrine) secretion of cargo proteins occurs. Hence, pancreatic amylase or lipase in the serum are standard clinical markers of this disease. Given the presence of multiple different pathways for secretion, the key issue is to understand the molecular interactions of proteins destined for each pathway, which cause sorting to the correct immature granule or tubule as it forms, or which cause retention of the protein in the granule as it matures.

The best characterized model for specific sorting of soluble cargo proteins involves a transmembrane sorting receptor protein (Fig. 2). The receptor protein is present in the TGN membrane and is able to interact with coat proteins on the cytosolic side of the vesicle bud, and can also bind luminal cargo proteins. The transmembrane receptor is localized to the vesicle bud by the appropriate coat proteins, and in turn, selects the correct cargo. Hence, a single protein serves to coordinate the identity of the vesicle with the luminal contents. This model is exemplified by mannose-6-phosphate sorting receptors (MPRs), which are type I transmembrane receptors present in the TGN (Ghosh et al. 2003). Both the cation-dependent (MPR300) and the cation-independent (MPR46) MPRs deliver lysosomal enzymes from the TGN to endosomes for subsequent transfer to lysosomes. The cytosolic tails of MPRs have specific binding sites for multiple adaptor proteins, including AP1, AP2, GGAs and PACs1 (reviewed in (Ghosh et al. 2003)). In addition, the portion of MPR in the lumen of the vesicle binds to mannose 6-phosphate tags on cargo proteins. The mannose 6-phosphate is a specific posttranslational modification on the N-glycans of over 60 of acid hydrolases which need to be transported to lysosomes. Failure of this sorting causes lysosomal sorting disease, Mucolipidosis type II alpha/beta (I-cell disease) (Ghosh et al. 2003). A similar sorting mechanism has been described for stabilin-1 which binds GGAs and serves as a sorting receptor for a chitinase-like enzyme.

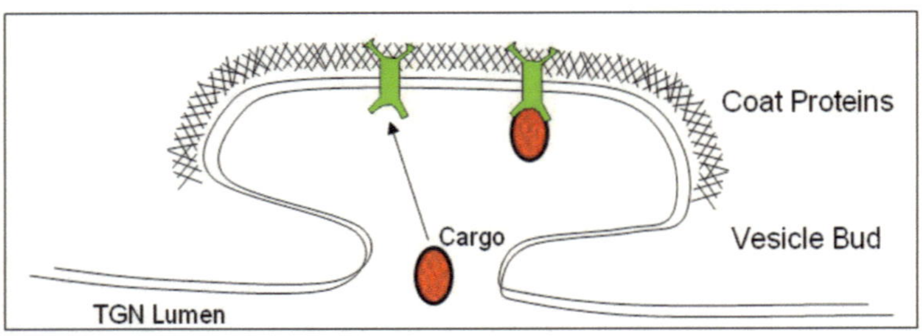

Fig. 2. Transmembrane sorting receptor model. The coat proteins localize a transmembrane sorting receptor, which collects the appropriate luminal cargo.

Various mechanisms have been suggested for sorting of secreted proteins into the regulated secretory pathway, as opposed to the constitutive secretory pathway or trafficking to intracellular targets (Dikeakos and Reudelhuber 2007; Park et al. 2008). Some proteins are secreted by the model discussed above. Phogrin (Ptprn2) is a type I transmembrane receptor present in the TGN and secretory granules of endocrine and neuroendocrine cells. Phogrin contains tyrosine and leucine motifs in the C-terminal (cytosolic) tail, which are important for localization to secretory granules, likely by interaction with coat proteins such as AP1

(Saito et al. 2011). Within the lumen, phogrin can bind to carboxypeptidase E (CPE), and contributes to sorting of the complex (Saito et al. 2011). Hence, information about the identity of the vesicle encoded by the coat proteins, is related through phogrin to determine the luminal cargo. Phogrin also has PI-phosphatase activity which may be important for regulating glucose-stimulated insulin secretion (Caromile et al. 2010).

Other sorting receptors have been reported in neural and endocrine cells. CPE itself binds the granule membrane. The C-terminus of CPE can span the membrane, although only 5 amino acid residues are cytosolic (Dhanvantari et al. 2002). Recycling of CPE from the plasmalemma requires ARF6 apparently due to direct binding to the coat protein. CPE acts as a sorting receptor for proopiomelanocortin (POMC) and proBDNF trafficking into the regulated secretion pathway (reviewed in (Park and Loh 2008)). Secretogranin III (SgIII) can also serve as a sorting receptor. Despite the absence of a transmembrane domain, SgIII binds to granule membranes or to cholesterol-rich liposomes, and anchors chromograninA (CgA) to the membrane (Hosaka et al. 2004). SgIII also interacts with CPE, POMC, and adrenomedullin (Hosaka et al. 2005; Han et al. 2008). It is unclear how SgIII is targeted to specific granules, but this may occur due to selective interactions with membrane rafts, or due to interaction with CPE and, indirectly, phogrin.

Dikeakos et al. (Dikeakos et al. 2007; Dikeakos and Reudelhuber 2007) have shown that a hydrophobic patch in short amphipathic alpha helices is sufficient to sort cargo proteins. Helical domains are implicated in sorting of somatostatin, CPE, prohormone convertase enzymes (PC1/3, PC2), and chromogranin A (CgA). The proposed mechanism for this sorting is that the hydrophobic patch of the helix embeds into the membrane of the forming granule (Dikeakos and Reudelhuber 2007). One of the sorting-competent helical domains (Hels13-5) bound liposomes composed of phosphatidylcholine (PC) and phosphatidylglycerol (PG) with a Kd=9.7 μM (Kitamura et al. 1999). This Hels13-5 peptide integrated into the non-polar layer to a variable extent depending on the pH of the liposome. No cholesterol was necessary for this interaction. In addition, the helical domain of PC1/3 was shown to directly interact with CHAPS detergent micelles (Dikeakos et al. 2009). Characterization of both natural and artificial helices which confer sorting of a cargo protein at the TGN is an important step forward, however, at this point it is unclear how the identity of the vesicle (determined by the cytosolic coat proteins) directs selective sorting of such cargo into the correct pathway.

For many years, aggregation of secreted proteins has been seen as critical to sorting into the regulated secretory pathway in neuroendocrine cells (Gorr et al. 2005). Dense-core secretory granules of the regulated pathway contain large aggregates comprising chromogranins secretogranins and other secreted proteins, and which are not present in the constitutive secretion pathway. Aggregation of granins and many hormones can be demonstrated *in vitro* in a Ca^{++} and pH-dependent fashion. The pH of the TGN is approximately 6.2. In AtT20 cells the pH decreases further as the granule matures, reaching pH 5.5 in mature secretory granules (Wu et al. 2001). The acidic pH and high Ca^{++} present in the regulated secretory pathway is essential for aggregation. This relatively non-specific interaction may allow trafficking of large aggregates of proteins even where only a few specific interactions with transmembrane sorting receptors are present. Nonetheless, it must be recognized that aggregation is a fairly ill-defined concept, and it will be necessary to characterize these interactions in order to determine how some cargo is excluded from the aggregate to be sorted into the constitutive pathway.

4. Sorting for regulated secretion in the parotid gland

The parotid salivary gland provides an excellent model for the study of regulated secretion of proteins. This gland has evolved to secrete copious amounts of specific proteins into the saliva. It secretes salivary proteins including amylase, Parotid Secretory Protein (PSP), a family of acidic (aPRP) and basic (bPRP) Proline-Rich Proteins, and less abundant proteins such as histatin and statherin (Helmerhorst and Oppenheim 2007). Secreted PSP has anti-bacterial activity which contributes to protection of the oral cavity (Gorr et al. 2011). Human PSP (SPLUNC2, BPIFA2) has been shown to be expressed in saliva as several isoforms due to alternative splicing of the mRNA (Bingle et al. 2009; Bingle et al. 2011). Another abundant salivary protein, amylase, initiates digestion of starch, and also adheres to oral bacteria and enamel. PRPs contribute to secretion of other cargo proteins (Venkatesh and Gorr 2002; Venkatesh et al. 2007), are part of the acquired dental pellicle, and also bind bacteria. These three proteins are the most abundant luminal cargo proteins within the secretory granule. In addition, hundreds of other proteins are secreted into saliva by the parotid gland, and have been cataloged by proteomic approaches (Denny et al. 2008). As these proteins move through the trans-Golgi network, they are each presumably sorted into the correct pathway for secretion. We previously reviewed the pathways of sorting and secretion in the parotid gland (Gorr et al. 2005). Secretion in the parotid, as with other exocrine cells, includes the major regulated pathway, a minor regulated pathway, apical and basolateral constitutive secretory pathways, and a constitutive-like secretory pathway (Perez et al. 2010). Similarly, endocrine cells have both regulated secretory and constitutive secretory pathways, in addition to pathways within the cell (Kim et al. 2006; Park and Loh 2008). Of the major salivary proteins, PSP is an excellent marker for the regulated secretory pathway. Western blot analysis of rat serum demonstrates that some fraction of salivary amylase is normally present in serum, however, PSP was undetectable (Venkatesh et al. 2007). This indicates that under normal conditions PSP is tightly sorted into the apical regulated secretory pathway, whereas a portion of amylase enters a basolateral pathway *in vivo*.

Salivary glands are being studied for their potential to produce and secrete therapeutic proteins from transgenes introduced to patients (Perez et al. 2010). Towards this goal, it is important to understand the molecular mechanisms that control parotid sorting and secretion, in order to regulate whether the transgenic protein is secreted by an apical regulated (exocrine) or basolateral (endocrine) pathway (Perez et al. 2010). Progress has been made in defining molecular interactions that affect sorting in some cell types; however, many of these mechanisms do not appear to be present in parotid acinar cells. As described above, the pH of the secretory granules of neuroendocrine and endocrine cells decreases during maturation from about 6.2 at the TGN to 5.5 - 5.0 in the mature granule (Wu et al. 2001; Kim et al. 2006). This acidic environment is important for sorting in PC12 and AtT20 cells, and is essential for protein aggregation (reviewed in (Kim et al. 2006)). However, in the parotid gland, the pH of the acinar cell granule increases from about pH 6.2 at the TGN to 6.8 or higher after maturation (Arvan and Castle 1986). Furthermore, granule cargo proteins from parotid acinar cells (amylase and PSP) are unable to aggregate even in the presence of Ca^{++} and low pH, whereas pancreatic exocrine granule proteins (used as a control) aggregate in a fashion similar to endocrine cells (Venkatesh et al. 2004). This indicates that sorting of amylase and PSP in the parotid gland have at least some important differences from the mechanisms described for neuro/endocrine cells. Therefore, we have investigated the molecular interactions of PSP in the secretory granule in an attempt to understand how sorting may be controlled (Venkatesh et al. 2011).

4.1 Parotid secretory protein (PSP) binds to secretory granule membranes

We analyzed rat parotid granule membranes by mass spectrometry with the goal of identifying integral membrane proteins that may be candidate sorting receptors in the parotid gland. Parotid gland homogenates were fractionated on sucrose gradients to isolate the secretory granules, which were osmotically lysed. Membranes were washed and further enriched by an additional sucrose gradient. Sucrose gradient-purified granule membranes were electrophoresed on polyacrylamide gels, and trypsinized peptides identified by MS/MS as described (Uriarte et al. 2008). Numerous integral or membrane-bound proteins were identified, including several involved in vesicle trafficking and cytoskeletal proteins, as expected. However, potential sorting receptors such as SgIII (Han et al. 2008) or carboxypeptidase E (Dhanvantari et al. 2002) were not identified in parotid membranes by this method. Nonetheless, one salivary cargo protein, PSP, was identified. Other abundant soluble cargo proteins such as amylase and acidic Proline-Rich Protein (PRP) were not detected. To confirm the binding of PSP to granule membranes, western blot analysis was done with independent preparations of purified and extensively washed granule membranes. This confirmed that PSP is selectively bound, whereas amylase and PRP are absent from purified membranes (Fig. 3) (Venkatesh et al. 2011). While this approach failed to identify candidate sorting receptor proteins, it did demonstrate that PSP is a good marker for interactions with the membrane. In other cell types, putative sorting proteins such as secretogranin III (Hosaka et al. 2004), carboxypeptidase E (Dhanvantari et al. 2002), PC1/3, and PC2 (Jutras et al. 2000) are also associated with isolated secretory granule membranes.

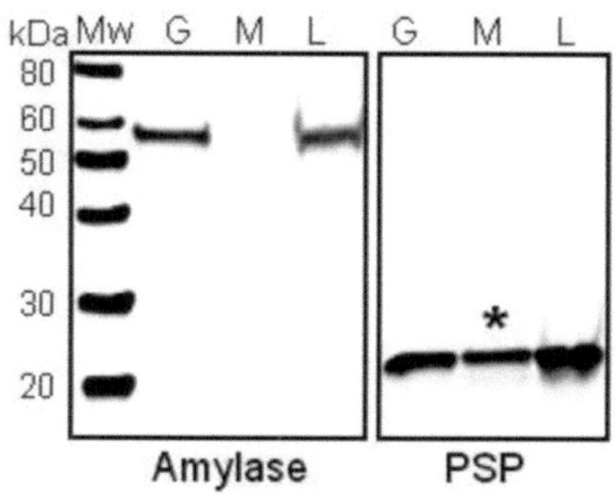

Fig. 3. Western blots of purified secretory granule membranes, probed with either anti-amylase or anti-PSP. Lanes contain either intact granules (G), purified granule membranes (M), or soluble cargo protein lysate (L). Equal proportions (0.5%) of each fraction was analyzed. The star indicates PSP in the purified membrane fraction. Mw: molecular size standards.

Given the existing models for sorting receptors, we tested whether PSP was bound to a sorting receptor protein in the secretory granule membrane. Numerous experiments were done attempting to crosslink PSP to a membrane protein; however, PSP never crosslinked into a specific membrane-dependent higher molecular weight band. Taking the opposite approach, we extensively digested parotid granule membranes with either trypsin or pronase to destroy all membrane-associated proteins, and subsequently found that exogenous PSP still binds quite effectively. These results indicated that PSP does not require a protein receptor for binding to the membrane. In contrast, exogenous amylase did not bind the trypsinized membranes, and was present in the unbound fraction only, emphasizing the specificity of the binding of PSP.

4.2 PSP binds specifically to phosphatidylinositol 3,4-bisphosphate

The binding of a cargo protein to the vesicle membrane is of great interest in defining the mechanisms of sorting. However, the ability of PSP to bind granule membranes in the absence of any sorting receptor protein ruled out the most common model for sorting. Several secreted proteins have been shown to interact with lipid microdomains (e.g., CPE, SgIII, PC1/3), presumably by relatively non-specific hydrophobic interactions (Park and Loh 2008; Dikeakos et al. 2007), whereas other classes of protein bind to a highly specific lipid headgroup (Di Paolo and De Camilli 2006). Therefore, we tested the ability of PSP to bind specific lipids, and to bind liposomes. Parotid secretory granules were isolated, and the lysate supernatant containing soluble cargo proteins was used in lipid-overlay assays to determine whether PSP or other salivary proteins (amylase or PRP) bind specific lipids (Dowler et al. 2000). Importantly, none of the cargo proteins bound directly to the most abundant membrane lipids (phosphatidylcholine, phosphatidylethanolamine, cholesterol, or sphingomyelin). Similarly, acidic PRP never bound any lipid spots, and amylase showed little or no binding. In contrast, PSP bound with remarkable selectivity to phosphatidylinositol phosphates (PtdInsPs), but did not bind to unphosphorylated PtdIns (Fig. 4). We observed decreased PSP binding to PtdInsPs at more acidic conditions, but clear binding was still present at pH 6.0. Hence, this interaction could contribute to sorting of PSP in the TGN where the pH is approximately 6.2 (Arvan and Castle 1986), and may also contribute to retention as the pH increases during granule maturation.

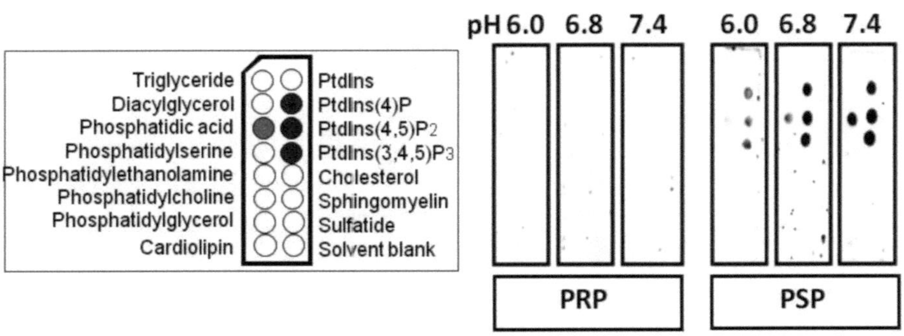

Fig. 4. PSP binds to phosphatidylinositol phosphates. Lipid strips (Echelon Biosciences) were incubated with parotid granule soluble lysate at 2 µg/ml (Venkatesh et al. 2011). Bound protein was detected with antibodies to PSP or acidic PRP. A schematic of the lipid strips is shown on the left (filled circles represent PSP binding).

The inability of PSP to bind unphosphorylated PtdIns suggested that specific interactions with the headgroup were required. Therefore, we compared the binding of PSP to a dilution series of each of the seven different phosphorylated forms of PtdInsPs. We found that native PSP binds 3- to 5-fold more to PtdIns(3,4)P_2 compared to PtdIns(4,5)P_2 or PtdIns(3,4,5)P_3, and 10-fold greater than PtdIns(3,5)P_2 or PtdIns(4)P (Fig. 5). PSP does not bind PtdIns(3)P, PtdIns(5)P, or PtdIns. Half-maximal binding of PSP was with approximately 35 pmoles PtdIns(3,4)P_2. Parallel blots failed to detect any bound amylase or acidic PRP, both of which are abundant in the granule lysates. This high degree of specificity indicates that PSP binds the head group of PtdInsPs, analogous to known PtdInsP-binding proteins (Di Paolo and De Camilli 2006). For example, PSP is more selective than the PH-domain protein DAPP1 which binds PtdIns(3,4)P_2 or PtdIns(3,4,5)P_3 with similar avidity (Dowler et al. 2000).

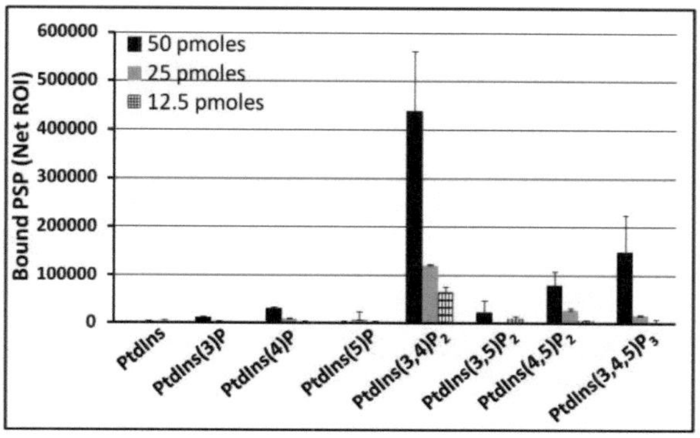

Fig. 5. Binding of PSP to phosphatidylinositol phosphates. PtdInsP array membranes spotted with serially diluted lipids were used to define the binding of native PSP. Membranes were blocked and probed with anti-PSP as described (Venkatesh et al. 2011).

The experiments described above use a secretory granule lysate as the source of PSP. This leaves open the possibility that PSP binds indirectly to PtdInsP_2. Therefore, PSP was expressed *in vitro* in rabbit reticulocyte lysates, and also was expressed in bacteria. Chloramphenicol acetyltransferase (CAT) was used as a negative control since CAT and rPSP are similar in size and also have similar acidic pI values. Rat PSP, and human PSP (human Splunc2, BPIFA2), each with a V5 tag, were translated *in vitro* and bound selectively to PtdIns(3,4)P_2 and PtdIns(4,5)P_2 demonstrating that this activity is conserved between species. CAT-V5 was unable to bind any of the lipids. Similarly, both human and rat PSP proteins were expressed in bacteria as glutathione-S-transferase (GST) fusion proteins having V5 tags. Again, GST-rPSP-V5 and GST-hPSP-V5 each bound strongly to PtdIns(3,4)P_2 and did not bind PtdIns, whereas GST-V5 had no binding activity (Venkatesh et al. 2011). Bacterially expressed and GST-affinity purified rPSP-V5 also preferentially binds PtdIns(3,4)P_2 (Fig. 6). The binding of *in vitro* synthesized, and bacterially expressed, rPSP and human PSP to PtdIns(3,4)P_2 demonstrates that the interaction is independent of other parotid granule proteins.

Pleckstrin homology domain proteins bind phosphoinositides with a moderate to high affinity (Vicinanza et al. 2008). We used bacterially expressed rPSP-V5 to determine the affinity of PSP for PtdIns(3,4)P_2. Bacterially expressed affinity-purified rPSP-V5 was incubated at 0.1 to 3.5 μg/ml with membranes spotted with 50 pmoles of PtdIns(3,4)P_2. Bound protein was detected with anti-V5 antibody, and the intensity of the signal used to calculate the amount of free and bound protein from a PSP-V5 standard curve, as described

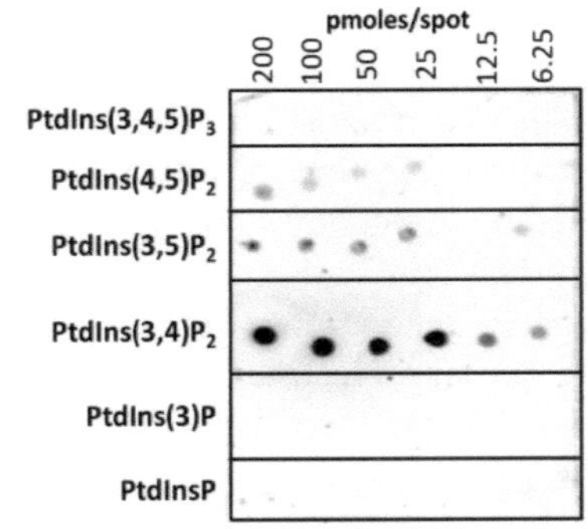

Fig. 6. Bacterially expressed rPSP-V5 binds to PtdIns(3,4)P_2. GST-PSP-V5 was affinity purified, and the rPSP-V5 was isolated separate from GST using PreScission protease. Protein overlay assays were performed using the purified rPSP-V5 (1 µg/ml) on nitrocelluose membranes spotted with lipids (200-6.25 pmoles/spot). Bound protein was detected with anti-V5 antibody. Bacterially expressed rPSP shows a similar pattern of specificity as native PSP.

(Venkatesh et al. 2011). The affinity was derived from the binding curve (Fig. 7). In three independent experiments, the binding affinity of PSP ranged from K_d= 1.85 x 10^{-10} to 3.9 x 10^{-11} M demonstrating a high affinity interaction. This is similar to the affinity of 5 x 10^{-9} M for TAPP1 binding PtdIns(3,4)P_2 measured by a similar method (Dowler et al. 2000). The affinity of p47phox for PtdIns(3,4)P_2 is reported as 3.8 x 10^{-8} (Karathanassis et al. 2002). In a direct comparison of bacterially expressed p47phox-V5 and rPSP-V5 we confirmed that PSP binds PtdIns(3,4)P_2 more strongly than p47phox.

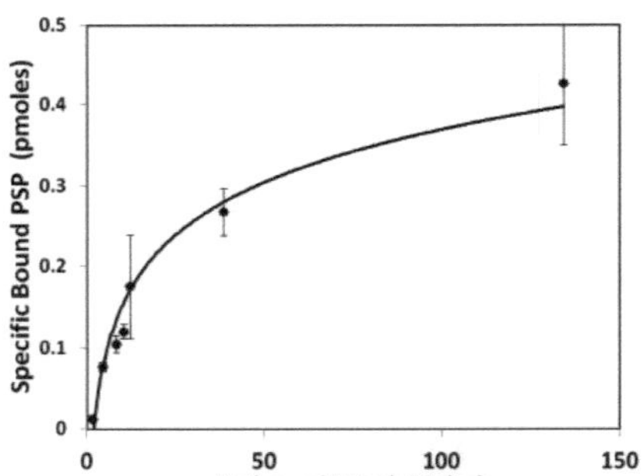

Fig. 7. Binding curve of PSP to PtdIns(3,4)P_2. Bacterially expressed affinity-purified rPSP-V5 was incubated at different concentrations to define binding to PtdIns(3,4)P_2 as described (Venkatesh et al. 2011).

We studied the interaction of PSP with lipid bilayers in liposomes for three reasons. Primarily, we wanted to determine if PSP binds specifically to PtdInsPs in intact membranes. In addition, it was important to test whether PSP can bind non-specifically to membranes due to hydrophobic interactions similar to helical peptides (Kitamura et al. 1999; Dikeakos et al. 2007) and to compare these two types of interaction. Synthetic liposomes were made by standard methods to contain phosphatidylcholine (PC), phosphatidylethanolamine (PE), and PtdIns (or PtdInsP) at a molar ratio of 77:20:3. We did not detect any interaction of PSP with liposomes comprising PC, PE, and unphosphorylated PtdIns (77:20:3), nor did it bind to liposomes with PtdIns(3,5)P_2 (Fig. 8). This suggests that PSP does not interact with membranes through the relatively non-specific hydrophobic interactions observed for amphipathic alpha helices, and shown to be important for sorting of certain cargo proteins (Kitamura et al. 1999; Dikeakos et al. 2007). Conversely, native rat PSP, in granule lysates, bound repeatably to lysosomes spiked with 3% PtdIns(3,4)P_2 or PtdIns(4,5)P_2 (Fig. 8), consistent with the previous results. Amylase present in the same granule lysates did not bind to any of these liposomes. PSP also bound to stabilized PIPosomes (from Echelon Biosciences Inc.) containing 5% PtdIns(3,4)P_2. These results show that native PSP binds PtdIns(3,4)P_2 in an intact lipid membrane. This binding is not detectably due to hydrophobic interactions of an alpha helical domain, does not require a different protein to act as a membrane tether or a transmembrane sorting receptor, nor does it require cholesterol-rich domains.

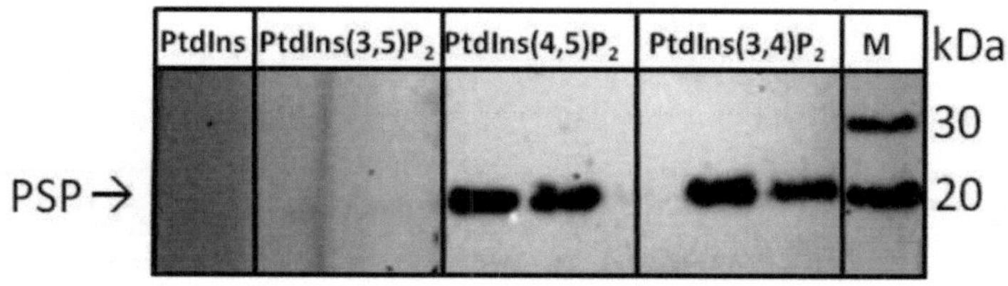

Fig. 8. PSP binds to PtdIns(3,4)P_2 and PtdIns(4,5)P_2 in lipid bilayers. Parotid granule extract was incubated with liposomes spiked with 3% either PtdIns, PtdIns(3,4)P_2, PtdIns(4,5)P_2 or PtdIns(3,5)P_2. After binding, liposomes were extensively washed and analyzed by SDS-PAGE and western blotting with anti-PSP. Each lane is a separate incubation. PSP bound to PtdIns(3,4)P_2 and PtdIns(4,5)P_2 only. M: molecular size markers.

4.3 The PSP family of BPI-fold proteins

The results described above suggest that PSP binds directly to the headgroup of PtdIns(3,4)P_2 in a membrane. It is of interest to compare this activity to related proteins. Following the newly developed nomenclature (Bingle et al. 2011), rat PSP/BPIFA2E is a member of the BPI-fold superfamily. This diverse superfamily includes bactericidal/permeability-increasing protein (BPI), LPS-binding protein (LBP), cholesteryl ester transfer protein (CETP), and phospholipid transfer protein (PLTP). In addition, the superfamily includes the BPIFA subfamily (previously referred to as the SPLUNC family, containing PSP), and the BPIFB subfamily (previously termed LPLUNCs). The history of cloning the PSP/Splunc/BPIFA family, and changes in the nomenclature have recently been described in depth (Bingle et al. 2011). An important observation is that proteins in this superfamily have quite divergent amino acid sequences. For example, the optimal possible alignment of rat PSP to rat BPI gives a sequence identity of only 19%. However, there is strong conservation of secondary structure and predicted tertiary structures across the superfamily. The crystal structure of BPI shows a hollow boomerang-shaped structure (Beamer et al. 1997). The two halves (domains) of the boomerang show clear similarity to each other at both the primary sequence level, and at the structural level. The Long PLUNC (BPIFB) subfamily maps across both the two domains, whereas PSP and the rest of the Short PLUNC (BPIFA) subfamily consists of only one domain. Many of these proteins have important roles that involve binding to lipids. For example, both BPI and LBP bind to lipopolysaccaride (LPS) by the lipid A region containing multiple acyl chains. The crystal structure of BPI shows the presence of two bound phosphatidylcholine molecules, each with the acyl chains deeply embedded in the hollow protein tube (Beamer et al. 1997). A similar model is found for CETP, and PLTP (Huuskonen et al. 1999; Qiu et al. 2007). This is distinctly different from PSP binding to lipids observed in our results. Using two different assays, PSP does not bind to phosphatidylcholine. Further, PSP requires critical interactions with the headgroup of the lipid, which are not apparent in BPI binding to phosphatidylcholine. Based on these differences, it does not appear that PSP binding to PtdIns(3,4)P_2 uses the same mechanism as the well-characterized binding of BPI to lipids. As an initial hypothesis, we predict that separate binding sites on PSP will be identified for binding PtdIns(3,4)P_2 and lipopolysaccaride. PSP could bind PtdIns(3,4)P_2 in membranes during trafficking, and subsequently use a different interaction to bind LPS in the saliva.

4.4 Phosphoinositides in parotid granule membranes

The observed binding of PSP to PtdIns(3,4)P_2 could support sorting of PSP; and in addition PSP could act as a membrane tether (or chaperone) to aid in sorting of other cargo. However, PtdIns(3,4)P_2 binding can mediate selective trafficking into granules only if the presence of PtdInsP_2 is somehow linked to the identity of the forming granule. The idea that a specific type of rare lipid may direct the sorting of cargo proteins is an entirely novel suggestion. However, it is just an extrapolation of the role of PtdInsPs on the other side of the membrane. In the following sections we address two key questions. Is PtdIns(3,4)P_2 present on parotid secretory granule membranes? Can PtdIns(3,4)P_2 cross to the luminal side of the granule membrane?

PtdInsPs have highly specific intracellular distributions, anchoring critical proteins to specific membranes (Graham and Burd 2011). PtdIns(3,4)P_2 could reasonably be present in the TGN since both PI3-kinase and PI4-kinase are bound at the TGN, and both PI-kinase activities are required for regulated secretion (Meunier et al. 2005; Low et al. 2010). However, PtdIns(3,4)P_2 has tended to be neglected in studies of PtdInsP distribution, so little information is available. Nonetheless, immunofluorescence of a transfected PtdIns(3,4)P_2-binding protein (TAPP1) showed clear localization to a Golgi-like structure adjacent to the nucleus (Hogan et al. 2004). We isolated parotid gland secretory granules, and methanolic extracts of purified granule membranes were spotted on PVDF and probed with antibodies to specific phosphoinositides. As expected, we detected PtdIns(4)P which is reported to be on vesicles and TGN of several cell types. In addition, strong immunoreactivity was observed for PtdIns(3,4)P_2 (Venkatesh et al. 2011). Standard curves of PtdIns(4)P and PtdIns(3,4)P_2 were used to calculate the amounts of each lipid. PtdIns(4)P is abundant in the TGN, however, we find that PtdIns(3,4)P_2 is present at a slightly higher amount than PtdIns(4)P in parotid granule membranes.

As a separate approach, we used immunofluorescence to localize PtdIns(3,4)P_2 within the parotid acinar cell. Anti-PSP labels the secretory granules, which collect near the center of the acinus, in the apical end of each cell, but did not label the basal ends. Anti-PtdIns(3,4)P_2 also labeled the apical end of parotid acinar cells, giving a similar pattern. Superimposing the images shows that PSP and PtdIns(3,4)P_2 co-localize to the secretory granules (Fig. 9; yellow and orange). Therefore, using either biochemical or histological methods, we consistently find that PtdIns(3,4)P_2 is present in parotid granule membranes.

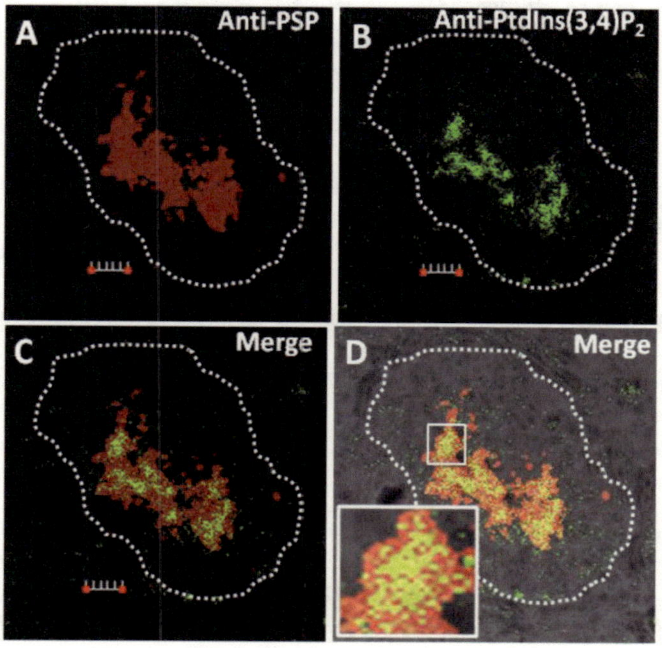

Fig. 9. PtdIns(3,4)P_2 co-localizes to secretory granules of parotid acinar cells. Rat parotid tissue sections were probed with anti-rPSP (red) and monoclonal anti-PtdIns(3,4)P_2 (green). A merge of the two images shows co-localization (C). The dotted line marks the boundary of the acinus. The boxed area of the Nomarski image (D) is enlarged in the inset. The scale bar is 5 μm.

4.5 Lipid translocases in parotid granule membranes

The observation that PtdIns(3,4)P_2 is present in parotid granule membranes supports our model that PSP binds the membrane by interacting with this lipid. However, the PI-kinases that produce PtdInsPs are located on the cytosolic leaflet of the TGN and granule membranes, yet PSP is present only inside the secretory granule.

The coat protein complex on budding vesicles of the TGN includes translocases which flip phospholipids to maintain lipid asymmetry, and can contribute to bending of membranes (van Meer and Sprong 2004; Natarajan et al. 2009; Contreras et al. 2010). Translocases (flippases) are reported on the TGN and post-Golgi vesicles, and are linked to vesicle budding (Muthusamy et al. 2009). Translocases on yeast TGN are involved in a clathrin-dependent pathway, vesicle bud formation, and membrane trafficking (Graham 2004; Daleke 2007; Natarajan et al. 2009). Translocase activity is also present on pig gastric parietal cell secretory vesicles (Suzuki et al. 1997), and adrenal chromaffin granules (Zachowski et al. 1989). Three types of lipid translocases have been described at the TGN or on post-Golgi vesicles. P-type ATPase translocases are present on the TGN and secretory granules (Suzuki et al. 1997), and are important for secretion (Muthusamy et al. 2009). Alternatively, the ATP-binding cassette (ABC) superfamily of transporters includes lipid translocases which can be found on the TGN, lysosomes, and secretory vesicles of lung type II cells (Stahlman et al. 2007). Similarly, the phospholipid scramblase family mediates bidirectional flipping of phospholipid across the membrane, and PLSCR1 is present on neutrophil secretory vesicle membranes (Frasch et al. 2004). Un-phosphorylated PtdIns can be translocated by a flippase, however, none of the translocase families have been tested for the ability to flip PtdInsPs.

Our data suggest that PtdIns(3,4)P_2 may be present on the luminal face of the parotid granule membrane as a binding site for PSP, however, it is unclear how it would get there. Therefore, intact rat parotid secretory granules were incubated with fluorescent NBD-tagged PtdInsP to measure flipping, according to (Natarajan and Graham 2006; Natarajan et al. 2009). In this method, added NBD-lipids integrate rapidly into the outer leaflet of the granule membrane on ice. After incubation at 37 °C to allow flipping, label remaining in the outer leaflet is quantitatively destroyed by addition of BSA and dithionite, however, NBD-lipids which had translocated to the luminal leaflet are protected by the membrane. We found that incubation for 1 hour at 37 °C allowed 15% of integrated NBD-PtdIns(3,4)P_2 to translocate to the protected inner leaflet (Fig. 10). Less than 2% of the PSP leaked from the granules after incubation at 37 °C for 1 h, indicating that the granules remained sealed during the assay.

Unphosphorylated PtdIns is translocated by a flippase at a rate similar to phosphatidylcholine (Vishwakarma et al. 2005). This provides us with a benchmark for comparison with the extent of flipping of PtdInsPs. In our experiments, approximately 10% of PtdIns was flipped to the inner leaflet of parotid granules, whereas 15-18% of PtdIns(4)P, PtdIns(3,4)P_2, or PtdIns(3,5)P_2 was translocated (Fig. 10). Hence, the flippase activity is relatively non-selective, and the phosphate groups do not inhibit translocation. Parotid granule membranes support flipping of phosphorylated forms to a greater extent than PtdIns. Taken together, our results demonstrate that PtdIns(3,4)P_2 is present in the granule membrane, and can flip to the inner face of the membrane. Further, the presence of PtdInsPs on the luminal leaflet of membranes raises the possibility that other intra-organelle proteins may be localized by PtdInsP anchors.

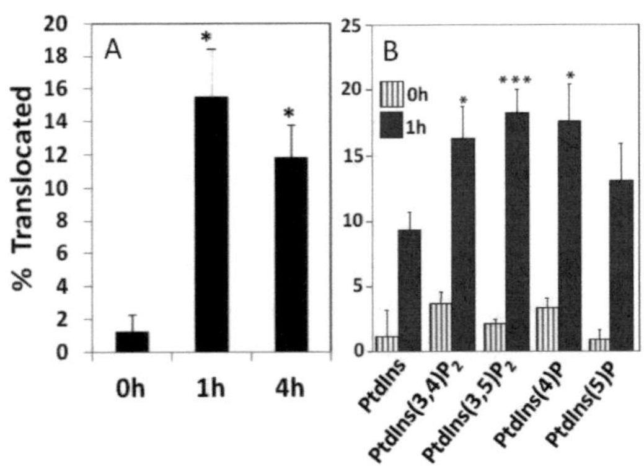

Fig. 10. A. Parotid secretory granules translocate PtdIns(3,4)P_2. Purified intact parotid secretory granules were incubated with fluorescent NBD-labeled PtdIns(3,4)P_2 to measure translocation, as described (Natarajan and Graham 2006; Venkatesh et al. 2011). Data show the amount of lipid flipped to the inner leaflet. Data are Mean±SE of 3 experiments in triplicate. *$p<0.01$ compared with 0h. B. Several PtdInsPs translocate parotid granule membranes. Data are the amount of flipped lipid at 0 or 1 hour. Data are Mean ± SE of 3-6 experiments in triplicate. *$p<0.05$, ***$p<0.01$ compared to PtdIns.

5. Conclusions: Lipids as sorting receptors

As discussed above, a central aspect of selective trafficking is ensuring that the vesicle targeting information encoded by the granule coat matrix directs the choice of which cargo proteins fill the lumen of the granule. Current models for selective sorting all rely on interactions with a transmembrane protein to convey that information, directly or indirectly. Our results with parotid secretory granules suggest the possibility of a variant of this model. Rather than a transmembrane sorting receptor, translocation (flipping) of a rare lipid, PtdIns(3,4)P_2, may convey the character of the coat matrix (Fig. 11).

Both PI3-kinase and PI4-kinase are present on the TGN or vesicles in several cell types, and create specific PtdInsPs in the outer leaflet of the membrane. In addition, lipid translocases, or flippase activity, has been reported on mammalian secretory granules. Our results demonstrate that PtdIns(3,4)P_2 is present in parotid secretory granule membranes. We observe that PtdIns(3,4)P_2 can translocate to the luminal bilayer of the granule membrane. Further, we find that PSP binds strongly to PtdIns(3,4)P_2 in the membrane. Since the translocase likely is recruited and localized by the coat proteins, we suggest that the translocase may create a local region of higher concentration of PtdIns(3,4)P_2 in the luminal leaflet within the forming vesicle bud, compared to other areas of the TGN. This may serve to localize PSP within the budding vesicle, thereby sorting it for secretion. The membrane-bound PSP may in turn act as a sorting chaperone for other cargo proteins. This hypothetical model has the advantage that it suggests specific interactions which can be tested for a role in sorting for secretion.

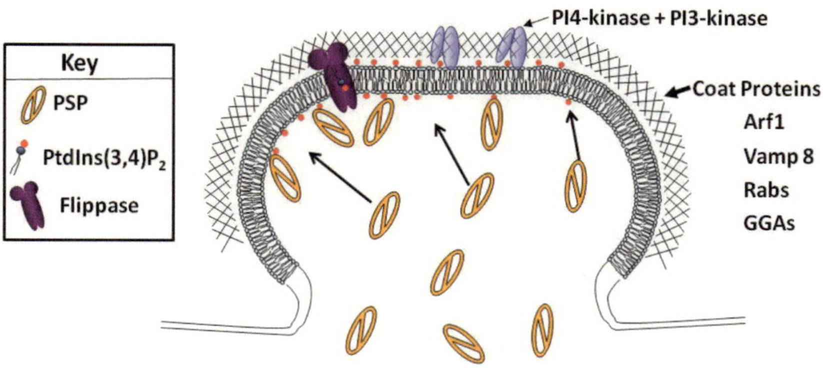

Fig. 11. A hypothetical model for how PtdIns(3,4)P_2 could mediate sorting of PSP.

6. Acknowledgments

Work described in this chapter was supported by NIH NIDCR grants DE012205 and DE019243.

7. References

Arvan, P. and Castle, J. (1986) Isolated secretion granules from parotid glands of chronically stimulated rats possess an alkaline internal pH and inward-directed H+ pump activity. J. Cell Biol. 103(4): 1257-1267.

Balla, A. and Balla, T. (2006) Phosphatidylinositol 4-kinases: old enzymes with emerging functions. Trends Cell Biol 16(7): 351-361.

Beamer, L.J., Carroll, S.F., et al. (1997) Crystal Structure of Human BPI and Two Bound Phospholipids at 2.4 Angstrom Resolution. Science 276(5320): 1861-1864.

Bingle, C.D., Seal, R.L., et al. (2011) Systematic nomenclature for the PLUNC/PSP/BSP30/SMGB proteins as a subfamily of the BPI fold-containing superfamily. Biochem Soc Trans 39(4): 977-983.

Bingle, L., Barnes, F.A., et al. (2009) Characterisation and expression of SPLUNC2, the human orthologue of rodent parotid secretory protein. Histochem Cell Biol 132(3): 339-349.

Caromile, L.A., Oganesian, A., et al. (2010) The Neurosecretory Vesicle Protein Phogrin Functions as a Phosphatidylinositol Phosphatase to Regulate Insulin Secretion. J. Biol. Chem. 285(14): 10487-10496.

Contreras, F., Sanchez-Magraner, L., et al. (2010) Transbilayer (flip-flop) lipid motion and lipid scrambling in membranes. FEBS Lett 584(9): 1779-1786.

Cullen, P.J. (2011) Phosphoinositides and the regulation of tubular-based endosomal sorting. Biochem Soc Trans 39(4): 839-850.

D'Angelo, G., Vicinanza, M., et al. (2008) The multiple roles of PtdIns(4)P - not just the precursor of PtdIns(4,5)P2. J. Cell Sci. 121(12): 1955-1963.

Daleke, D.L. (2007) Phospholipid flippases. J Biol Chem 282(2): 821-825.

De Matteis, M. and Luini, A. (2011) Mendelian disorders of membrane trafficking. N Engl J Med 365(10): 927-938.

De Matteis, M.A., Di Campli, A., et al. (2005) The role of the phosphoinositides at the Golgi complex. Biochim Biophys Acta 1744(3): 396-405.

De Matteis, M.A. and Luini, A. (2008) Exiting the Golgi complex. Nat Rev Mol Cell Biol 9(4): 273-284.

Denny, P., Hagen, F.K., et al. (2008) The proteomes of human parotid and submandibular/sublingual gland salivas collected as the ductal secretions. J Proteome Res 7(5): 1994-2006.

Dhanvantari, S., Arnaoutova, I., et al. (2002) Carboxypeptidase E, a prohormone sorting receptor, is anchored to secretory granules via a C-terminal transmembrane insertion. Biochemistry 41(1): 52-60.

Di Paolo, G. and De Camilli, P. (2006) Phosphoinositides in cell regulation and membrane dynamics. Nature 443(7112): 651-657.

Dikeakos, J.D., Di Lello, P., et al. (2009) Functional and structural characterization of a dense core secretory granule sorting domain from the PC1/3 protease. Proc Natl Acad Sci U S A 106(18): 7408-7413.

Dikeakos, J.D., Lacombe, M.J., et al. (2007) A hydrophobic patch in a charged alpha-helix is sufficient to target proteins to dense core secretory granules. J Biol Chem 282(2): 1136-1143.

Dikeakos, J.D. and Reudelhuber, T.L. (2007) Sending proteins to dense core secretory granules: still a lot to sort out. J Cell Biol 177(2): 191-196.

Dowler, S., Currie , R., et al. (2000) Identification of pleckstrin-homology-domain-containing proteins with novel phosphoinositide-binding specificities. Biochem J 351(Pt 1): 19-31.

Folsch, H., Mattila, P.E., et al. (2009) Taking the scenic route: biosynthetic traffic to the plasma membrane in polarized epithelial cells. Traffic 10(8): 972-981.

Frasch, S.C., Henson, P.M., et al. (2004) Phospholipid flip-flop and phospholipid scramblase 1 (PLSCR1) co-localize to uropod rafts in formylated Met-Leu-Phe-stimulated neutrophils. J Biol Chem 279(17): 17625-17633.

Fujita-Yoshigaki, J., Katsumata, O., et al. (2006) Difference in distribution of membrane proteins between low- and high-density secretory granules in parotid acinar cells. Biochem Biophys Res Commun 344(1): 283-292.

Ghosh, P., Dahms, N.M., et al. (2003) Mannose 6-phosphate receptors: new twists in the tale. Nat Rev Mol Cell Biol 4(3): 202-212.

Godi, A., Di Campli, A., et al. (2004) FAPPs control Golgi-to-cell-surface membrane traffic by binding to ARF and PtdIns(4)P. Nat Cell Biol 6(5): 393-404.

Gorr, S.-U., Venkatesh, S.G., et al. (2005) Parotid Secretory Granules: Crossroads of Secretory Pathways and Protein Storage. J Dent Res 84(6): 500-509.

Gorr, S.U., Abdolhosseini, M., et al. (2011) Dual host-defence functions of SPLUNC2/PSP and synthetic peptides derived from the protein. Biochem Soc Trans 39(4): 1028-1032.

Graham, T.R. (2004) Flippases and vesicle-mediated protein transport. Trends Cell Biol 14(12): 670-677.

Graham, T.R. and Burd, C.G. (2011) Coordination of Golgi functions by phosphatidylinositol 4-kinases. Trends Cell Biol 21(2): 113-121.

Guerriero, C.J., Lai, Y., et al. (2008) Differential sorting and Golgi export requirements for raft-associated and raft-independent apical proteins along the biosynthetic pathway. J Biol Chem 283(26): 18040-18047.

Han, L., Suda, M., et al. (2008) A large form of secretogranin III functions as a sorting receptor for chromogranin A aggregates in PC12 cells. Mol Endocrinol 22(8): 1935-1949.

Helmerhorst, E.J. and Oppenheim, F.G. (2007) Saliva: a Dynamic Proteome. J Dent Res 86(8): 680-693.

Hogan, A., Yakubchyk, Y., et al. (2004) The phosphoinositol 3,4-bisphosphate-binding protein TAPP1 interacts with syntrophins and regulates actin cytoskeletal organization. J Biol Chem 279(51): 53717-53724.

Hosaka, M., Suda, M., et al. (2004) Secretogranin III binds to cholesterol in the secretory granule membrane as an adapter for chromogranin A. J Biol Chem 279(5): 3627-3634.

Hosaka, M., Watanabe, T., et al. (2005) Interaction between secretogranin III and carboxypeptidase E facilitates prohormone sorting within secretory granules. J. Cell Sci. 118(20): 4785-4795.

Huuskonen, J., Wohlfahrt, G., et al. (1999) Structure and phospholipid transfer activity of human PLTP: analysis by molecular modeling and site-directed mutagenesis. J. Lipid Res. 40(6): 1123-1130.

Jutras, I., Seidah, N.G., et al. (2000) A predicted alpha -helix mediates targeting of the proprotein convertase PC1 to the regulated secretory pathway. J Biol Chem 275(51): 40337-40343.

Karathanassis, D., Stahelin, R., et al. (2002) Binding of the PX domain of p47(phox) to phosphatidylinositol 3,4-bisphosphate and phosphatidic acid is masked by an intramolecular interaction. EMBO J 21(19): 5057-5068.
Kim, T., Gondre-Lewis, M.C., et al. (2006) Dense-core secretory granule biogenesis. Physiology (Bethesda) 21: 124-133.
Kitamura, A., Kiyota, T., et al. (1999) Morphological behavior of acidic and neutral liposomes induced by basic amphiphilic alpha-helical peptides with systematically varied hydrophobic-hydrophilic balance. Biophys J 76(3): 1457-1468.
Lacy, P. and Stow, J.L. (2011) Cytokine release from innate immune cells: association with diverse membrane trafficking pathways. Blood 118(1): 9-18.
Lang, T. (2007) SNARE proteins and 'membrane rafts'. J Physiol 585(Pt 3): 693-698.
Lara-Lemus, R., Liu, M., et al. (2006) Lumenal protein sorting to the constitutive secretory pathway of a regulated secretory cell. J. Cell Sci. 119(9): 1833-1842.
Low, P.C., Misaki, R., et al. (2010) Phosphoinositide 3-kinase delta regulates membrane fission of Golgi carriers for selective cytokine secretion. J Cell Biol 190(6): 1053-1065.
Meunier, F.A., Osborne, S.L., et al. (2005) Phosphatidylinositol 3-kinase C2alpha is essential for ATP-dependent priming of neurosecretory granule exocytosis. Mol Biol Cell 16(10): 4841-4851.
Muthusamy, B.P., Natarajan, P., et al. (2009) Linking phospholipid flippases to vesicle-mediated protein transport. Biochim Biophys Acta 1791(7): 612-619.
Natarajan, P. and Graham, T. (2006) Measuring translocation of fluorescent lipid derivatives across yeast Golgi membranes. Methods 39(2): 163-168.
Natarajan, P., Liu, K., et al. (2009) Regulation of a Golgi flippase by phosphoinositides and an ArfGEF. Nat Cell Biol 11(12): 1421-1426.
Nicol, L., McNeilly, JR, et al. (2004) Differential secretion of gonadotrophins: investigation of the role of secretogranin II and chromogranin A in the release of LH and FSH in LbetaT2 cells. J. Mol. Endocrinol. 32(2): 467-480.
Park, J.J., Cawley, N.X., et al. (2008) Carboxypeptidase E cytoplasmic tail-driven vesicle transport is key for activity-dependent secretion of peptide hormones. Mol Endocrinol 22(4): 989-1005.
Park, J.J., Gondre-Lewis, M.C., et al. (2011) A distinct trans-Golgi network subcompartment for sorting of synaptic and granule proteins in neurons and neuroendocrine cells. J. Cell Sci. 124(5): 735-744.
Park, J.J. and Loh, Y.P. (2008) How peptide hormone vesicles are transported to the secretion site for exocytosis. Mol Endocrinol 22(12): 2583-2595.
Perez, P., Rowzee, A.M., et al. (2010) Salivary epithelial cells: an unassuming target site for gene therapeutics. Int J Biochem Cell Biol 42(6): 773-777.
Puri, N. and Roche, P.A. (2006) Ternary SNARE complexes are enriched in lipid rafts during mast cell exocytosis. Traffic 7(11): 1482-1494.
Qiu, X., Mistry, A., et al. (2007) Crystal structure of cholesteryl ester transfer protein reveals a long tunnel and four bound lipid molecules. Nat Struct Mol Biol 14(2): 106-113.
Reefman, E., Kay, J.G., et al. (2010) Cytokine secretion is distinct from secretion of cytotoxic granules in NK cells. J Immunol 184(9): 4852-4862.
Roth, M.G. (2004) Phosphoinositides in Constitutive Membrane Traffic. Physiol Rev 84(3): 699-730.

Saito, N., Takeuchi, T., et al. (2011) Luminal interaction of phogrin with carboxypeptidase E for effective targeting to secretory granules. Traffic 12(4): 499-506.

Santiago-Tirado, F.H. and Bretscher, A. (2011) Membrane-trafficking sorting hubs: cooperation between PI4P and small GTPases at the trans-Golgi network. Trends Cell Biol 21(9): 515-525.

Shilo, B.Z. and Schejter, E.D. (2011) Regulation of developmental intercellular signalling by intracellular trafficking. EMBO J 30(17): 3516-3526.

Simons, K. and Sampaio, J.L. (2011) Membrane organization and lipid rafts. Cold Spring Harb Perspect Biol 3(10): a004697.

Stahlman, M.T., Besnard, V., et al. (2007) Expression of ABCA3 in developing lung and other tissues. J Histochem Cytochem 55(1): 71-83.

Stanley, A.C. and Lacy, P. (2010) Pathways for cytokine secretion. Physiology (Bethesda) 25(4): 218-229.

Suzuki, H., Kamakura, M., et al. (1997) The Phospholipid Flippase Activity of Gastric Vesicles. J. Biol. Chem. 272(16): 10429-10434.

Trucco, A., Polishchuk, R., et al. (2004) Secretory traffic triggers the formation of tubular continuities across Golgi sub-compartments. Nat Cell Biol 6(11): 1071-1081.

Uriarte, S.M., Powell, D.W., et al. (2008) Comparison of Proteins Expressed on Secretory Vesicle Membranes and Plasma Membranes of Human Neutrophils. J. Immunol. 180(8): 5575-5581.

van Meer, G. and Sprong, H. (2004) Membrane lipids and vesicular traffic. Curr Opin Cell Biol 16(4): 373-378.

Venkatesh, S.G., Cowley, D.J., et al. (2004) Differential aggregation properties of secretory proteins that are stored in exocrine secretory granules of the pancreas and parotid glands. Am J Physiol Cell Physiol 286(2): C365-371.

Venkatesh, S.G. and Gorr, S.U. (2002) A sulfated proteoglycan is necessary for storage of exocrine secretory proteins in the rat parotid gland. Am J Physiol Cell Physiol 283(2): C438-445.

Venkatesh, S.G., Goyal, D., et al. (2011) Parotid secretory protein binds phosphatidylinositol (3,4) bisphosphate. J Dent Res 90(9): 1085-1090.

Venkatesh, S.G., Tan, J., et al. (2007) Isoproterenol increases sorting of parotid gland cargo proteins to the basolateral pathway. Am J Physiol Cell Physiol 293(2): C558-565.

Vicinanza, M., D'Angelo, G., et al. (2008) Function and dysfunction of the PI system in membrane trafficking. EMBO J 27(19): 2457-2470.

Vieira, O.V., Verkade, P., et al. (2005) FAPP2 is involved in the transport of apical cargo in polarized MDCK cells. J Cell Biol 170(4): 521-526.

Vishwakarma, R.A., Vehring, S., et al. (2005) New fluorescent probes reveal that flippase-mediated flip-flop of phosphatidylinositol across the endoplasmic reticulum membrane does not depend on the stereochemistry of the lipid. Org Biomol Chem 3(7): 1275-1283.

Wang, C.C., Shi, H., et al. (2007) VAMP8/endobrevin as a general vesicular SNARE for regulated exocytosis of the exocrine system. Mol Biol Cell 18(3): 1056-1063.

Wang, Y.J., Wang, J., et al. (2003) Phosphatidylinositol 4 Phosphate Regulates Targeting of Clathrin Adaptor AP-1 Complexes to the Golgi. Cell 114(3): 299-310.

Weixel, K.M., Blumental-Perry, A., et al. (2005) Distinct Golgi populations of phosphatidylinositol 4-phosphate regulated by phosphatidylinositol 4-kinases. J Biol Chem 280(11): 10501-10508.

Wilson, C., Venditti, R., et al. (2011) The Golgi apparatus: an organelle with multiple complex functions. Biochem J 433(1): 1-9.

Wu, M.M., Grabe, M., et al. (2001) Mechanisms of pH Regulation in the Regulated Secretory Pathway. J. Biol. Chem. 276(35): 33027-33035.

Zachowski, A., Henry, J., et al. (1989) Control of transmembrane lipid asymmetry in chromaffin granules by an ATP-dependent protein. Nature 340(6228): 75-76.

4

Molecular Machinery Regulating Exocytosis

T. Shandala, R. Kakavanos-Plew, Y.S. Ng,
C. Bader, A. Sorvina, E.J. Parkinson-Lawrence,
R.D. Brooks, G.N. Borlace, M.J. Prodoehl and D.A. Brooks
*Mechanisms in Cell Biology and Diseases Research Group,
School of Pharmacy and Medical Sciences, Sansom Institute for Health Research,
University of South Australia
Australia*

1. Introduction

Exocytosis is the major intracellular route for the delivery of proteins and lipids to the plasma membrane and the means by which vesicular contents are released into the extracellular space. The anterograde trafficking of vesicles to the plasma membrane is vital for membrane expansion during cell division; cell growth and migration; the delivery of specialised molecules to establish cell polarity; cell-to-cell communication; neurotransmission and the secretion of response factors such as hormones, cytokines and antimicrobial peptides. There are two major trafficking routes in eukaryotic cells, which are referred to as constitutive and regulated (Ory & Gasman, 2011). Constitutive exocytosis involves the steady state delivery of secretory carrier vesicles from the endoplasmic reticulum via the Golgi apparatus to the plasma membrane (Lacy & Stow, 2011). Regulated or granule-mediated exocytosis involves a specific trigger, usually a burst of intracellular calcium following an extrinsic stimulus. This system is utilized for secretion in neuronal cells and other specialist secretory cells, such as neuroendocrine, endocrine and exocrine cells (Burgoyne & Morgan, 2003; Jolly & Sattentau, 2007; Lacy & Stow, 2011). Regulated exocytosis enables a rapid response from a subpopulation of vesicles already primed and competent for fusion (Manjithaya & Subramani, 2011; Nickel & Seedorf, 2008; Nickel, 2010). Regulated exocytosis is also used for polarised traffic of vesicular membrane and cargo to specific spatial landmarks and this is particularly important during times of dramatic change in cell morphology, such as cell division, cell motility, phagocytosis and axonal outgrowth.

Regulated exocytosis involves the shuttling of carrier vesicles between vesicular compartments, as they are transported towards the plasma membrane. Each step in this process requires the fission of a vesicle from a donor compartment. This carrier vesicle is then targeted/trafficked to an acceptor compartment where docking and fusion takes place, and the cargo is either unloaded or further processed (Bonifacino & Glick, 2004). These fission and fusion steps are repeated until the cargo reaches the plasma membrane (Bonifacino & Glick, 2004). This sequential trafficking of secretory vesicles is orchestrated by a complex set of molecular machinery including: small GTPases of the Ral, Rab and Rho subfamilies that regulate the processes of vesicle formation, traffic and fusion; the exocyst complex for vesicle assembly and membrane tethering; and soluble *N*-ethylmaleimide-sensitive factor attachment protein receptor (SNARE) proteins for vesicular fusion. At the target domain of the plasma membrane, cross-talk between the exocyst complex and SNARE proteins culminates in vesicle-to-plasma membrane fusion, and thereby delivery of membrane proteins and luminal cargo. There are many post-translational modifications of the vesicular machinery that facilitate exocytosis. These include the addition of lipid moieties to increase membrane binding affinity, the switching of GTPase activity by nucleotide exchange factors, phosphorylation, and ubiquitination. Phosphorylation is of particular importance as it incorporates the vesicular trafficking machinery into a circuit of cellular signaling cascades. This chapter focuses on the process of exocytosis and the regulatory role that post-translational modification has on the exocytic machinery. Because the small GTPases and the exocyst complex have multiple inter-connected functions during vesicle formation, trafficking and fusion, we have focused discussion here to the final steps of the exocytic process, which occur in close proximity to the plasma membrane.

2. The exocyst complex and vesicle interaction with the plasma membrane

The exocyst is a scaffolding complex that is required for the final steps of regulated, and constitutive exocytosis (Hsu, *et al.*, 2004). The exocyst complex is attached to the cytosolic face of the exocytic vesicular membrane, and tethers the vesicle to specific domains of the plasma membrane (Brymora, *et al.*, 2001; X. W. Chen, *et al.*, 2011a; Fukai, *et al.*, 2003; Inoue, *et al.*, 2003; Li, *et al.*, 2007; Moskalenko, *et al.*, 2002) (Figure 1). The pioneering studies of the early 1990's discovered that there are six yeast secretion (Sec) proteins; Sec3, Sec5, Sec6, Sec8, Sec10 and Sec15; and two exocyst (Exo) subunit proteins; Exo70 and Exo84, which form the exocyst complex (TerBush, *et al.*, 1996). The constituents of the exocyst complex are conserved between yeast and mammals (He & Guo, 2009) and there are striking structural and topological similarities in the C-terminal domains of Sec6, Sec15, Exo70 and Exo84, despite there being less than 10% sequence identity between the individual proteins. These C-terminal domains consist of multiple rod-like helical bundles, which appear to be evolutionarily related molecular scaffolds that have diverged to create functionally distinct exocyst proteins (Sivaram, *et al.*, 2006). The interaction between these helical structures may create the framework that is necessary for the assembly of the exocyst complex (Munson & Novick, 2006).

There is some evidence that the exocyst complex may be present as distinct sub-complexes on vesicular and plasma membranes. In yeast, two members of the complex are associated with the plasma membrane; Sec3 and Exo70, while in mammals only Exo70 appears to be found on the plasma membrane (He, *et al.*, 2007; He & Guo, 2009; Inoue, *et al.*, 2003; J. Liu, *et al.*, 2007). It is likely that the membrane localisation of Sec3 and Exo70 controls targeting of secretory vesicles to distinct domains of the plasma membrane, thereby defining the sites of active exocytosis and membrane growth during cell migration and cytokinesis (Liu & Guo, 2011). It has been suggested that the Sec3 and Exo70 plasma membrane complex also contains Sec5, Sec8 and Sec6, while Exo84, Sec10 and Sec15 are complexed to the vesicle membrane (Moskalenko, *et al.*, 2003). By binding to the vesicular membrane, Sec15 initiates the assembly of the vesicular exocyst sub-complex, while Sec3 and Exo70 mediate assembly of the plasma membrane sub-complex. Sec3 relies on a Rho-mediated targeting

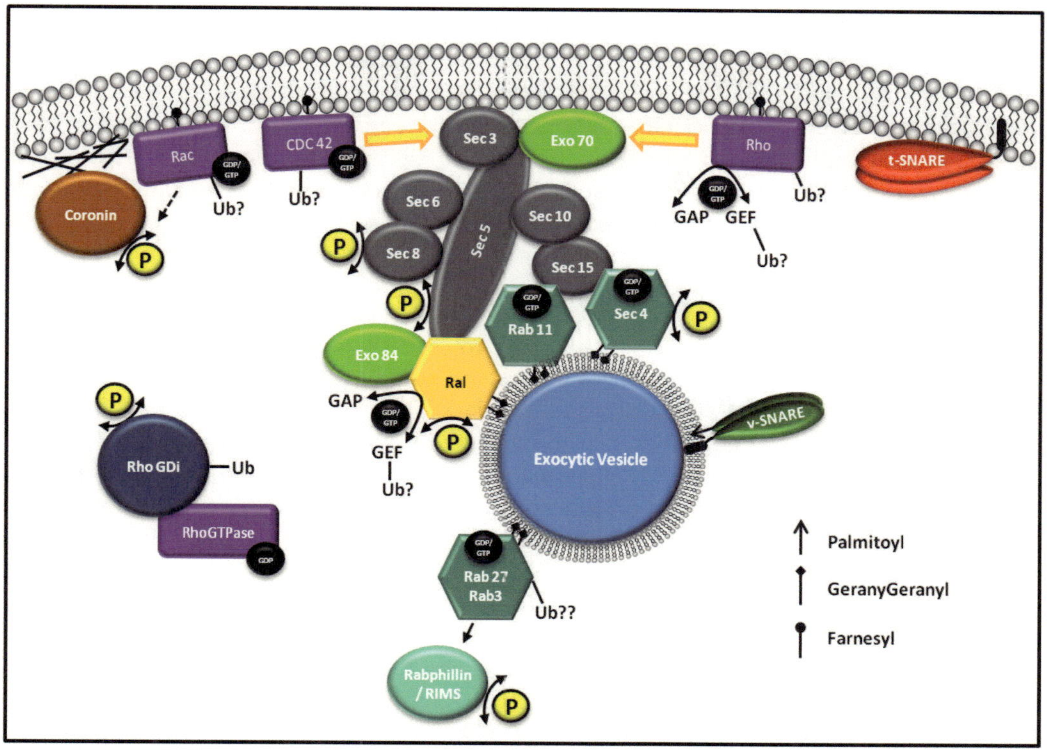

Fig. 1. Post-translational regulation of exocytic vesicle tethering via the exocyst complex

mechanism for its plasma membrane localization (He, *et al.*, 2007; Moskalenko, *et al.*, 2002; H. Wu, *et al.*, 2010), which is distinct from the Rab-dependent targeting of Sec15 to the vesicular membrane (Guo, *et al.*, 1999; Langevin, *et al.*, 2005; S. Wu, *et al.*, 2005; Zhang, *et al.*, 2004). Co-assembly of these two exocyst sub-complexes to form the entire complex is governed by Ral-GTPase via its interaction with Sec5 (Hohlfeld, 1990). Prior to membrane fusion, SNAREs (e.g. Sec1, Sro7p and Sro77p) interact with the exocyst complex (via Sec6, Exo84) to facilitate fusion between the vesicle and plasma membranes (Morgera, *et al.*, 2012; Zhang, *et al.*, 2005).

3. Small GTPases as regulators of exocytosis

While the exocyst complex has a clear role in exocytosis, the factors promoting the final orchestration of exocytosis are yet to be characterized. Emerging data highlights that small GTPases of the Ras super-family, including the Ras homologous (Rho), Ras-associated binding proteins (Rabs), adenosine ribosylation factors (Arfs), and Ras-like proteins (Ral) subfamilies, are involved in regulating distinct steps during exocytosis, some of which are mediated via interaction with the exocyst (reviewed in (Csepanyi-Komi, *et al.*, 2011; Hutagalung & Novick, 2011; Segev, 2011)). Thus, there appears to be stage-specific requirements for small GTPase subfamily members during exocytosis (Figure 1).

The unique feature of the small GTPase superfamily (G-protein family) is the presence of a 20 kDa, catalytic domain (Bourne, *et al.*, 1991; Pai, *et al.*, 1990). Through guanosine nucleotide-dependent conformational transitions within their G-protein domain (Pereira-Leal & Seabra, 2000), these GTPases act as molecular switches; cycling between the inactive GDP bound form and a GTP-bound active form, the process which regulates the activity of downstream effectors. This activity switch can be triggered by a variety of intracellular stimuli, most notably calcium ions (Khvotchev, *et al.*, 2003; Zajac, *et al.*, 2005). The current dogma suggests that guanosine-triphosphate-dependent activation is essential for Rho, Ral and Rab relocation to target membranes, which then triggers their function.

3.1 Rab GTPases and vesicular tethering

The Rab family of small GTPases is defined by the presence of at least one of five characteristic Rab motifs, with the RabF1 motif frequently positioned within the effector binding domain, and the RabF2 motif usually in the GTPase domain (Pereira-Leal & Seabra, 2000). Recent bioinformatic analysis of the Rab family using the "Rabifier" and "RabDB" tools have uncovered an interesting phenomenon, namely the highly dynamic evolution of this family, with a significant expansion and specialization of the Rabs involved in the secretory pathway (Diekmann, *et al.*, 2011). The repertoire of secretion-related Rabs includes 14 subfamilies, which co-evolved with *Metazoan* multicellularity and may reflect either unique roles in tissue-specific membrane trafficking events or restricted trafficking of specialist vesicles (Diekmann, *et al.*, 2011). The animal-specific subfamilies have purported roles in the regulation of secretion (e.g. Rab3, Rab26, Rab27, Rab33, Rab37, Rab39), while there are also Golgi-specific Rabs (Rab30, Rab33, Rab34, Rab43) and Rabs relating to the traffic to (Rab43) and from (Rab10) the Golgi. Rab proteins usually play positive roles in anterograde membrane trafficking, but the exact nature of their involvement (in vesicle budding, biogenesis, transport, docking, priming and fusion) depends on the particular pathway, and is yet to be defined. One of the most evolutionary conserved proteins, Rab11, appears to be associated with both constitutive and regulated secretory pathways, as shown in mammalian and insect cells (Chen, *et al.*, 1998; Shandala, *et al.*, 2011; Urbe, *et al.*, 1993; Ward, *et al.*, 2005).

There have been mechanistic links established between some Rab proteins and components of the exocyst complex. For example, the interaction of Sec15 with Rab proteins appears to be essential for the tethering of exocyst components to designated membranes. In yeast, the small Rab GTPase Sec4 (orthologous to mammalian Rab10) may bring Sec15 to the vesicular membrane (Guo, *et al.*, 1999), which is an essential step in the tethering and assembly of the exocyst complex (Zajac, *et al.*, 2005). Metazoan Sec15 is a known effector of Rab11 (S. Wu, *et al.*, 2005; X. M. Zhang, *et al.*, 2004). Through its C-terminal domain, *Drosophila* Sec15 can interact with Rab11 (and is found co-localized with Rab11 in *Drosophila* photoreceptor and sensory organ precursor cells (Jafar-Nejad, *et al.*, 2005; S. Wu, *et al.*, 2005)), as well as with Rab3, Rab8, and Rab27 (S. Wu, *et al.*, 2005). The functional relationship between Sec15 and Rab11 also exists in mammalian cells (Langevin, *et al.*, 2005; Zhang, *et al.*, 2004), where the interaction with Rab11 is involved in sequestering the exocyst complex to the endosome recycling compartment. Interestingly, there is a functional co-dependence of Rab11 and Sec15, where the loss of Sec15 function affects the intracellular localisation of Rab11, and mimics a phenotype of abnormal Rab11 function. Evidence of this interaction can be observed during the dramatic changes that occur in photoreceptor cell development (S. Wu, *et al.*, 2005). More specifically, the mutant *Sec15* phenotype involves impaired trafficking from recycling endosomes to the plasma membrane, and restricts cargo trafficking, e.g. DE-Cadherin in *Drosophila* (Langevin, *et al.*, 2005).

Rabs have important interactive functions at different stages of exocytosis. Protein sorting in recycling endosomes depends upon the function of the small GTPase Rab4, a close homologue of Rab11 (Li, *et al.*, 2008; Ward, *et al.*, 2005). Rab3 also has a role in anterograde traffic between the trans-Golgi network and recycling endosomes (Mohrmann, *et al.*, 2002; van der Sluijs, *et al.*, 1992). The Rab small GTPases, Sec4 in yeast and Rab11 in metazoans, facilitate trafficking of secretory vesicles carrying Sec15-exocyst components from recycling endosomes to the plasma membrane (Langevin, *et al.*, 2005). Sec15 does not appear to interact with mammalian Rab4a, and therefore does not function as a Rab4 effector (Zhang, *et al.*, 2004). This suggests a unique role for Rab11 in the final delivery of exocyst carrying secretory vesicles, to the plasma membrane (Chen, *et al.*, 1998; Shandala, *et al.*, 2011; Urbe, *et al.*, 1993; Ward, *et al.*, 2005). Sec15 does however interact with Rab3, a closely related homologue of Rab11, both of which appear to play a critical role in: secretory vesicle biogenesis; docking and priming of specialised secretory vesicles; delivery of synaptic and dense core vesicles to the active zone of exocytosis; and in maintaining a primed pool of vesicles available for rapid release (Schonn, *et al.*, 2010; S. Wu, *et al.*, 2005). Thus, the loss of Rab3 led to a reduction in the total number of synaptic vesicles as well as the number recruited to the active zone of the neural synapse (Gracheva, *et al.*, 2008). Similarly, there was a reduction in the number of dense core vesicles docked at the plasma membrane in adrenal chromaffin cells, isolated from a mouse quadruple knockout lacking all four Rab3 A to D paralogues (Schonn, *et al.*, 2010). An increased number of docked dense core vesicles was observed in PC12 and in adrenal chromaffin cells following Rab3 overexpression and this correlated with a strong inhibition of secretion (Holz, *et al.*, 1994; Johannes, *et al.*, 1994). Interestingly, there is evidence of some functional redundancy between Rab3 and its closest homologues, Rab27A and Rab27B, which are involved in the delivery of vesicles near the exocytic site (Fukuda, 2008; Gomi, *et al.*, 2007; Ostrowski, *et al.*, 2010). Studies of melanosome dynamics have indicated that Rab27A has a role in vesicular recruitment and this is mediated by its interaction with a specific effector called Melanophilin, which in turn binds an actin motor protein, MyosinVa (Hume & Seabra, 2011; Seabra & Coudrier, 2004). There appears to be a further functional divergence of Rabs, where Rab27A and Rab27B control different steps of the secretion pathway (Ostrowski, *et al.*, 2010). Rab25, a close homologue of Rab11 with a different C-terminus, shows co-localization with Rab11 in exocytic/recycling vesicle membranes of some cells, and may function as a tissue specific tethering factor (Calhoun & Goldenring, 1997; Khandelwal, *et al.*, 2008).

Recent studies have implicated Rabs in the movement of transport vesicles from their site of formation to their site of fusion, and several Rabs have been linked to specific microtubule- or actin-based motor proteins (Hammer & Wu, 2002; Lapierre, *et al.*, 2001) The role of Rabs in docking of secretory vesicles to the plasma membrane is mediated by their effectors (Fukuda, 2008). Thus, the small GTPases of the Rab family, through interplay with their specialist effector molecules, cooperatively target secretory vesicles from the trans-Golgi network (TGN) to the plasma membrane, and facilitate their docking at the active site of exocytosis (Fukuda, 2008; Orlando & Guo, 2009). This poses the question: what is the molecular mechanism defining the plasma membrane docking sites?

3.2 Rho GTPases and assembly of the plasma membrane exocyst complex

The most highly conserved and best studied members of the Rho family, Rho1/A, Rac1 and Cdc42, play a crucial role in tethering and fusion of vesicles during regulated exocytosis (Ory & Gasman, 2011; Ridley, 2006; Williams, *et al.*, 2009). Most Rho GTPases transiently localize at the plasma membrane, after being targeted to specific phosphoinositide-containing sub-domains. On the one hand, the Rho GTPases, Rho1/A and Rac1, are thought to regulate secretion by remodelling microtubules and the membrane-associated actin cytoskeleton (Ory & Gasman, 2011; Williams, *et al.*, 2009). On the other hand, recent findings have implicated yeast Cdc42, Rho1/A and Rho3/C and mammalian TC10 in actin-independent regulation of exocytosis by anchoring the plasma membrane exocyst components, Sec3 and Exo70, to specific plasma membrane microdomains (Bendezu & Martin, 2011; Guo, *et al.*, 2001; He, *et al.*, 2007; He & Guo, 2009; Inoue, *et al.*, 2003; J. Liu, *et al.*, 2007; Moskalenko, *et al.*, 2002; Novick & Guo, 2002; Wu & Brennwald, 2010; H. Wu, *et al.*, 2010; Xiong, *et al.*, 2012; Zajac, *et al.*, 2005). However, in this case, the normal functioning of the Sec3 and Exo70 plasma membrane exocyst components is a prerequisite for the correct localization of cell polarity regulators such as Cdc42 (Zajac, *et al.*, 2005). This might be due to the fact that Sec3 and Exo70 could independently bind to phosphatidylinositol-4,5-bisphosphate (PI(4,5)P2), via their C-terminal D domain, thereby forming a plasma membrane targeting patch for exocytic proteins (He, *et al.*, 2007; He & Guo, 2009; Inoue, *et al.*, 2003; J. Liu, *et al.*, 2007). Moreover, Exo70 was found to be directly associated with type Iγ phosphatidylinositol phosphate kinase (PIPKIγ), which facilitates the generation of a PI(4,5)P2 phospholipid microdomain and recruitment of Exo70 to the plasma membrane (Xiong, *et al.*, 2012). Thus, the cooperation between the Sec3 and Exo70 exocyst components and Rho small GTPases defines competent sites for exocytosis. The next question is: how are secretory vesicles targeted to these sites?

3.3 Ral small GTPases and the exocyst complex

Ras-like (Ral) small GTPase was first discovered in human platelet cells in 1991 (van der Meulen, *et al.*, 1991), where its association with dense granules suggested a potential regulatory role in the release of storage contents from these granules (Mark, *et al.*, 1996). Subsequently, Ral small GTPases were linked to exocytosis in neural, epithelial, endothelial endocrinal, and pancreatic tissues (Hazelett, *et al.*, 2011; Lopez, *et al.*, 2008; Moskalenko, *et al.*, 2003; Polzin, *et al.*, 2002; Rondaij, *et al.*, 2004; Rondaij, *et al.*, 2008; Takaya, *et al.*, 2007). The two mammalian Ral homologues RalA and RalB share 85% protein sequence identity, and are well conserved in evolution (van Dam & Robinson, 2006). The bulk of the Ral protein comprises a conserved nucleotide phosphate-binding motif (Marchler-Bauer, *et al.*, 2011; van Dam & Robinson, 2006). RalA, but not RalB, contains a short amphipathic helix that binds the Ca^{2+}-sensing protein Calmodulin, conferring Ca^{2+}-dependent activation of RalA during regulated exocytosis (van Dam & Robinson, 2006). Ral functions as an essential component of the cellular machinery, regulating the post-Golgi processing of secretory vesicle membrane, via activation of the exocyst complex (X. W. Chen, *et al.*, 2011a; Feig, 2003; Kawato, *et al.*, 2008; Mark, *et al.*, 1996; Mott, *et al.*, 2003). It has been suggested that GTP-bound Ral, through its effectors Sec5 and Exo84, brings together the plasma and vesicular membrane exocyst subunits, as the loss of RalA, or mutation of its effector binding motif, leads to the disassembly of the exocyst complex (Moskalenko, *et al.*, 2003). Activation of the exocyst complex is initiated by the binding of Ral to Sec5 and Exo84 (Mott, *et al.*, 2003). This is followed by the assembly of the full octameric exocyst complex at the interface of the vesicular and plasma membranes (Moskalenko, *et al.*, 2003). Thus, the interaction of the exocyst complex with Ral is an essential step in anchoring secretory vesicles to the exocytosis-competent microdomains of the plasma membrane (Angus, *et al.*, 2003; Fukai, *et al.*, 2003; Jin, *et al.*, 2005; Mark, *et al.*, 1996; Moskalenko, *et al.*, 2002).

The functional interaction of Ral and the exocyst complex is highlighted by their co-involvement in multiple exocytic processes. The exocyst complex has well-established roles in anterograde trafficking of membrane receptors from recycling endosomes (Langevin, *et al.*, 2005; Xiong, *et al.*, 2012; Yeaman, *et al.*, 2004); membrane delivery in cell growth (Chernyshova, *et al.*, 2011; Genre, *et al.*, 2011); and the translocation of glucose transporters in response to insulin (Ljubicic, *et al.*, 2009; Lopez, *et al.*, 2008). Each of these processes has been linked to a functional requirement for a member of the Ral protein family. RalA is required for establishing neuronal cell polarity (Lalli, 2009), and the regulation of readily releasable pools of synaptic vesicles (Lee, *et al.*, 2002; Li, *et al.*, 2007). In the epithelium, RalB is required for delivery of membrane to the dynamic leading edge of migrating cells (Rosse, *et al.*, 2006); while RalA is involved in polarised delivery of the membrane protein, E-Cadherin, to the basolateral surface of epithelial cells (Shipitsin & Feig, 2004). Exocytosis of vesicular content, such as hormones, chemokines, enzymes, and adhesion molecules from Weibel-Palade bodies (endothelial cell-specific storage organelles), occurs in response to a specific agonist that requires Ral regulation (Kim, *et al.*, 2010; Rondaij, *et al.*, 2008). RalA is required in glucose regulation where it mediates insulin secretion from pancreatic cells (Ljubicic, *et al.*, 2009; Lopez, *et al.*, 2008), and translocation of the glucose transporter GLUT4 in adipocytes (Chen, *et al.*, 2007). Ral is also required for dense granule secretion from platelets (Kawato, *et al.*, 2008) and cell growth and migration, all of which have been shown to be reliant on Ral for lipid raft trafficking to the plasma membrane (Balasubramanian, *et al.*, 2010; Spiczka & Yeaman, 2008).

Given that multiple GTPases regulate the assembly of the full octameric exocyst complex, which is necessary for vesicular tethering to the site of fusion, the assembly of the exocyst complex might represent the integration of various cellular signaling pathways that ensure tight control of exocytosis (Sugihara, *et al.*, 2002).

4. Exocyst, SNARE complexes and membrane fusion machinery

The exocyst mediated tethering of secretory vesicles to specific sites of the plasma membrane precedes the assembly of SNARE complexes and membrane fusion (He & Guo, 2009; Novick & Guo, 2002) (Figures 1 & 2). In the early 1980s, Rothman and colleagues used an *in vitro* trafficking assay to identify the soluble factors; *N*-ethylmaleimide-sensitive factor (NSF) and Soluble NSF Attachment Protein (SNAP) (Balch, *et al.*, 1984). This was followed by isolation of their membrane receptors termed SNAREs (for SNAP receptors) (Sollner, *et al.*, 1993). SNAREs were initially isolated from mammalian brain cells, as factors crucial for vesicle fusion–mediated release of neurotransmitters at synapses. It soon became evident that SNAREs are involved in most, if not all vesicular fusion events (Malsam, *et al.*, 2008).

SNAREs comprise evolutionarily conserved families of membrane-associated proteins (including the Synaptobrevin/vesicle associated membrane protein (VAMP), Syntaxin and SNAP, families), which are characterized by the presence of a 60–70 amino acid long SNARE motifs located, in most cases, immediately adjacent to a C-terminal trans-membrane anchor (Weimbs, *et al.*, 1997). In some cases (e.g. Sec9), this trans-membrane motif is absent, and the membrane binding is mediated by lipid modifications or by the presence of domains capable of binding lipid head groups (Grote, *et al.*, 2000). A coiled-coil structure formed by SNARE motifs is known to mediate protein-protein interactions between different SNAREs, and this is believed to create the membrane curvature required for membrane fusion (Groffen, *et al.*, 2010). It has been suggested that the SNARE-induced membrane curvature becomes fusion competent when decorated by two Ca^{2+} sensors, Synaptotagmin1 and double C2 domain (DOC2) protein (Groffen, *et al.*, 2010; Hui, *et al.*, 2009).

SNAREs act at all levels of the secretory pathway, although individual family members tend to be compartment-specific and thus, are thought to contribute to the specificity of docking and fusion events (Gerst, 1999). The exocyst complex may facilitate vesicular targeting to the plasma membrane by direct interaction with SNARE proteins, as is the case with the exocyst component Sec6 and the target t-SNAREs Sec1 and Sec9 (Morgera, *et al.*, 2012). In some cases this interaction may be indirect and mediated by adaptor proteins. For example, the yeast exocyst component Exo84 is capable of interacting with t-SNAREs though the WD40 domain adaptor proteins Sro7p and Sro77p (Sivaram, *et al.*, 2005; Zhang, *et al.*, 2005). In addition, during the process of regulated exocytosis involved in neurotransmitter release, a unique set of components (Synaptotagmin, Complexin, Munc13, RIM, - also called the trans-SNARE complex), assemble between v-SNAREs on the transport vesicle and t-SNAREs on the fusion target, to ensure efficient membrane fusion (Li & Chin, 2003). Some elements of this SNARE complex can also act as negative regulators of membrane fusion: as is the case during sperm exocytosis, where α-SNAP prevents docking of the acrosome by sequestering monomeric Syntaxin (Rodriguez, *et al.*, 2011).

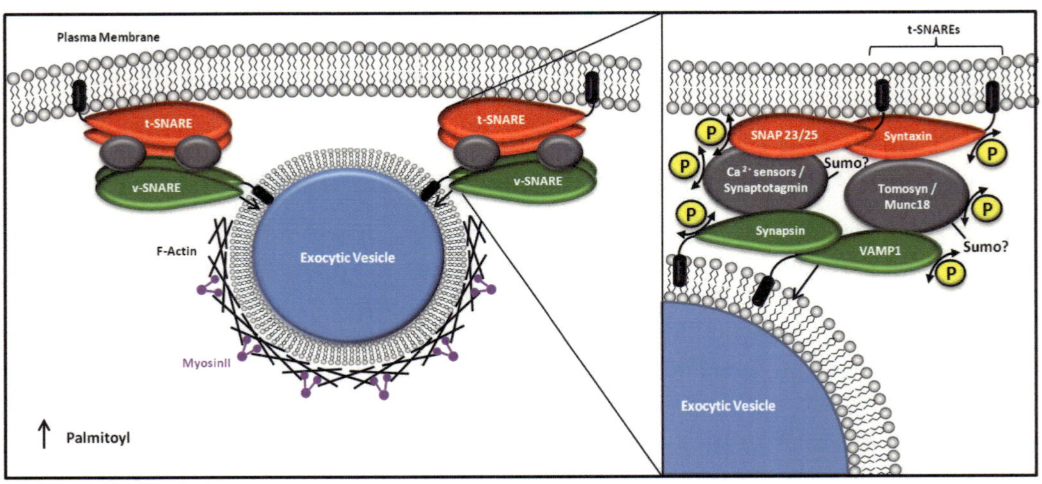

Fig. 2. Post-translational regulation of docking and fusion at the plasma membrane

5. Post-translational modifications that regulate components of the exocytic machinery

5.1 Regulation of GTPases: Activity by nucleotide exchange

As mentioned above, the small GTPases involved in exocytosis are regulated by switching between an active (GTP-bound) and inactive (GDP-bound) state (Ali & Seabra, 2005; Barr & Lambright, 2010; Csepanyi-Komi, *et al.*, 2011; Stenmark, 2009; Uno, *et al.*, 2010) (Figure 1). This exchange between two states is regulated by guanine nucleotide exchange factors (GEFs) and GTPase activating proteins (GAPs) (Csepanyi-Komi, *et al.*, 2011). GEF proteins activate GTPases by a mechanism involving the destabilization of the interaction between the GEF and guanosine di-phosphate (GDP), resulting in the fast release of the GDP and generation of a nucleotide-free GTPase-GEF intermediary complex (Liao, *et al.*, 2008). As guanosine tri-phosphate (GTP) is much more abundant in the cell than GDP (Gamberucci, *et al.*, 1994), GTP will preferentially bind to the nucleotide binding pocket of the GTPase inducing dissociation of the GEF, leaving the GTP-loaded GTPase ready to perform its function. The subsequent hydrolysis of the GTP by the GTPase provides the energy required for the GAP enzyme to accomplish its specific task. In some cases, the membrane affinity of GEFs and GAPs has been linked to the binding modules of the Phox-homology domain (PX domain in TCGAP, a protein with intrinsic GAP activity towards Rho GTPases as shown *in vitro*), and the Pleckstrin-homology domain (PH domain in Trio, a GEF for Rho GTPases). These domains target the proteins to specific phosphoinositide-containing sub-domains of intracellular membranes (Chiang, *et al.*, 2003; Hyvonen, *et al.*, 1995; Kutateladze, 2007; van Rijssel, *et al.*, 2012; Zhou, *et al.*, 2003).

Once activated by GEFs, the small GTPases (Rho, Rab, Arf, and Ral) translocate to their destination membrane in order to carry out their specific function (Csepanyi-Komi, *et al.*, 2011). In exocytosis, the GEFs identified so far control the function of members of three of the GTPase families, the Rabs, Rals, and the Rhos. In the budding yeast *S. cerevisiae*, the Rab GTPase Sec4 and its GEF Sec2 were found to be associated with the polarized transport of vesicles from the TGN to the site of bud formation at the plasma membrane. Sec2 catalyzes the guanine nucleotide exchange and subsequent activation of Sec4, a key mediator of the docking and fusion steps (Dong, *et al.*, 2007; Medkova, *et al.*, 2006; Ortiz, *et al.*, 2002). The Rab GTPase Rab3a performs a similar role in mammalian neuroendocrine cells, with its function being required for docking and fusion of hepatocyte growth factor granules in chromaffin cells, as demonstrated by the use of mutated forms of the protein (Holz, *et al.*, 1994; Luo, *et al.*, 2001; Macara, 1994). The GEF for Rab3a, called GRAB, has been shown to be essential for the targeting of secretory vesicles to the plasma membrane (Luo, *et al.*, 2001; Sato, *et al.*, 2007). The small Rho GTPase Cdc42 is involved in remodeling of the actin cytoskeleton of the cell including some of the events that occur in Ca^{2+} mediated regulated exocytosis. Recent studies in pancreatic β-cell islets, using a combination of *in vitro* binding assays and RNAi silencing screens in cell culture, have identified Cool-1/β-pix as a GEF for Cdc42 (Kepner, *et al.*, 2011). A number of GEF proteins have been proposed to mediate Ral function in exocytosis, which include Ral guanine nucleotide dissociation stimulator (RalGDS) (Albright, *et al.*, 1993) and RalGDS-like 2 (Rgl2)/RalGDS-like factor (Rlf) (Wolthuis, *et al.*, 1996). The activity of RalGEFs, are themselves regulated by the Ras GTPase protein. Active GTP-bound Ras facilitates activation by binding to RalGEFs (Spaargaren & Bischoff, 1994), and delivering them to Ral in the plasma membrane. RalGDS has been shown to be required for the activation of RalA dependent exocytosis from Weibel-Palade bodies in endothelial cells and the secretion of insulin in β-pancreatic cells (Kim, *et al.*, 2010; Ljubicic, *et al.*, 2009; Rondaij, *et al.*, 2008), while calcium induced exocytosis from endosomes is predicted to be regulated by Rgl2/Rlf - mediated activation of Ral (Takaya, *et al.*, 2007). Thus, the regulation of Ral dependent exocytosis may be at the nexus of a number of upstream signaling pathways. Often GEFs will display a certain level of promiscuity, being capable of activating more than one GTPase family member. For example, the Rab3a GEF GRAB is also capable of activating Rab27a (Baisamy, *et al.*, 2005; Baisamy, *et al.*, 2009). This might reflect a need for simultaneous activation of more than one GTPase in order to coordinate the progression of exocytosis. Moreover, individual GTPases may be regulated by multiple GEFs (for example Cdc42 is regulated by Cool-1/β-pix (Kepner, *et al.*, 2011), FGD1 (Olson, *et al.*, 1996), FGD3 (Pasteris, *et al.*, 2000), and PEM-2 (Reid, *et al.*, 1999)), although this may reflect a mechanism for the tissue-specific activation of these uniformly expressed proteins.

When driven by the intrinsic enzymatic activity of small GTPases, the hydrolysis of GTP to GDP is a slow process (Crechet & Parmeggiani, 1986; Liao, *et al.*, 2008; Trahey & McCormick, 1987), but this can be stimulated by GTPase activating proteins (GAPs) (Csepanyi-Komi, *et al.*, 2011; Trahey & McCormick, 1987). In yeast for example, once GTP-bound Sec4 reaches the plasma membrane, its GAPs, Msb3 and Msb4, promote hydrolysis of the GTP to switch it to its GDP-bound inactive form and translocate it away from the plasma membrane (Gao, *et al.*, 2003). Yeast cells with deletions of Msb3 and Msb4 display defective exocytosis as indicated by the accumulation of abnormally large numbers of vesicles within the cells and a reduction in secretion (Gao, *et al.*, 2003). As is the case with GEFs, there is also a certain promiscuity of RhoGAP activity. For example, the neuron specific GAP Nadrin has been demonstrated to have activity towards Rho1/A, Cdc42, and Rac1 (Harada, *et al.*, 2000). In doing so, Nadrin appears to modulate the actin related function of Rho GTPases, as its overexpression in NIH3T3 fibroblast cells resulted in abnormal reorganization of the actin stress fibers (Harada, *et al.*, 2000). Biochemical studies have identified Rab3GAP1/Rab3GAP2 as a GAP for Rab3, and EV15 as a GAP for Rab11, but the physiological relevance of their function is yet to be fully defined (Csepanyi-Komi, *et al.*, 2011; Dabbeekeh, *et al.*, 2007; Fukui, *et al.*, 1997). The RalGAP complex (RGC), which is made up of RalGAP complex 1 (RGC1) and a catalytic subunit RalGAP complex 2 (RGC2) has been identified for its role in Ral dependent exocytosis of GLUT4 (Chen & Saltiel, 2011). RCG interacts with RalA under steady state conditions, acting as its inhibitor. Insulin induced activation of the PI3-kinase/Akt pathways leads to Akt-mediated phosphorylation of the RGC2 catalytic subunit, disengaging it from the GTPase and reversing the inhibitory effect of RGC on RalA activity. This allows RalA to associate with the exocyst and facilitate exocyst complex assembly (Chen & Saltiel, 2011).

After GTPases complete their function on the membrane, they are returned to the cytosol in an inactive GDP-bound form, terminating the exocytic process and preventing inappropriate activity of the enzyme and its effectors. This sequestration and inactivation is mediated by members of the guanine dissociation inhibitor (GDI) family. Notably, GDIs have dual modes of GTPase binding. One domain of the protein binds to and occludes the nucleotide binding site of the GTPase (Gosser, *et al.*, 1997; Grizot, *et al.*, 2001; Hoffman, *et al.*, 2000; Keep, *et al.*, 1997; Longenecker, *et al.*, 1999; Scheffzek, *et al.*, 2000), preventing the release of the nucleotide and interaction with GTPase effector proteins. The second domain of the protein forms an immunoglobin-like hydrophobic pocket into which the prenyl group of the GTPase can be inserted (see section 5.2), preventing it from interacting with membranes (Gosser, *et al.*, 1997; Grizot, *et al.*, 2001; Hoffman, *et al.*, 2000; Keep, *et al.*, 1997; Longenecker, *et al.*, 1999; Scheffzek, *et al.*, 2000). This dual mode of interaction suggests that GDIs could play a negative regulatory role in the GTPase cycle; holding a GTPase in an inactive GDP-bound state, while still bound to the membrane via a lipid moiety, then transporting it away from the membrane after the lipid moiety is engaged (DerMardirossian & Bokoch, 2005; Johnson, *et al.*, 2009). This has been demonstrated most clearly in studies examining the mechanism of action of RhoGDI on Cdc42. In solution, RhoGDI exhibits an equal affinity to GDP or GTP bound Cdc42 (Nomanbhoy & Cerione, 1996). However, when Cdc42 is membrane associated, RhoGDI shows a much higher affinity to GDP bound Cdc42 (Johnson, *et al.*, 2009). During its normal activation/hydrolysis/inactivation cycle, Cdc42 translocates from membrane bound to freely solvent GDI-bound states. In the latter state, the immunoglobin-like hydrophobic pocket of RhoGDI prevents Cdc42 from re-associating with the plasma membrane while the rest of the protein prevents the release of GDP from Cdc42 (Johnson, *et al.*, 2009).

GDIs have been shown to have important regulatory roles during exocytosis. In yeast, once Sec4 is converted from a GTP to a GDP-bound form, it is removed from the plasma membrane by the GDI, Sec19 (Garrett, *et al.*, 1994). The loss of Sec19 in mutant yeast strains leads to the accumulation of Sec4 at the site of budding, indicating a block of Sec4-GDP recycling (Zajac, *et al.*, 2005). Also, members of the rat Rab GDI family, GDI α and β, were demonstrated to block nucleotide exchange of both Rab3 and Rab11 in *in vitro* assays (Nishimura, *et al.*, 1994), and RNAi mediated knockdown of RhoGDI in pancreatic β-cells was found to result in a defect in Rac1/Cdc42 mediated insulin secretion (Kowluru & Veluthakal, 2005; Wang & Thurmond, 2010).

In general, GEFs function in large membrane associated complexes, and serve to translate signals from membrane bound receptors to specific members of the small GTPase superfamily (reviewed in (Csepanyi-Komi, *et al.*, 2011)). This mechanistic link ensures that the activation of small GTPases is dependent on major cellular signaling events. GEFs also provide a link between the exocyst complex and the vesicle. This is demonstrated by the interaction between the GTPase Sec4 and the exocyst component Sec15, an interaction that relies upon the activation of Sec4 by its GEF Sec2 (Guo, *et al.*, 1999). Phosphorylation also plays a key role in regulating the activity of the GEFs, GAPs and GDIs and examples of this regulatory role in exocytosis will be discussed in section 5.3.

5.2 Lipid modifications as a mean to control intracellular localization of proteins

The covalent attachment of lipids to exocytic proteins provides an important mechanism for regulating their location and in doing so, their activity. In exocytosis, two major lipid modifications, prenylation and palmitoylation (Yalovsky, *et al.*, 1999), facilitate the targeting of proteins to the plasma membrane and intracellular membrane compartments. Although somewhat controversial, modification by specific lipid moieties appears to play a role in controlling the strict compartmentalisation of the modified protein to specialist intracellular organelles (Figure 1&2).

5.2.1 Prenylation

Protein prenylation involves the attachment of either of two isoprenoids (C15 farnesyl or the C20 geranylgeranyl) to cysteine residues at the C terminus of the protein (Seabra, 1998;

Sebti, 2005; Zhang & Casey, 1996). The choice of isoprenoid in Rho proteins is dictated by the amino acid composition of a conserved *CAAX* motif, where C is cysteine, *A* is an aliphatic amino acid residue and *X* represents any amino acid (Seabra, 1998; Sebti, 2005; . Zhang & Casey, 1996). However, in Rho1/A, Rho2/B, Rac1, Rac2, Cdc42 and Ral, where *X* is a leucine residue, the protein becomes a substrate for geranylgeranyl transferase I (GGTase I), which attaches a 20-carbon (geranylgeranyl) isoprenoid moiety (Berzat, *et al.*, 2006; Matsubara, *et al.*, 1997). In Rho proteins where the *X* is a methionine, serine, alanine or glutamine (e.g. Rho2/B), a 15-carbon (farnesyl) isoprenoid will be attached by the enzyme farnesyl transferase (FTase) (Adamson, *et al.*, 1992; Zhang & Casey, 1996). There is some evidence to suggest that prenylation triggers the cleavage of the *AAX* peptide by Rce1 (Ras-converting enzyme 1) endoprotease and the addition of a methyl group to the prenylated cysteine residue (catalyzed by isoprenylcysteine-*O*-carboxyl methyl transferase (Icmt)), which enhances membrane association of the prenylated protein (Roberts, *et al.*, 2008; Sebti & Der, 2003).

In contrast to the Rho small GTPases, the C terminal peptide motif for the majority of Rab proteins contains two cysteine residues (e.g. *XXXCC, XCCXX, XXCXC,* CC*XXX, or XXCCX*), both of which can be prenylated resulting in mono- or di-geranylgeranylated proteins (Pereira-Leal & Seabra, 2000, 2001). This modification is catalyzed by geranylgeranyl transferase II (GGTase II, also called Rab geranylgeranyl transferase, RabGGTase), acting in conjunction with a chaperone Rab escort protein (human REP/yeast Yip-1) (Anant, *et al.*, 1998; Andres, *et al.*, 1993; Calero, *et al.*, 2003; Seabra, *et al.*, 1992).

5.2.1.1 Prenylation of Rho GTPases

Our understanding of the role of lipid modification in protein targeting to a particular membrane comes from studies using either specific enzyme inhibitors or genomic mutants (Mitin, *et al.*, 2012). Rho2/B GTPase is known to undergo modification with both farnesyl (Rho2/B-F) and geranylgeranyl groups (Rho2/B-GG). In one study, treatment of HeLa cells with FTase inhibitors resulted in a loss of Rho2/B association with the plasma membrane, although association with the endosomal membranes remained unaffected (Lebowitz, *et al.*, 1997). This suggested that the farnesyl moiety was specific for Rho2/B targeting to the plasma membrane and that the geranylgeranyl moiety controlled targeting to the endosomal membranes. However, for experiments performed in mouse embryonic fibroblasts or yeast cells, Rho1/A and Rho2/B (normally found on the cell surface and in cytoplasmic vesicles) accumulated in the nucleus when treated with a GGTase-I inhibitor (Rho1/A) or a combination of a GGTase-1 and FTase inhibitor (Rho2/B), presumably because they were unable to traffic to their target membranes (Roberts, *et al.*, 2008). In this study, treatment with FTase inhibitor alone had a minimal impact on the membrane localisation of Rho2/B (Roberts, *et al.*, 2008). Although conflicting, the results of these inhibitor studies can be explained by the inhibitors exhibiting a competitive effect on the substrate proteins and thereby partially affecting both enzymes (Lobell, *et al.*, 2002). These results also suggest a potential functional redundancy between farnesyl and geranylgeranyl isoprenoids (Sjogren, *et al.*, 2007), which is further complicated by the suggested equilibrium between the two modifications of Rho2/B (Lebowitz, *et al.*, 1997). Evidence for functional redundancy between the isoprenoids comes from studies using mouse embryonic fibroblasts (MEFs) genetically depleted for GGTase-I function (Sjogren, *et al.*, 2007). Loss of GGTase-1 in MEFs isolated from mutant mice resulted in reduced actin polymerization leading to impaired cell migration. Transfection of these MEFs with plasmid vectors, driving the expression of farnesylated Rho1/A and Cdc42, restored the actin cytoskeleton, and the cells' ability to migrate and proliferate. This phenotypic rescue could be due to similar functionality of these two variants, as well as compensatory up-regulation of farnesylated variants, similar to that observed in COS cells, where treatment with an FTase inhibitor led to the accumulation of the geranylgeranylated form Rho2/B-GG (Du, *et al.*, 1999; Lebowitz, *et al.*, 1997).

The role of post-translational prenylation for the appropriate targeting of the lipid modified Rho protein to intracellular membranes has been further clarified by the mutational analysis of the enzyme recognition *CAAX*-motif as well as of the respective enzymes. Changing the *CAAX*-motif cysteine to a serine, or deleting the AAX component, rendered members of the Rho GTPase family unable to be prenylated and disrupted their ability to associate with the appropriate membranes (Winter-Vann & Casey, 2005). In addition, the loss of the prenylation enzyme GGTase-I in murine bone marrow macrophages (in *Pggt1b*$^{fl/fl}$ mutants) resulted in high levels of the activated Rho small GTPases, Cdc42, and Rho1/A, which predominantly accumulated in the cytoplasm, and Rac1 which was found preferentially associated with the plasma membrane. Unexpectedly, this was correlated with increased secretion of the pro-inflammatory cytokines IL-1β and TNF-α and hyper-activation of macrophages, resulting in erosive arthritis in GGTase-I deficient mice (Khan, *et al.*, 2011) . The authors proposed that this may be due to the lack of isoprenyl moieties in these small GTPases interfering with their recognition by endogenous regulatory proteins, such as RhoGDIs, which are responsible for GTPase inactivation and sequestration from the plasma membrane to the cytoplasm (Berzat, *et al.*, 2006). While similar studies in yeast have suggested that prenylation may have a regulatory role in the targeting of vesicles to specific sites on the plasma membrane (such as the site of yeast budding), this is yet to be fully explored (H. Wu, *et al.*, 2010).

5.2.1.2 Prenylation of Rab GTPases

To date, there have been no effective inhibitors identified for RabGGTase (El Oualid, *et al.*, 2005), so the functional studies of Rab GTPase prenylation have been carried out using genetic manipulation, either of the enzyme recognition sites for the Rab proteins, or involving genomic RabGGTase knock-out mutants. Experiments in yeast have demonstrated that the mode of geranylgeranylation can alter which organelle membrane the Rab proteins associate with (Calero, *et al.*, 2003). For instance, the wild-type di-geranylgeranylated yeast Rab proteins, YPT1 and Sec4, localize to the Golgi stack and the yeast bud tip respectively (Calero, *et al.*, 2003). In yeast cells expressing Rab proteins, carrying a C-terminal peptide with one cysteine mutated, mono-geranylated YPT1 and Sec4 were often mis-localized to a reticular structure (possibly the endoplasmic reticulum) or to the cytoplasm (Calero, *et al.*, 2003). Similarly, Rab27a, which is normally associated with melanosomes in mammals, became mis-localized when one of the cysteine residues in the C-terminal peptide motif was mutated (Gomes, *et al.*, 2003). Genomic mutations in the catalytic subunit of RabGGTase (in *gunmetal* (*gm/gm*) mutant mice) resulted in reduced prenylation and vesicular membrane association of Rab27a and reduced Rab11 prenylation in mouse platelets and melanocytes (Detter, *et al.*, 2000; Zhang, *et al.*, 2002). The *gunmetal* mutant showed a phenotype indicative of defective exocytosis in affected cells; defective killing capability by cytotoxic T-cells (Stinchcombe, *et al.*, 2001), hypopigmentation, macrothrombocytopenia and reduced platelet synthesis leading to extended bleeding time/excessive bleeding (Detter, *et al.*, 2000; Novak, *et al.*, 1995; Swank, *et al.*, 1993). Secretion defects in resorptive osteoclasts and bone forming osteoblasts were also observed in this mutant, resulting in imbalanced bone homeostasis (Taylor, *et al.*, 2011). Both of these cell populations exhibited problems with anterograde vesicular trafficking. In osteoclasts, plasma membrane delivery of proteases and ion pumps was defective, and in osteoblasts there was a reduced mineralisation potential, possibly due to a failure to deliver vesicles containing bone-matrix (Taylor, *et al.*, 2011). The underlying molecular cause is likely to be under-prenylation of many RabGTPases, including Rab3d, and possibly Rab27a.

Together, these *in vitro* and *in vivo* data indicate a significant role for prenylation in the function and activity of small GTPases of the Rho and Rab families, most likely through stable localization in the target membrane. Interestingly, there is a lipid-sensitive step at the secretory vesicle and plasma membrane fusion stage, which is proposed to be mediated by geranylgeranyl modification of SNAREs (Grote, *et al.*, 2000; Tong, *et al.*, 2009). In yeast, the exocytic SNARE Sec9 contains a geranylgeranyl lipid anchor (GG). When SNARE-GG is overexpressed, this inhibits exocytosis at a stage after SNARE complex assembly, possibly due to the generated membrane curvature blocking the merger of the secretory vesicle and plasma membrane. This inhibitory effect could be partially rescued by changing the curvature, by for example inserting an inverted cone–shaped lipid (e.g. lysophosphatidylcholine).

5.2.2 Palmitoylation

In contrast to prenylation, palmitoylation is reversible, allowing for a cycle of palmitoylation/depalmitoylation to occur in either a constitutive or regulated manner, such as in response to a specific extracellular signal (Fukata & Fukata, 2010). The palmitoylated proteins do not share an easily recognisable palmitate-binding motif, and palmitoylation can occur at the N-terminal, C-terminal or mid regions of proteins (Aicart-Ramos, *et al.*, 2011; Baekkeskov & Kanaani, 2009; Resh, 2004). Palmitoylation often occurs at cysteine residues (S-palmitoylation), where a 16-carbon saturated fatty acid (palmitate) is added via a thioester linkage with the aid of the Golgi localized palmitoyl acyltransferase (PAT), an enzyme belonging to a DHHC (Asp-His-His-Cys) protein family. Although there is no known consensus sequence for palmitoylation, a pre-existing prenylation may promote the attachment of a palmitoyl motif. For instance, impaired Rac1 prenylation caused by mutation of cysteine to serine in the *CAAX* motif (*SAAX*), also prevents the incorporation of [H^3]-palmitic acid at other sites on the Rac1 protein. This leads to the mis-localization of Rac1 to the cytosol and nucleus, instead of the plasma membrane (Navarro-Lerida, *et al.*, 2011). Depalmitoylation is carried out by an acyl protein thioesterase (APT), which releases protein from membranes in the cell periphery to traffic back to the Golgi membranes (Baekkeskov & Kanaani, 2009).

Disruption of palmitoylation prevents anterograde traffic to the target membrane. For example, the treatment of COS-7 cells and mouse embryonic fibroblasts (MEFs) with the palmitoylation inhibitor 2-bromo-palmitate (2BP) consistently led to an accumulation of GFP-tagged Rac in the perinuclear region, resulting in a partial loss from the plasma membrane and an exclusion from the nucleus (areas where the wild-type protein normally resides) (Navarro-Lerida, *et al.*, 2011). In addition, this treatment abolished the attachment of a palmitate moiety to the protein in living cells (blocking incorporation of [H^3]-palmitate).

Furthermore, mutation of the conserved amino acid residue at position 178 in Rac1 from cysteine to serine (Rac1^{C178S}) prevented the incorporation of radiolabeled [H^3]-palmitate and plasma membrane localisation of GFP-tagged Rac1. The loss of Rac1 palmitoylation in Rac1^{C178S} mutants affected cell motility, with mutants showing reduced cell spreading and delays in wound closure in scratch wound-healing assays (Navarro-Lerida, *et al.*, 2011). In *Drosophila* palmitoyl transferase (Huntington-interacting protein 14, *hip14*) null mutants, the resulting loss of palmitoylation significantly impaired neurotransmitter release from neuromuscular junctions isolated from 3rd instar larvae as indicated by fluorescent dye uptake and electrophysiology studies (Ohyama, *et al.*, 2007). The *hip14* mutant phenotype was linked to the mis-localization of two exocytic components, cysteine string protein (CSP) and SNAP25 at neural synapses, due to the loss of palmitoylation.

The equilibrium between palmitoylation and depalmitoylation is essential for the dynamic association of proteins with the plasma membrane, and for recycling back to the Golgi. A deficiency in the depalmitoylation enzyme palmitoyl thioesterase1, in *PPT1*-knockout neurons, evoked the retention of the palmitoylated SNARE proteins VAMP2 and SNAP25 on synaptic vesicles, making neural synapses non-responsive to depolarisation signals and stopping synaptic vesicles from releasing their cargo (Kim, *et al.*, 2008). In humans, the essential role for depalmitoylation in the recycling of vesicular proteins was confirmed by the early onset neurodegenerative pathology detected in brain tissue from patients with infantile neuronal ceroid lipofuscinosis, a disease linked to a mutation in *PPT1* gene (Mitchison, *et al.*, 1998; Vesa, *et al.*, 1995). This indicated that palmitoylation, is as essential as prenylation for targeting of these proteins to the plasma membrane. Notably, palmitoylation favors association of proteins with lipid rafts (Delint-Ramirez, *et al.*, 2011), which are specialized microdomains of plasma membrane containing distinctive arrangement of lipids and signaling molecules (Simons & Ikonen, 1997). It is this preferential association with lipid rafts which may bring palmitoylated proteins into close proximity with specific signaling cascades.

5.2.3 Multiple lipid modifications are associated with spatial and functional diversity in the Rho GTPase family

It has been previously proposed that prenylated *CAAX* proteins associate initially with the endomembrane prior to trafficking to the plasma membrane, and this process requires a secondary targeting motif (Choy, *et al.*, 1999). Now there is emerging evidence that modification with different lipid moieties could localize a protein to different microdomains within the plasma membrane, thereby diversifying the functions within a protein family. In *Saccharomyces cerevisiae*, Cdc42 and Rho3 are found at distinct sites on the plasma membrane, with Cdc42 restricted to the tip of the bud in a budding yeast cell and Rho3 found across the plasma membrane with only a slight enrichment at the bud (Wu & Brennwald, 2010). These differences in localisation correlate with the unique functions of these two proteins in regulating polarized exocytosis and overall cell polarity (Wu & Brennwald, 2010). Curiously, both Cdc42 and Rho3 contain the C-terminal *CAAX* prenylation motif, with Cdc42 known to be geranylgeranylated and Rho3 predicted to be farnesylated (Moores, *et al.*, 1991). However, Rho3 contains a long N-terminal extension with a cysteine residue that can be palmitoylated by the Ankyrin repeat-containing protein (Akr1) palmitoyl transferase, whereas Cdc42 does not (Wu & Brennwald, 2010). In yeast mutation studies, the diverse localization and functional differences for Rho3 and Cdc42 are imparted by the N-terminal region, which contains palmitoylated residues (Wu & Brennwald, 2010).

Another member of the Rho family of GTPases is the Wnt-regulated Cdc42 homolog-1 (Wrch-1). Although sharing functional properties with Cdc42, it is also present on some intracellular membranes, which is distinct from Cdc42. Wrch-1 contains an unusual CCFV C-terminal motif, (Berzat, *et al.*, 2005). This non canonical CCFV motif of Wrch-1 more readily incorporated [^{3}H]-palmitate rather than isoprenyl moieties ([^{3}H]-FPP or [^{3}H]-GGPP). In addition, the plasma membrane localization of Wrch-1 was found to be more sensitive to a palmitoylation inhibitor (2-Bromo-palmitate) as opposed to prenyl trasferase inhibitors (GGTase I and/or FTase). Mutation of the second cysteine residue (CCFV to CSFV) demonstrated that the second cysteine of this non-conventional motif was palmitoylated, as CSFV-mutant Wrch-1 resulted in a failure to incorporate [^{3}H]palmitate, and HA-tagged mutant Wrch-1protein was re-distributed from the plasma membrane to the cytosol and nucleus (Berzat, *et al.*, 2005). Palmitoylation of the *CCFV*-motif on Wrch-1, in addition to other possible sequences within its C terminus, may serve as a secondary membrane targeting signal, which may define a distinct Wrch-1 plasma membrane localization, and thus account for its divergent roles in cell physiology.

In conclusion, prenylation, palmitoylation and depalmitoylation might be interdependent processes, and the precise control over these lipid modifications is critical for traffic to and from the plasma membrane and or the shuttling membrane and content between intracellular compartments (Fukata, *et al.*, 2006). The importance of this precise control is borne out by the many examples of defects in post-translational lipid modifications that are linked to neurodegenerative diseases, such as Schizophrenia, X-linked mental retardation, Huntington disease, as well as colorectal and bladder cancers (Buff, *et al.*, 2007; Fukata & Fukata, 2010; Williams, 1991). Despite significant progress in identifying the enzymes involved in post-translational lipid modification, their specific protein targets and the relationship between lipid modifications and membrane targeting remain unclear. Recent advances in proteomics together with improved bioinformatics prediction algorithms (CSS-Palm, PrePS) have shed some light on these lipid related post-translational modifications in some proteins (Adamson, *et al.*, 1992; Fukata, *et al.*, 2006; Kang, *et al.*, 2008; Shmueli, *et al.*, 2010), although they don't provide.

5.3 Phosphorylation as a means of controlling protein activity

In the past decade, a significant regulatory role has been revealed for post-translational phosphorylation as a means of controlling exocytosis. Phosphorylation involves the covalent addition of a phosphate group to a specific protein residue (namely serine, threonine, tyrosine or histidine), and is catalysed by one of the many protein kinases found within the cell (Alberts, *et al.*, 2002). Phosphorylation can affect a protein in two ways; conformational change that can alter the proteins activity through an allosteric affect; and the attached phosphate can form part of a recognition domain that facilitates protein-protein interactions. These types of modifications are transient, and are often followed by de-phosphorylation, where the phosphate can be removed by the catalytic activity of a specific phosphatase. The analysis of protein activities and interactions that are controlled by phosphorylation and de-phosphorylation has been empowered by technologies such as: genetic manipulation in animal models; the generation and expression of phosphomimetic mutant proteins; and the generation of phosphospecific antibodies. This has facilitated the functional study of individual proteins as targets of particular protein kinases, whilst uncovering the significance of phosphorylation/dephosphorylation events in the regulation of exocytosis, such as in: synaptic transmission and cell plasticity (Amin, *et al.*, 2008; Barclay, *et al.*, 2003; Boczan, *et al.*, 2004), neuronal morphogenesis (Chernyshova, *et al.*, 2011), insulin secretion (Butelman, 1990; Sugawara, *et al.*, 2009; Wang & Thurmond, 2010), insulin stimulated GLUT4 transport (Aran, *et al.*, 2011; X. W. Chen, *et al.*, 2011a; Sano, *et al.*, 2011), mast cell and platelet degranulation (Fitzgerald & Reed, 1999; Foger, *et al.*, 2011), exocytosis of factors required for neutrophil adhesion (Fu, *et al.*, 2005), acrosomal exocytosis in sperm (Castillo Bennett, *et al.*, 2010; Zarelli, *et al.*, 2009) and lung surfactant exocytosis (Gerelsaikhan, *et al.*, 2011). It has now become obvious that phospho-regulation of exocytosis is a complex and dynamic process implicated at almost all points along the exocytic route, from recruitment and transport of vesicles to their ultimate fusion at the plasma membrane (Figure 1&2).

5.3.1 Phospho-regulation of small GTPases, their effectors and regulators

Recruitment and tethering of secretory vesicles that are destined to fuse with the plasma membrane involves GEFs, GAPs and small GTPases, as well as their effector proteins, all of which have been reported to undergo phospho-regulation by a number of different kinases. Phosphorylation appears to add an additional level of regulation to small GTPases beyond the GTP/GDP cycle (see 5.1 section above) and lipid modification/s (see 5.2 section above) modulating the intracellular localization of the protein and its function.

Aikawa and Martin identified a role for the plasma membrane bound GTPase Arf6 in specifying exocytosis competent plasma membrane microdomains (Aikawa & Martin, 2003). This process depends on the site-specific localization of the phospholipid kinase phosphatidylinositol 4-phosphate 5-kinase (PIP_5K), which itself is regulated by phosphorylation, and is stimulated by agonists and cell stresses. Arf6 regulates various cellular functions by activation of PIP_5K including the exocytosis of insulin and dense core vesicles, and neurotransmitter release (Funakoshi, *et al.*, 2011). Arf6 interacts with the dephosphorylated form of PIP_5Kc in PC12 cells when stimulated by depolarization of the cell (Aikawa & Martin, 2003). This interaction activates PIP_5Kc resulting in increased levels of the lipid messenger phosphatidylinositol 4,5-biphosphate (PI(4,5)P) at the plasma membrane, triggering exocytosis of dense core vesicles (Aikawa & Martin, 2003; Funakoshi, *et al.*, 2011).

Altered localization due to phosphorylation has been shown for the Rab3A and Rab27A effector, Rabphilin. Rabphilin, together with Rab3A and Rab27A, has been implicated in the modulation of the docking step for dense core and synaptic vesicles, destined for exocytosis (Deak, *et al.*, 2006; Lin, *et al.*, 2007; Tsuboi & Fukuda, 2006). The modulation of this docking step has recently been attributed to the interaction of Rabphilin with the plasma membrane SNARE protein, SNAP25 (Tsuboi, *et al.*, 2007; Tsuboi, 2009). Rabphilin can be phosphorylated by the kinases PKA, PKC and CaMKII. Further investigation revealed that calcium-mediated phosphorylation of non-vesicle bound Rabphilin at Ser234 and Ser 274 reduced its affinity for the vesicular membrane. Therefore, phosphorylation of Rabphilin modulates its membrane localization during synaptic transmission (Foletti, *et al.*, 2001; Lonart & Sudhof, 2001).

Recently, studies of pancreatic, colorectal, and other cancers identified RalGEFs as key effectors of Ras signaling cascades (Feldmann, *et al.*, 2010; Neel, *et al.*, 2011). Phosphorylation of constitutively active RalA (RalA^{G23V}) at a conserved C-terminal S194 residue, by Aurora A kinase, promotes collagenI-induced cell motility and anchorage-independent growth in MDCK epithelial cells (J. C. Wu, *et al.*, 2005). This phosphorylation caused RalA relocation from the plasma membrane to endomembranes, where RalA associated preferentially with its effector Ral binding protein 1 RalBP1/RLIP76, thereby promoting cell motility during human cancer cell metastasis (Lim, *et al.*, 2010; Z. Wu, *et al.*, 2010). Similarly, phosphorylation of RalB by PKC at the C-terminal S198 residue led to its translocation from the plasma membrane to the perinuclear region of the cell. S198 phosphorylation of a constitutively activated G23V RalB mutant enhanced tumor growth in experimental lung metastasis of T24 or UMUC3 human bladder cancer cells (Wang, *et al.*, 2010).

RalA and RalB are present on synaptic vesicles, platelet granules and glucose transporting GLUT4 vesicles, and are involved in the assembly of the octameric exocyst complex. During insulin stimulated GLUT4 vesicle transport, phosphorylation of the RalGAP complex by protein kinase Akt2 (downstream of PI3-Kinase) relieves its inhibitory effect on RalA activity, and thereby increases GLUT4 exocytosis at the plasma membrane (Chen, *et al.*, 2011b; Chen & Saltiel, 2011). Intriguingly, phosphorylated RalGAP complex was shown to be capable of binding RalA *in vitro*, and the authors suggested that in cells this phosphorylation may act through sequestering RalGAP in the cytoplasm, away from site of RalA action (X. W. Chen, *et al.*, 2011b). Once activated, GTP-bound RalA protein interacts with Sec5 protein to assemble vesicular and plasma membrane subunits of the exocyst complex. Protein kinase C (PKC) then promotes exocytosis though the phosphorylation of the G-protein binding domain of Sec5 to disengage RalA from the exocyst, allowing GLUT4 exocytosis in adipocytes (X. W. Chen, *et al.*, 2011a). Phosphorylation has been implicated in the modulated function of another exocyst subunit, Sec8 (B. H. Kim, *et al.*, 2011). The FGF receptor mediated tyrosine phosphorylation of Sec8 was required for efficient recruitment of the exocyst complex to growth cones of growing neurites. However, the exact mechanism by which phosphorylation modulates Sec 8 function still remains to be elucidated.

In a similar scenario to the RalGAP complex, phosphorylation of the RabGAP, AS160, relieves its inhibitory activity on Rab10. Insulin activates the PI3 kinase/Akt signaling pathway where Akt phosphorylates AS160 and inhibits its GAP activity, leading to increased levels of active Rab10-GTP and triggering movement of GLUT4 containing vesicles to the plasma membrane ready for fusion (Sano, *et al.*, 2011). Vasopressin-stimulated activation of Akt leads to phosphorylation of AS160 at Ser473 in kidney collecting duct cells, and this phosphorylation enhances the translocation of the Aquaporin water channel (AQP2) to the apical membrane (H. Y. Kim, *et al.*, 2011). Although it is clear that phosphorylation plays a role in modulating GAP function, its role in modulating GEF function remains unclear and there are relatively few examples in the literature (Schmidt & Hall, 2002). However, phosphorylation of the GEF Vav by Src and Syk tyrosine kinases, has been shown to result in stimulation of its catalytic activity (Aghazadeh, *et al.*, 2000). Phosphorylation of Tyr174 induced a conformational change in Vav's N-terminal region, subsequently allowing access to Rac GTPase.

Rho GTPases, particularly Rac2, also play a crucial role in actin remodeling in response to pathogen invasion, regulating neutrophil degranulation, and the release of antimicrobial mediators. A recent proteomic analysis of Rac2-mediated degranulation in neutrophils identified several actin-binding proteins as novel downstream effectors of this pathway, including Coronin1A. This Rac2 function is mediated by phosphorylation of Coronin1A at Thr418, after secretagogue-stimulation (Eitzen, *et al.*, 2011). Phosphorylation of Coronin1A at Ser2 has also been implicated in modulating degranulation in mast cells (Foger, *et al.*, 2011).

Thus, specific phosphorylation events modulate the engagement and disengagement of small GTPases with their regulators and effector proteins, adding an additional tier of temporal and spatial control over this vesicular machinery.

5.3.2 Phospho-regulation of SNARE complex assembly

There is emerging evidence to suggest that phosphorylation plays a significant role in the regulation of the final steps of exocytosis through the membrane bound tethering of SNARE family proteins and their accessory proteins (Figure 1&2). For example, phosphorylation modulates the association of SNARE accessory proteins with v- and t-SNAREs adding another level of regulation to SNARE complex formation.

A number of phosphorylation events have been found to regulate the interaction of the SNARE accessory proteins Sec1/Munc18c with SNARE proteins. In most, but not all systems, Munc18 has an inhibitory role in SNARE complex formation, and phosphorylation of the Munc18 isoforms by different kinases alleviates this inhibitory role. Exocytosis of Weibel-Palade bodies (WPBs) by endothelial cells in response to inflammatory mediators, such as Thrombin, is important for initiating the adhesion of neutrophils. Thrombin, acting in conjunction with an influx of intracellular calcium, activates PKCα, resulting in the phosphorylation of Munc18c and Syntaxin4. This leads to the dissociation of the inhibitory complex, allowing SNARE complex formation and fusion of WPBs at the plasma membrane (Fu, *et al.*, 2005). In another cell system, phosphorylation of Munc18c at Tyr521 by the insulin receptor tyrosine kinase in pancreatic β-cells was shown to inhibit its interaction with Syntaxin4 and VAMP2 (Aran, *et al.*, 2011). Similarly, in neurotransmitter release, phosphorylation of Munc18a at Ser 306 and Ser313 by PKC reduces its affinity for Syntaxin and the phosphorylation state of Ser 313 alone was sufficient to alter the dynamics of vesicle release events (Barclay, *et al.*, 2003). In support of the importance of Ser313 phosphorylation, the phosphomimetic mutant S313D increased vesicle docking and enhanced the ready releasable vesicle pool in adrenal chromaffin cells (Nili, *et al.*, 2006). In contrast to these examples of Munc18 as a negative regulator of SNARE complex formation, there have been reports of Munc18 promoting fusion by stabilising SNARE complex formation. In parietal epithelial cells, gastric acid exocytosis involved Cdk5 phosphorylation of Munc18b at Thr572, which enhanced its association with the Syntaxin3 - SNAP25 complex, and facilitated vesicle docking and fusion (Y. Liu, *et al.*, 2007). This functional difference might be attributed to the distinct modes by which Munc18 binds with SNARE proteins. In its inhibitory role, Munc18 binding to the SNARE motif of monomeric Syntaxin disables SNARE complex formation. In its fusion promoting role, the N-terminal domain of Munc18 binds to the N-terminal Habc-domain of Syntaxin, which allows it to fold back onto the SNARE complex. When v-SNARE proteins are present, Munc18 exhibits a binding capacity that enables it to clasp the SNARE complex and promote fusion (reviewed in (Burgoyne, *et al.*, 2009; Sudhof & Rothman, 2009)). It is obvious that phosphorylation is an important modulator of Munc18 and that Munc18 activity can differ depending on the isoform and cell system involved.

In addition to promoting Munc18 and Syntaxin interaction, Cdk5 can also phosphorylate the filamentous protein Septin5 at Ser327 and Ser161, with the former site resulting in dissociation of Septin5 from Syntaxin at the synapse (Amin, *et al.*, 2008). It has been suggested that Septin5 filaments may act as tethers that prevent access of exocytic vesicles to the membrane, until there is appropriate signaling to fuse and release their contents (Beites, *et al.*, 2005). Phosphorylation of the Ser327 and Ser161 sites can affect Septin-Septin filament assembly (Amin, *et al.*, 2008), and Cdk5 phosphorylation of Septin5 results in its dissociation from the vesicle associated protein Synapsin. This may play an important role in modulating vesicle interaction with the SNARE complex (Amin, *et al.*, 2008).

Syntaphilin, a Syntaxin1 clamp that controls SNARE complex assembly, competes with SNAP25 for Syntaxin1 binding, and inhibits SNARE complex formation by regulating the availability of free Syntaxin during synaptic vesicle exocytosis. PKA phosphorylation of Syntaphilin at Ser43 in PC12 cells modulates its interaction with Syntaxin1 and annuls its inhibitory affect on SNARE complex formation and exocytosis (Boczan, *et al.*, 2004). Another SNARE regulatory protein that interacts with Syntaxin1 and inhibits SNARE complex formation is Tomosyn (Fujita, *et al.*, 1998). PKA phosphorylation of Tomosyn at Ser724 reduces its association with Syntaxin1 and promotes SNARE complex formation increasing the size of the readily releasable pool of synaptic vesicles in neurons (Baba, *et al.*, 2005). In contrast, phosphorylation of Syntaxin1 by the Rho-associated coiled-coiled forming kinase (ROCK) at Ser14, positively regulates the Syntaxin1-Tomosyn association by increasing Syntaxin1 affinity for Tomosyn. This inhibits SNARE complex formation during neurite extension in hippoccampal cultured neurons (Sakisaka, *et al.*, 2004).

Protein phosphorylation of the membrane fusion machinery plays a central role sperm activation in keeping acrosomal exocytosis on hold until the sperm contacts the egg. N-ethylmaleimide-sensitive factor 5 (NSF5), which is necessary for the disassembly of fusion incompetent cis-SNARE complexes, and the calcium sensor SynaptotagminVI, are phosphorylated in resting sperm and inactive. Dephosphorylation of NSF5 by protein-tyrosine phosphatase PTP1B results in its activation and allows disassembly of cis-SNARE complexes, a requirement for subsequent fusion competent trans-SNARE complex assembly and membrane fusion (Zarelli, *et al.*, 2009). Similarly dephosphorylation of SynaptotagminVI by the calcium-dependent phosphatase Calcineurin, was shown to be a requirement for acrosomal exocytosis (Castillo Bennett, *et al.*, 2010). Snapin is a SNARE-binding protein that enhances the association of Synaptotagmin with the SNARE complex. Phosphorylation of Snapin at Ser50 by PKA enhances its interaction with SNAP25 and resulted in increased binding of Synaptotagmin to the SNARE complex in chromaffin cells (Chheda, *et al.*, 2001). In contrast to Snapin, phosphorylation of the synaptic vesicle membrane protein CSP (a cysteine string protein) by PKA (on Ser10) inhibited its interaction with Syntaxin and Synaptotagmin (Evans, *et al.*, 2001; Evans & Morgan, 2002). More recently synapse specific localisation of phosphorylated CSP was shown in rat brain suggesting a role in synapse specific regulation of neurotransmitter release (Evans & Morgan, 2005).

There are clearly a multitude of SNARE accessory proteins that regulate SNARE complex assembly, particularly in the central nervous system. Phosphorylation/dephosphorylation modulates their association with SNARE complex proteins and this involves a number of signaling pathways.

5.3.3 Phospho-regulation beyond fusion

A recent and elegant *intra-vital* study demonstrated a role for filamentous actin and non-muscle Myosin2 in the secretion of cargo from vesicles. In the salivary glands of transgenic mice, real-time live imaging of secretory vesicles showed that shortly after a vesicle docks at the plasma membrane it becomes coated with actin filaments (Masedunskas, *et al.*, 2011). Furthermore, it was found that to release the content of these vesicles, Myosin2a and 2b recruitment and activity were required to provide the contractile force necessary to complete fusion of the vesicle with the plasma membrane (Masedunskas, *et al.*, 2011). Similar observation has been reported for the secretion of von Willebrand factor from human endothelial cells (Nightingale, *et al.*, 2011).

Phospho-regulation of the non-muscle associated actin molecular motor, Myosin2, plays a role in modulating its function at the post fusion level during exocytosis. Inhibition of phosphorylation by the myosin light chain kinase (MLCK) at sites Tyr18 and Ser19 of Myosin2, slowed down the opening of the fusion pore, during the release of catecholamines and peptide transmitters in chromaffin cells (Doreian, *et al.*, 2008; Doreian, *et al.*, 2009; Neco, *et al.*, 2008). Similarly, phosphorylation at Ser 19 of Myosin2 by MLCK was shown to be necessary for maintaining the opening of the fusion pore in pancreatic cells (Bhat & Thorn, 2009). In addition to Myosin2, phosphorylation by PKC of Myristoylated alanine-rich C-kinase substrate (MARCKS), another actin-associated protein, has been implicated in regulating the activity dependent rearrangement of the actin cytoskeleton. Phosphorylation of both MARCKS and Myosin2 have been implicated in modulating the transition from an omega kiss-and-run mode of exocytosis involving a narrow fusion pore, to the full granule collapse mode involving fusion pore expansion (Doreian, *et al.*, 2009).

5.3.4 Multiple phosphorylation sites modulated by single or multiple kinases

There is increasing evidence of multiple phosphorylation sites within individual proteins, which are phospho-regulated by single or multiple kinases/phosphatases. This brings to light the intriguing possibility of a convergence of different signaling events on key components of the vesicular machinery. The dynamic interplay of phosphorylation events can alter a proteins' physiological function. An example of this is phosphorylation of the Rab11 effector, Rab11 interacting protein (Rip11), where phosphorylation at Ser 357 by PKA modulates the recruitment of insulin granules to the plasma membrane (Sugawara, *et al.*, 2009). In addition, a non-PKA dependent phosphorylation at a serine/threonine site by an as yet unidentified kinase was shown to be important for its role in apical membrane recycling in MDCK cells (Prekeris, *et al.*, 2000; Sugawara, *et al.*, 2009). Phosphorylation of SNAP25 by the kinases PKA (at Thr138) and PKC (at Ser187) respectively modulated the size of the releasable vesicle pool and rate of refilling after the pools were emptied in chromaffin cells (Nagy, *et al.*, 2004). Similarly, differential phosphorylation of the synaptic vesicle protein Synapsin has been implicated in modulating its various roles including neurotransmitter release, vesicle clustering, maintaining the reserve pool, and vesicle delivery to the active zones. These processes are regulated via a dynamic phospho-regulation cycle which involves multiple phosphorylation sites and several kinases including cAMP-dependent protein kinase A, PKA (at site 1 (Ser9)) (Angers, *et al.*, 2002; Menegon, *et al.*, 2006), Ca2+/calmodulin-dependent kinase CaMKII and VI (at sites 1, 2 and 3 (Ser9, Ser566 and ser603)) (Chi, *et al.*, 2003), mitogen-activated kinase MAPK (sites 4, 5, 6 and 7 (ser62, Ser67 and Ser549 and Ser551)) (Chi, *et al.*, 2003; Giachello, *et al.*, 2010), and tyrosine kinase Src (site 8 (Tyr301)) (Messa, *et al.*, 2010). Phosphorylation on serine residues upon activation of PKA, CaMK and MAPK signaling pathways, promotes the dissociation of Synapsin from synaptic vesicles and/or the actin network which results in trafficking of synaptic vesicles from the reserve pool to the ready releasable pool for exocytosis (Chi, *et al.*, 2003; Giachello, *et al.*, 2010; Menegon, *et al.*, 2006). In contrast, Src kinase-mediated phosphorylation of Synapsin enhances its oligomerization and increases its association with synaptic vesicles and the cytoskeleton, stimulating the re-clustering of recycled vesicles and subsequent recruitment to the reserve pool (Messa, *et al.*, 2010). In addition, phosphorylation at different Synapsin sites can occur concurrently through the selectivity of kinase/phosphatase activation, which is dependant on the stimulus and the signaling pathways implicated. For example, in synaptosomal preparations, calcium entry stimulated bidirectional phospho-regulation of Synapsin involving phosphorylation at CaMKII dependent and PKA dependent sites and dephosphorylation at MAPK/Calcineurin sites (Cesca, *et al.*, 2010; Jovanovic, *et al.*, 2001; Yamagata, *et al.*, 2002). Synapsin phospho-regulation involving multiple signaling pathways allows Synapsin to control synaptic vesicle mobilisation and trafficking.

Phosphorylation at different sites within an individual protein can also modulate a proteins' function through the differential kinetics of phosphorylation/dephosphorylation events, as in the case of Rabphilin. Rab3A recruits Rabphilin to synaptic vesicles where it can undergo phosphorylation during membrane depolarisation stimulated calcium influx. Rabphilin Ser234 and Ser274 are phosphorylated by different kinases (PKA for the former site and PKA, PKC, CaMKII for the latter) and can show distinct regulatory effects on Rabphilin during synaptic transmission, depending upon the extent of phosphorylation at these two sites and the kinetics of dephosphorylation after stimulus removal (Foletti, *et al.*, 2001).

Regulation of a proteins' function by phosphorylation may occur in a hierarchical sequence. For instance, RhoGDI undergoes sequential phosphorylation in response to glucose stimulation in β-cells (Wang & Thurmond, 2010). The first phosphorylation event involves Tyr156 and coincides with RhoGDI-Cdc42 dissociation. This is then followed by Ser101/ Ser174 phosphorylation coinciding with RhoGDI-Rac1 dissociation. The sequential phosphorylation of RhoGDI allows differential temporal activation of the Rho GTPases, Cdc42 and Rac1, during insulin secretion from pancreatic β-cells. (Wang & Thurmond, 2010). The yeast tethering Rab GTPase, Sec4, has multiple phosphorylation sites (Ser8, Ser11, Ser201 and Ser204), and phosphorylation of the N-terminal serines (Ser8, Ser11) prevents binding to its effector, the Sec15 exocyst subunit, and hinders polarised exocytosis (Heger, *et al.*, 2011). The authors also identified protein phosphatase 2A (when containing the regulatory subunit Cdc55) as the phosphatase responsible for alleviating the inhibitory affect of Sec4 phosphorylation (Heger, *et al.*, 2011). Structural analysis of Sec4 suggests a clustering and physical proximity of the N- and C-terminal phosphorylation sites. This has led the authors to postulate that the impact of the N-terminal serines can be modulated by phosphorylation at the C-terminal serines and may involve phosphorylation at these sites in a hierarchical manner (Heger, *et al.*, 2011), although further studies are required to verify this assertion.

Finally, the unique combination of tandem phosphorylated sites within a protein can act as recognition sites for phosphoprotein-binding proteins, such as 14-3-3. An increasing number of proteins are being identified with tandem sites that can act as 14-3-3 dimer binding sites

(Chen, *et al.*, 2011). These sites can be phosphorylated by distinct protein kinases and the combination of phosphorylated sites can alter the effect of 14-3-3 on the protein target. 14-3-3 has been shown to bind to the Rab GAPs, AS160 (Ser341 and Thr642) and TBC1D1 (Ser237 and Thr596), which are involved in the regulation of GLUT4 trafficking to the plasma membrane, and 14-3-3s' interaction with these two proteins occurs in response to insulin and energy stress respectively (reviewed (Chen, *et al.*, 2011)). 14-3-3 binding sites have been identified on Rab3A effectors - the Rab3A interacting molecules Rim1 and Rim2 and Rabphilin3 (Sun, *et al.*, 2003). The physiological relevance of the 14-3-3 interaction with these proteins in neuroendocrine exocytosis and synaptic transmission is still being investigated.

Exocytosis requires the orchestrated actions of distinct exocytic machinery that is governed by multiple signaling pathways, and this occurs in a temporally and spatially regulated manner that is dependent on the types of cells and their stimuli. Phospho-regulation of the exocytic machinery is one of the mechanisms by which the cell coordinates the function of these proteins during exocytosis and it is implicated at each stage of the process. It is clear from the literature that the proteins implicated in exocytosis, which contain multiple phosphorylation sites, can be differentially modulated by single or multiple kinases. With the identification of an increasing number of phosphorylation targets and the elucidation of the precise functional roles of the exocytic machinery, the physiological significance of these phospho-regulation events may be determined. A huge task lays ahead to delineate the functional role of each phosphorylation site for all the exocytosis players and then integrate this information into a comprehensive model that can define the signaling pathways that are responsible for modulating these events.

6. Ubiquitin and small ubiquitin-like modifier in exocytosis

6.1 Ubiquitination

Post-translational modification with ubiquitin has also been recognised as an important sorting signal on cargo transported by the endosomal network (specifically as a signal for internalisation), particularly at the late endosome (LE)/multivesicular body (MVB) and at the trans-Golgi apparatus. For example, at the LE/MVB, the proteins that make up the endosomal sorting complex required for transport (ESCRT) machinery (ESCRT I and ESCRT II) are known to contain ubiquitin-binding domains that enable them to recognise ubiquitinated cargo proteins and sort them into internalised vesicles destined for lysosomal degradation or for secretion events (reviewed in (Hurley, 2010)).

Ubiquitin has an established role in regulating protein relocation and targeted destruction at the proteasome (see (Hershko & Ciechanover, 1998; Hershko, 2005) for some excellent reviews). The three Rho GTPases, Rho1/A, Rac1, and Cdc42 have now been demonstrated to be ubiquitinated and degraded under certain stimuli (de la Vega, *et al.*, 2011) (Figure 1&2). Furthermore, both Rac1 and Cdc42 ubiquitination and protein levels are increased when cells are treated with protease inhibitors (Doye, *et al.*, 2006). Inactive Rho1/A was shown to be ubiquitinated by the E3 ubiquitin ligase Smurf1 and degraded in migrating Mv1Lu epithelial cells (H. R. Wang, *et al.*, 2003). Ubiquitination of Rho1/A may be required to prevent the Rho1/A mediated formation of actin stress fibres at the leading edge of migrating cells, and to allow the Cdc42 and Rac1 mediated dynamic actin rearrangement necessary for anterograde delivery of membranes to the leading edge of migrating cells (Y. Wang, *et al.*, 2003). This site specific ubiquitination/degradation of Rho1/A appears to be restricted to the lamellipodia and filopodia of the leading edge, where Smurf1 is recruited through atypical protein kinase C zeta (aPKC)-mediated phosphorylation (H. R. Wang, *et al.*, 2003). The latter is activated by the Cdc42/Rac1 polarity complex (H. R. Wang, *et al.*, 2003). Hence the polarised exocytosis during cell migratory activity is regulated by a hierarchy of post-translational modifications, where ubiquitination of Rho1/A appears to be important for switching between the competing actin modifying activities of Rho1/A and Cdc42/Rac1, which in turn controls phosphorylation dependent ubiquitination activity of Smurf, and thereby Rho degradation. A further level of complexity is added by the down-regulation of Rho1/A in migrating cells by another E3 ligase, the Cul3/BACURD complex (Chen, *et al.*, 2009). The Cul3/BACURD complex is a ring finger E3 ubiquitin ligase complex that has been shown to ubiquitinate Rho1/A in a diverse range of organisms from human cell lines (293T and HeLa fibroblasts), insect cells (*Drosophila melanogaster* S2 cells) and amphibians (*Xenopus laevis* embryos). Depletion of the Cul3 and BACURD ligase complex by siRNA results in defective migration of HeLa cells and mouse embryonic fibroblasts, and in embryonic abnormalities resulting from defective cell migration in *Xenopus* embryos (Chen, *et al.*, 2009). There is also evidence for the ubiquitination of Rac1 by the ubiquitin E3 ligase POSH2 but the purpose of this ubiquitination is not yet clear (Karkkainen, *et al.*, 2010). A proteasomal degradation resistant and thus constitutively active mutant of Rac1 (Rac1b), is found in colorectal and breast cancer tumour cells (Jordan, *et al.*, 1999; Schnelzer, *et al.*, 2000). Interestingly, RNAi mediated silencing of this mutant results in a failure of cancer cells to undergo an epithelial to mesenchymal transition (Radisky, *et al.*, 2005) suggesting a role for ubiquitination of Rac1 in controlling cell motility (Visvikis, *et al.*, 2008).

Rho GTPases can also be regulated by the ubiquitination of their GEF activators. Activation of Rho1/A via ubiquitination of PDZ-RhoGEF, was found to be initiated by Cul3/KLHL20 (Lin, *et al.*, 2011). Likewise, Cdc42 can be activated via ubiquitination of its GEF, hPEM-2 (Yamaguchi, *et al.*, 2008). It is yet to be established whether Smurf-mediated regulation these two Rho GTPases occurs in a coordinated manner. In addition to Smurf, Cdc42 activity could be regulated by ubiquitination and or proteasomal degradation of its GEFs, FGD1 and FGD3 by E3 ligase $SCF^{FWD1/\beta\text{-}TrCP}$ (Hayakawa, *et al.*, 2005; Hayakawa, *et al.*, 2008). There is an interesting interplay between the regulatory effects of ubiquitination and phosphorylation. At the leading edge of migrating cells, $SCF^{FWD1/\beta\text{-}TrCP}$ ligase recognises only forms of GEFs inactivated by GSK-3β kinase phosphorylation; the latter kinase could in turn be inactivated by aPKC-mediated phosphorylation (Etienne-Manneville & Hall, 2003; H. R. Wang, *et al.*, 2003). Therefore, phosphorylation by aPKC appears to be at the nexus of regulation of ubiquitination of small Rho GTPases, promoting degradation in Rho1/A and preventing it in CDC42. Finally, two of the small Rab GTPase GEFs, Rabex5 and Rabring7 (Xu, *et al.*, 2010; Yan, *et al.*, 2010) are known to have ubiquitin E3 ligase activity (Sakane, *et al.*, 2007), and Rabex5 cellular localisation is regulated by its ability to bind a ubiquitin signal (Mattera, *et al.*, 2006). As yet there are no known Rab proteins that are themselves ubiquitinated.

Ubiquitination of the negative regulators of small Rho GTPases, RhoGDIs, has also been shown. RhoGDI is ubiquitinated by the E3 ubiquitin ligase GRAIL which, while not resulting in its proteasomal degradation, did appear to increase the stability the RhoGDI protein (Su, *et al.*, 2006). This results in sequestration of Rho molecules in the cytosol, blocking their activation and initiation of the Rho signaling pathway, and thereby impairing cytoskeletal polarization or actin polymerization. It is yet to be defined why, in the context of GRAIL-mediated ubiquitination, RhoGDI inhibition is restricted to Rho1/A, but not Rac1 or Cdc42 (Su, *et al.*, 2006).

6.2 Sumoylation

Sumo is a small ubiquitin-like modifier that, like ubiquitin, can be covalently attached to a protein via an internal lysine residue on the target and can serve to modify its function (for recent reviews see (Wang & Dasso, 2009; Wilkinson & Henley, 2010)). Sumoylation is emerging an additional level of control over the proteins that regulate exocytosis (Figure 2). At least two SNARE proteins are believed to be sumoylated. Sumoylation of the SNARE accessory protein, Tomosyn, relieves its inhibitory effect on SNARE complex assembly, and thereby on exocytosis (Williams, *et al.*, 2011). In response to Ca^{2+} signaling, sumoylation is known to inhibit exocytosis of insulin granules following their docking at the plasma membrane, and this is most likely to occur through SynaptotagminVII (Dai, *et al.*, 2011). In addition, the Rho GTPase Rac1 was found to be sumoylated in response to hepatocyte growth factor stimulation of a number of cell lines (HEK293T, MDCKII, HeLa, and Cos7 cells (Castillo-Lluva, *et al.*, 2010)). Sumoylation of Rac1 resulted in sustained activation of Rac1 which promoted the formation of lamellipodia and cell motility (Castillo-Lluva, *et al.*, 2010). Sumoylation as a post-translational modification of the proteins in the exocytic pathway is a new field of research and will undoubtedly be found to regulate many more of these proteins.

7. Concluding remarks

Here we have illustrated that there is a complex array of specialist molecular machinery that is used to control each step in the process of exocytosis. Emerging evidence suggests that there is a highly organised regulatory network required to achieve control of exocytosis. This involves the post-translational modification of the vesicular machinery and membrane associated proteins that orchestrate exocytosis; including the addition of lipid moieties, phosphorylation, and ubiquitination and sumoylation. These post-translational modifications are responsible for mediating protein intracellular localization, protein-protein interactions, complex assembly, and ultimately protein function. The dynamics and precision of exocytosis often require multiple modifications of a single protein in order to tightly control temporal/spatial function. Moreover, to ensure the harmonious reaction of the cell to a specific stimulation, these post-translational modifications respond to a variety of cell-type specific signaling events. The challenge facing researchers in this field is to investigate the cross-talk between different modifications in the context of a specific signal, and to determine how these are coordinated with other cellular functions. Thus, it is tempting to speculate about an even higher point of control in the regulation of exocytosis, involving proteins that recognise post-translational modifications and facilitate appropriate functional interaction.

8. References

Adamson, P., Marshall, C. J., Hall, A. & Tilbrook, P. A. (1992). Post-translational modifications of p21rho proteins. *The Journal of biological chemistry*. Vol. 267, No.28 (October 1992), pp.20033-20038, 0021-9258

Aghazadeh, B., Lowry, W. E., Huang, X. Y. & Rosen, M. K. (2000). Structural basis for relief of autoinhibition of the Dbl homology domain of proto-oncogene Vav by tyrosine phosphorylation. *Cell*. Vol. 102, No.5 (September 2000), pp.625-633, 0092-8674

Aicart-Ramos, C., Valero, R. A. & Rodriguez-Crespo, I. (2011). Protein palmitoylation and subcellular trafficking. *Biochimica et biophysica acta*. Vol. 1808, No.12 (December 2011), pp.2981-2994, 0006-3002

Aikawa, Y. & Martin, T. F. (2003). ARF6 regulates a plasma membrane pool of phosphatidylinositol(4,5)bisphosphate required for regulated exocytosis. *The Journal of cell biology*. Vol. 162, No.4 (August 2003), pp.647-659, 0021-9525

Alberts, B., Johnson, A., Lewis, J., Raff, M., Roberts, K. & Walter, P. (2002). *Molecular Biology of the Cell* (fourth edition), Garland Science, ISBN-10: 0-8153-3218-1, *New York*.

Albright, C. F., Giddings, B. W., Liu, J., Vito, M. & Weinberg, R. A. (1993). Characterization of a guanine nucleotide dissociation stimulator for a ras-related GTPase. *The EMBO journal*. Vol. 12, No.1 (January 1993), pp.339-347, 0261-4189

Ali, B. R. & Seabra, M. C. (2005). Targeting of Rab GTPases to cellular membranes. *Biochemical Society transactions*. Vol. 33, No.Pt 4 (August 2005), pp.652-656, 0300-5127

Amin, N. D., Zheng, Y. L., Kesavapany, S., Kanungo, J., Guszczynski, T., Sihag, R. K., Rudrabhatla, P., Albers, W., Grant, P. & Pant, H. C. (2008). Cyclin-dependent kinase 5 phosphorylation of human septin SEPT5 (hCDCrel-1) modulates exocytosis. *The Journal of neuroscience*. Vol. 28, No.14 (April 2008), pp.3631-3643, 0270-6474

Anant, J. S., Desnoyers, L., Machius, M., Demeler, B., Hansen, J. C., Westover, K. D., Deisenhofer, J. & Seabra, M. C. (1998). Mechanism of Rab geranylgeranylation: formation of the catalytic ternary complex. *Biochemistry*. Vol. 37, No.36 (September 1998), pp.12559-12568, 0006-2960

Andres, D. A., Seabra, M. C., Brown, M. S., Armstrong, S. A., Smeland, T. E., Cremers, F. P. & Goldstein, J. L. (1993). cDNA cloning of component A of Rab geranylgeranyl transferase and demonstration of its role as a Rab escort protein. *Cell*. Vol. 73, No.6 (June 1993), pp.1091-1099, 0092-8674

Angers, A., Fioravante, D., Chin, J., Cleary, L. J., Bean, A. J. & Byrne, J. H. (2002). Serotonin stimulates phosphorylation of Aplysia synapsin and alters its subcellular distribution in sensory neurons. *The Journal of neuroscience*. Vol. 22, No.13 (July 2002), pp.5412-5422, 0270-6474

Angus, S. P., Solomon, D. A., Kuschel, L., Hennigan, R. F. & Knudsen, E. S. (2003). Retinoblastoma tumor suppressor: analyses of dynamic behavior in living cells reveal multiple modes of regulation. *Molecular and cellular biology*. Vol. 23, No.22 (November 2003), pp.8172-8188, 0270-7306

Aran, V., Bryant, N. J. & Gould, G. W. (2011). Tyrosine phosphorylation of Munc18c on residue 521 abrogates binding to Syntaxin 4. *BMC biochemistry*. Vol. 12 (May 2011), pp.19, 1471-2091

Baba, T., Sakisaka, T., Mochida, S. & Takai, Y. (2005). PKA-catalyzed phosphorylation of tomosyn and its implication in Ca2+-dependent exocytosis of neurotransmitter. *The Journal of cell biology*. Vol. 170, No.7 (September 2005), pp.1113-1125, 0021-9525

Baekkeskov, S. & Kanaani, J. (2009). Palmitoylation cycles and regulation of protein function (Review). *Molecular membrane biology*. Vol. 26, No.1 (January 2009), pp.42-54, 0968-7688

Baisamy, L., Jurisch, N. & Diviani, D. (2005). Leucine zipper-mediated homo-oligomerization regulates the Rho-GEF activity of AKAP-Lbc. *The Journal of biological chemistry*. Vol. 280, No.15 (April 2005), pp.15405-15412, 0021-9258

Baisamy, L., Cavin, S., Jurisch, N. & Diviani, D. (2009). The ubiquitin-like protein LC3 regulates the Rho-GEF activity of AKAP-Lbc. *The Journal of biological chemistry*. Vol. 284, No.41 (October 2009), pp.28232-28242, 0021-9258

Balasubramanian, N., Meier, J. A., Scott, D. W., Norambuena, A., White, M. A. & Schwartz, M. A. (2010). RalA-exocyst complex regulates integrin-dependent membrane raft exocytosis and growth signaling. *Current biology : CB*. Vol. 20, No.1 (January 2010), pp.75-79, 0960-9822

Balch, W. E., Dunphy, W. G., Braell, W. A. & Rothman, J. E. (1984). Reconstitution of the transport of protein between successive compartments of the Golgi measured by the coupled incorporation of N-acetylglucosamine. *Cell*. Vol. 39, No.2 Pt 1 (Deceber 1984), pp.405-416, 0092-8674

Barclay, J. W., Craig, T. J., Fisher, R. J., Ciufo, L. F., Evans, G. J., Morgan, A. & Burgoyne, R. D. (2003). Phosphorylation of Munc18 by protein kinase C regulates the kinetics of exocytosis. *The Journal of biological chemistry*. Vol. 278, No.12 (March 2003), pp.10538-10545, 0021-9258

Barr, F. & Lambright, D. G. (2010). Rab GEFs and GAPs. *Current opinion in cell biology*. Vol. 22, No.4 (August 2010), pp.461-470, 0955-0674

Beites, C. L., Campbell, K. A. & Trimble, W. S. (2005). The septin Sept5/CDCrel-1 competes with alpha-SNAP for binding to the SNARE complex. *The Biochemical journal*. Vol. 385, No.Pt 2 (January 2004), pp.347-353, 0264-6021

Bendezu, F. O. & Martin, S. G. (2011). Actin cables and the exocyst form two independent morphogenesis pathways in the fission yeast. *Molecular biology of the cell*. Vol. 22, No.1 (January 2011), pp.44-53, 1059-1524

Berzat, A. C., Buss, J. E., Chenette, E. J., Weinbaum, C. A., Shutes, A., Der, C. J., Minden, A. & Cox, A. D. (2005). Transforming activity of the Rho family GTPase, Wrch-1, a Wnt-regulated Cdc42 homolog, is dependent on a novel carboxyl-terminal palmitoylation motif. *The Journal of biological chemistry*. Vol. 280, No.38 (September 2005), pp.33055-33065, 0021-9258

Berzat, A. C., Brady, D. C., Fiordalisi, J. J. & Cox, A. D. (2006). Using inhibitors of prenylation to block localization and transforming activity. *Methods in enzymology*. Vol. 407, (June 2006), pp.575-597, 0076-6879

Bhat, P. & Thorn, P. (2009). Myosin 2 maintains an open exocytic fusion pore in secretory epithelial cells. *Molecular biology of the cell*. Vol. 20, No.6 (March 2009), pp.1795-1803, 1059-1524

Boczan, J., Leenders, A. G. & Sheng, Z. H. (2004). Phosphorylation of syntaphilin by cAMP-dependent protein kinase modulates its interaction with syntaxin-1 and annuls its inhibitory effect on vesicle exocytosis. *The Journal of biological chemistry*. Vol. 279, No.18 (April 2003), pp.18911-18919, 0021-9258

Bonifacino, J. S. & Glick, B. S. (2004). The mechanisms of vesicle budding and fusion. *Cell*. Vol. 116, No.2 (January 2004), pp.153-166, 0092-8674

Bourne, H. R., Sanders, D. A. & McCormick, F. (1991). The GTPase superfamily: conserved structure and molecular mechanism. *Nature*. Vol. 349, No.6305 (January 1991), pp.117-127, 0028-0836

Brymora, A., Valova, V. A., Larsen, M. R., Roufogalis, B. D. & Robinson, P. J. (2001). The brain exocyst complex interacts with RalA in a GTP-dependent manner: identification of a novel mammalian Sec3 gene and a second Sec15 gene. *The Journal of biological chemistry*. Vol. 276, No.32 (August 2001), pp.29792-29797, 0021-9258

Buff, H., Smith, A. C. & Korey, C. A. (2007). Genetic modifiers of *Drosophila* palmitoyl-protein thioesterase 1-induced degeneration. *Genetics*. Vol. 176, No.1 (May 2007), pp.209-220, 0016-6731

Burgoyne, R. D. & Morgan, A. (2003). Secretory granule exocytosis. *Physiological reviews*. Vol. 83, No.2 (April 2003), pp.581-632, 0031-9333

Burgoyne, R. D., Barclay, J. W., Ciufo, L. F., Graham, M. E., Handley, M. T. & Morgan, A. (2009). The functions of Munc18-1 in regulated exocytosis. *Annals of the New York Academy of Sciences*. Vol. 1152, (January 2009), pp.76-86, 0077-8923

Butelman, E. R. (1990). The effect of NMDA antagonists in the radial arm maze task with an interposed delay. *Pharmacology, biochemistry, and behavior*. Vol. 35, No.3 (March 1990), pp.533-536, 0091-3057

Calero, M., Chen, C. Z., Zhu, W., Winand, N., Havas, K. A., Gilbert, P. M., Burd, C. G. & Collins, R. N. (2003). Dual prenylation is required for Rab protein localization and function. *Molecular biology of the cell*. Vol. 14, No.5 (May 2003), pp.1852-1867, 1059-1524

Calhoun, B. C. & Goldenring, J. R. (1997). Two Rab proteins, vesicle-associated membrane protein 2 (VAMP-2) and secretory carrier membrane proteins (SCAMPs), are present on immunoisolated parietal cell tubulovesicles. *The Biochemical journal*. Vol. 325, No. Pt 2, (July 1997), pp.559-564, 0264-6021

Castillo-Lluva, S., Tatham, M. H., Jones, R. C., Jaffray, E. G., Edmondson, R. D., Hay, R. T. & Malliri, A. (2010). SUMOylation of the GTPase Rac1 is required for optimal cell migration. *Nature cell biology*. Vol. 12, No.11 (November 2010), pp.1078-1085, 1465-7392

Castillo Bennett, J., Roggero, C. M., Mancifesta, F. E. & Mayorga, L. S. (2010). Calcineurin-mediated dephosphorylation of synaptotagmin VI is necessary for acrosomal exocytosis. *The Journal of biological chemistry*. Vol. 285, No.34 (August 2010), pp.26269-26278, 0021-9258

Cesca, F., Baldelli, P., Valtorta, F. & Benfenati, F. (2010). The synapsins: key actors of synapse function and plasticity. *Progress in neurobiology*. Vol. 91, No.4 (August 2010), pp.313-348, 0301-0082

Chen, S., Synowsky, S., Tinti, M. & MacKintosh, C. (2011). The capture of phosphoproteins by 14-3-3 proteins mediates actions of insulin. *Trends in endocrinology and metabolism: TEM*. Vol. 22, No.11 (November 2011), pp.429-436, 1043-2760

Chen, W., Feng, Y., Chen, D. & Wandinger-Ness, A. (1998). Rab11 is required for trans-golgi network-to-plasma membrane transport and a preferential target for GDP dissociation inhibitor. *Molecular biology of the cell*. Vol. 9, No.11 (November 1998), pp.3241-3257, 1059-1524

Chen, X. W., Leto, D., Chiang, S. H., Wang, Q. & Saltiel, A. R. (2007). Activation of RalA is required for insulin-stimulated Glut4 trafficking to the plasma membrane via the exocyst and the motor protein Myo1c. *Developmental cell*. Vol. 13, No.3 (September 2007), pp.391-404, 1534-5807

Chen, X. W., Leto, D., Xiao, J., Goss, J., Wang, Q., Shavit, J. A., Xiong, T., Yu, G., Ginsburg, D., Toomre, D., Xu, Z. & Saltiel, A. R. (2011a). Exocyst function is regulated by effector phosphorylation. *Nature cell biology*. Vol. 13, No.5 (May 2011), pp.580-588, 1465-7392

Chen, X. W., Leto, D., Xiong, T., Yu, G., Cheng, A., Decker, S. & Saltiel, A. R. (2011b). A Ral GAP complex links PI 3-kinase/Akt signaling to RalA activation in insulin action. *Molecular biology of the cell*. Vol. 22, No.1 (January 2011), pp.141-152, 1059-1524

Chen, X. W. & Saltiel, A. R. (2011). Ral's engagement with the exocyst: breaking up is hard to do. *Cell cycle*. Vol. 10, No.14 (July 2011), pp.2299-2304, 1551-4005

Chen, Y., Yang, Z., Meng, M., Zhao, Y., Dong, N., Yan, H., Liu, L., Ding, M., Peng, H. B. & Shao, F. (2009). Cullin mediates degradation of RhoA through evolutionarily conserved BTB adaptors to control actin cytoskeleton structure and cell movement. *Molecular cell*. Vol. 35, No.6 (September 2009), pp.841-855, 1097-2765

Chernyshova, Y., Leshchyns'ka, I., Hsu, S. C., Schachner, M. & Sytnyk, V. (2011). The neural cell adhesion molecule promotes FGFR-dependent phosphorylation and membrane targeting of the exocyst complex to induce exocytosis in growth cones. *The Journal of neuroscience*. Vol. 31, No.10 (March 2011), pp.3522-3535, 0270-6474

Chheda, M. G., Ashery, U., Thakur, P., Rettig, J. & Sheng, Z. H. (2001). Phosphorylation of Snapin by PKA modulates its interaction with the SNARE complex. *Nature cell biology*. Vol. 3, No.4 (April 2001), pp.331-338, 1465-7392

Chi, P., Greengard, P. & Ryan, T. A. (2003). Synaptic vesicle mobilization is regulated by distinct synapsin I phosphorylation pathways at different frequencies. *Neuron*. Vol. 38, No.1 (April 2003), pp.69-78, 0896-6273

Chiang, S. H., Hwang, J., Legendre, M., Zhang, M., Kimura, A. & Saltiel, A. R. (2003). TCGAP, a multidomain Rho GTPase-activating protein involved in insulin-stimulated glucose transport. *The EMBO journal*. Vol. 22, No.11 (June 2003), pp.2679-2691, 0261-4189

Choy, E., Chiu, V. K., Silletti, J., Feoktistov, M., Morimoto, T., Michaelson, D., Ivanov, I. E. & Philips, M. R. (1999). Endomembrane trafficking of ras: the CAAX motif targets proteins to the ER and Golgi. *Cell*. Vol. 98, No.1 (July 1999), pp.69-80, 0092-8674

Crechet, J. B. & Parmeggiani, A. (1986). Characterization of the elongation factors from calf brain. 3. Properties of the GTPase activity of EF-1 alpha and mode of action of kirromycin. *European journal of biochemistry: FEBS*. Vol. 161, No.3 (December 1986), pp.655-660, 0014-2956

Csepanyi-Komi, R., Levay, M. & Ligeti, E. (2011). Small G proteins and their regulators in cellular signalling. *Molecular and cellular endocrinology*. Vol. 353, No.1-2 (April 2011), 10-20, 0303-7207

Dabbeekeh, J. T., Faitar, S. L., Dufresne, C. P. & Cowell, J. K. (2007). The EVI5 TBC domain provides the GTPase-activating protein motif for RAB11. *Oncogene*. Vol. 26, No.19 (April 2007), pp.2804-2808, 0950-9232

Dai, X. Q., Plummer, G., Casimir, M., Kang, Y., Hajmrle, C., Gaisano, H. Y., Manning Fox, J. E. & MacDonald, P. E. (2011). SUMOylation regulates insulin exocytosis downstream of secretory granule docking in rodents and humans. *Diabetes*. Vol. 60, No.3 (March 2011), pp.838-847, 0012-1797 de la Vega, M., Burrows, J. F. & Johnston, J. A. (2011). Ubiquitination: Added complexity in Ras and Rho family GTPase function. *Small GTPases*. Vol. 2, No.4 (July 2011), pp.192-201, 2154-1248

Deak, F., Shin, O. H., Tang, J., Hanson, P., Ubach, J., Jahn, R., Rizo, J., Kavalali, E. T. & Sudhof, T. C. (2006). Rabphilin regulates SNARE-dependent re-priming of synaptic vesicles for fusion. *The EMBO Journal*. Vol. 25, No.12 (June 2006), pp.2856-2866, 0261-4189

Delint-Ramirez, I., Willoughby, D., Hammond, G. V., Ayling, L. J. & Cooper, D. M. (2011). Palmitoylation targets AKAP79 protein to lipid rafts and promotes its regulation of calcium-sensitive adenylyl cyclase type 8. *The Journal of biological chemistry*. Vol. 286, No.38 (September 2011), pp.32962-32975, 0021-9258

DerMardirossian, C. & Bokoch, G. M. (2005). GDIs: central regulatory molecules in Rho GTPase activation. *Trends in cell biology*. Vol. 15, No.7 (July 2005), pp.356-363, 0962-8924

Detter, J. C., Zhang, Q., Mules, E. H., Novak, E. K., Mishra, V. S., Li, W., McMurtrie, E. B., Tchernev, V. T., Wallace, M. R., Seabra, M. C., Swank, R. T. & Kingsmore, S. F. (2000). Rab geranylgeranyl transferase alpha mutation in the gunmetal mouse reduces Rab prenylation and platelet synthesis. *Proceedings of the National Academy of Sciences of the United States of America*. Vol. 97, No.8 (April 2000), pp.4144-4149, 0027-8424

Diekmann, Y., Seixas, E., Gouw, M., Tavares-Cadete, F., Seabra, M. C. & Pereira-Leal, J. B. (2011). Thousands of rab GTPases for the cell biologist. *PLoS computational biology*. Vol. 7, No.10 (October 2011), pp.e1002217, 1553-734X

Dong, G., Medkova, M., Novick, P. & Reinisch, K. M. (2007). A catalytic coiled coil: structural insights into the activation of the Rab GTPase Sec4p by Sec2p. *Molecular cell*. Vol. 25, No.3 (February 2007), pp.455-462, 1097-2765

Doreian, B. W., Fulop, T. G. & Smith, C. B. (2008). Myosin II activation and actin reorganization regulate the mode of quantal exocytosis in mouse adrenal chromaffin cells. *The Journal of neuroscience*. Vol. 28, No.17 (April 2008), pp.4470-4478, 0270-6474

Doreian, B. W., Fulop, T. G., Meklemburg, R. L. & Smith, C. B. (2009). Cortical F-actin, the exocytic mode, and neuropeptide release in mouse chromaffin cells is regulated by myristoylated alanine-rich C-kinase substrate and myosin II. *Molecular biology of the cell*. Vol. 20, No.13 (July 2009), pp.3142-3154, 1059-1524

Doye, A., Boyer, L., Mettouchi, A. & Lemichez, E. (2006). Ubiquitin-mediated proteasomal degradation of Rho proteins by the CNF1 toxin. *Methods in enzymology*. Vol. 406, (n. d.), pp.447-456, 0076-6879

Du, W., Lebowitz, P. F. & Prendergast, G. C. (1999). Cell growth inhibition by farnesyltransferase inhibitors is mediated by gain of geranylgeranylated RhoB. *Molecular and cellular biology*. Vol. 19, No.3 (March 1999), pp.1831-1840, 0270-7306

Eitzen, G., Lo, A. N., Mitchell, T., Kim, J. D., Chao, D. V. & Lacy, P. (2011). Proteomic analysis of secretagogue-stimulated neutrophils implicates a role for actin and actin-interacting proteins in Rac2-mediated granule exocytosis. *Proteome science*. Vol. 9, (November 2011), pp.70, 1477-5956

El Oualid, F., van den Elst, H., Leroy, I. M., Pieterman, E., Cohen, L. H., Burm, B. E., Overkleeft, H. S., van der Marel, G. A. & Overhand, M. (2005). A combinatorial approach toward the generation of ambiphilic peptide-based inhibitors of protein:geranylgeranyl transferase-1. *Journal of combinatorial chemistry.* Vol. 7, No.5 (September 2005), pp.703-713, 1520-4766

Etienne-Manneville, S. & Hall, A. (2003). Cell polarity: Par6, aPKC and cytoskeletal crosstalk. *Current opinion in cell biology.* Vol. 15, No.1 (February 2003), pp.67-72, 0955-0674

Evans, G. J., Wilkinson, M. C., Graham, M. E., Turner, K. M., Chamberlain, L. H., Burgoyne, R. D. & Morgan, A. (2001). Phosphorylation of cysteine string protein by protein kinase A. Implications for the modulation of exocytosis. *The Journal of biological chemistry.* Vol. 276, No.51 (December 2001), pp.47877-47885, 0021-9258

Evans, G. J. & Morgan, A. (2002). Phosphorylation-dependent interaction of the synaptic vesicle proteins cysteine string protein and synaptotagmin I. *The Biochemical journal.* Vol. 364, No.Pt 2 (June 2002), pp.343-347, 0264-6021

Evans, G. J. & Morgan, A. (2005). Phosphorylation of cysteine string protein in the brain: developmental, regional and synaptic specificity. *The European journal of neuroscience.* Vol. 21, No.10 (May 2005), pp.2671-2680, 0953-816X

Feig, L. A. (2003). Ral-GTPases: approaching their 15 minutes of fame. *Trends in cell biology.* Vol. 13, No.8 (August 2003), pp.419-425, 0962-8924

Feldmann, G., Mishra, A., Hong, S. M., Bisht, S., Strock, C. J., Ball, D. W., Goggins, M., Maitra, A. & Nelkin, B. D. (2010). Inhibiting the cyclin-dependent kinase CDK5 blocks pancreatic cancer formation and progression through the suppression of Ras-Ral signaling. *Cancer research.* Vol. 70, No.11 (June 2010), pp.4460-4469, 0008-5472

Fitzgerald, M. L. & Reed, G. L. (1999). Rab6 is phosphorylated in thrombin-activated platelets by a protein kinase C-dependent mechanism: effects on GTP/GDP binding and cellular distribution. *The Biochemical journal.* Vol. 342, No.Pt 2 (September 1999), pp.353-360, 0264-6021

Foger, N., Jenckel, A., Orinska, Z., Lee, K. H., Chan, A. C. & Bulfone-Paus, S. (2011). Differential regulation of mast cell degranulation versus cytokine secretion by the actin regulatory proteins Coronin1a and Coronin1b. *The Journal of experimental medicine.* Vol. 208, No.9 (August 2011), pp.1777-1787, 0022-1007

Foletti, D. L., Blitzer, J. T. & Scheller, R. H. (2001). Physiological modulation of rabphilin phosphorylation. *The Journal of neuroscience.* Vol. 21, No.15 (August 2001), pp.5473-5483, 0270-6474

Fu, J., Naren, A. P., Gao, X., Ahmmed, G. U. & Malik, A. B. (2005). Protease-activated receptor-1 activation of endothelial cells induces protein kinase Calpha-dependent phosphorylation of syntaxin 4 and Munc18c: role in signaling p-selectin expression. *The Journal of biological chemistry.* Vol. 280, No.5 (February 2005), pp.3178-3184, 0021-9258

Fujita, Y., Shirataki, H., Sakisaka, T., Asakura, T., Ohya, T., Kotani, H., Yokoyama, S., Nishioka, H., Matsuura, Y., Mizoguchi, A., Scheller, R. H. & Takai, Y. (1998). Tomosyn: a syntaxin-1-binding protein that forms a novel complex in the neurotransmitter release process. *Neuron.* Vol. 20, No.5 (May 1998), pp.905-915, 0896-6273

Fukai, S., Matern, H. T., Jagath, J. R., Scheller, R. H. & Brunger, A. T. (2003). Structural basis of the interaction between RalA and Sec5, a subunit of the sec6/8 complex. *The EMBO journal*. Vol. 22, No.13 (July 2003), pp.3267-3278, 0261-4189

Fukata, Y., Bredt, D. S. & Fukata, M. (2006). Protein Palmitoylation by DHHC Protein Family, In *The Dynamic Synapse: Molecular Methods in Ionotropic Receptor Biology*, Kittler, J. T., Moss, S. J., Boca Raton (FL): CRC Press, ISBN-13: 978-0-8493-1891-7.

Fukata, Y. & Fukata, M. (2010). Protein palmitoylation in neuronal development and synaptic plasticity. *Nature reviews. Neuroscience*. Vol. 11, No.3 (March 2010), pp.161-175, 1471-003X

Fukuda, M. (2008). Regulation of secretory vesicle traffic by Rab small GTPases. *Cellular and molecular life sciences: CMLS*. Vol. 65, No.18 (September 2008), pp.2801-2813, 1420-682X

Fukui, K., Sasaki, T., Imazumi, K., Matsuura, Y., Nakanishi, H. & Takai, Y. (1997). Isolation and characterization of a GTPase activating protein specific for the Rab3 subfamily of small G proteins. *The Journal of biological chemistry*. Vol. 272, No.8 (February 1997), pp.4655-4658, 0021-9258

Funakoshi, Y., Hasegawa, H. & Kanaho, Y. (2011). Regulation of PIP5K activity by Arf6 and its physiological significance. *Journal of cell physiology*. Vol. 226, No.4 (April 2011), pp.888-895, 0021-9541

Gamberucci, A., Innocenti, B., Fulceri, R., Banhegyi, G., Giunti, R., Pozzan, T. & Benedetti, A. (1994). Modulation of Ca2+ influx dependent on store depletion by intracellular adenine-guanine nucleotide levels. *The Journal of biological chemistry*. Vol. 269, No.38 (September 1994), pp.23597-23602, 0021-9258

Gao, X. D., Albert, S., Tcheperegine, S. E., Burd, C. G., Gallwitz, D. & Bi, E. (2003). The GAP activity of Msb3p and Msb4p for the Rab GTPase Sec4p is required for efficient exocytosis and actin organization. *The Journal of cell biology*. Vol. 162, No.4 (August 2003), pp.635-646, 0021-9525

Garrett, M. D., Zahner, J. E., Cheney, C. M. & Novick, P. J. (1994). GDI1 encodes a GDP dissociation inhibitor that plays an essential role in the yeast secretory pathway. *The EMBO journal*. Vol. 13, No.7 (April 1994), pp.1718-1728, 0261-4189

Genre, A., S, I., M, F., A, F., V, Z., T, B. & P, B. (2011). Multiple Exocytotic Markers Accumulate at the Sites of Perifungal Membrane Biogenesis in Arbuscular Mycorrhizas. *Plant & cell physiology*. Vol. 53, No.1 (December 2011), 0032-0781

Gerelsaikhan, T., Chen, X. L. & Chander, A. (2011). Secretagogues of lung surfactant increase annexin A7 localization with ABCA3 in alveolar type II cells. *Biochimica et biophysica acta*. Vol. 1813, No.12 (December 2011), pp.2017-2025, 0006-3002

Gerst, J. E. (1999). SNAREs and SNARE regulators in membrane fusion and exocytosis. *Cellular and molecular life sciences: CMLS*. Vol. 55, No.5 (May 1999), pp.707-734, 1420-682X

Giachello, C. N., Fiumara, F., Giacomini, C., Corradi, A., Milanese, C., Ghirardi, M., Benfenati, F. & Montarolo, P. G. (2010). MAPK/Erk-dependent phosphorylation of synapsin mediates formation of functional synapses and short-term homosynaptic plasticity. *Journal of cell science*. Vol. 123, No.Pt 6 (March 2010), pp.881-893, 0021-9533

Gomes, A. Q., Ali, B. R., Ramalho, J. S., Godfrey, R. F., Barral, D. C., Hume, A. N. & Seabra, M. C. (2003). Membrane targeting of Rab GTPases is influenced by the prenylation motif. *Molecular biology of the cell*. Vol. 14, No.5 (May 2003), pp.1882-1899, 1059-1524

Gomi, H., Mori, K., Itohara, S. & Izumi, T. (2007). Rab27b is expressed in a wide range of exocytic cells and involved in the delivery of secretory granules near the plasma membrane. *Molecular biology of the cell*. Vol. 18, No.11 (November 2007), pp.4377-4386, 1059-1524

Gosser, Y. Q., Nomanbhoy, T. K., Aghazadeh, B., Manor, D., Combs, C., Cerione, R. A. & Rosen, M. K. (1997). C-terminal binding domain of Rho GDP-dissociation inhibitor directs N-terminal inhibitory peptide to GTPases. *Nature*. Vol. 387, No.6635 (June 1997), pp.814-819, 0028-0836

Gracheva, E. O., Hadwiger, G., Nonet, M. L. & Richmond, J. E. (2008). Direct interactions between *C. elegans* RAB-3 and Rim provide a mechanism to target vesicles to the presynaptic density. *Neuroscience letters*. Vol. 444, No.2 (October 2008), pp.137-142, 0304-3940

Grizot, S., Faure, J., Fieschi, F., Vignais, P. V., Dagher, M. C. & Pebay-Peyroula, E. (2001). Crystal structure of the Rac1-RhoGDI complex involved in nadph oxidase activation. *Biochemistry*. Vol. 40, No.34 (August 2001), pp.10007-10013, 0006-2960

Groffen, A. J., Martens, S., Diez Arazola, R., Cornelisse, L. N., Lozovaya, N., de Jong, A. P., Goriounova, N. A., Habets, R. L., Takai, Y., Borst, J. G., Brose, N., McMahon, H. T. & Verhage, M. (2010). Doc2b is a high-affinity Ca2+ sensor for spontaneous neurotransmitter release. *Science*. Vol. 327, No.5973 (March 2010), pp.1614-1618, 0036-8075

Grote, E., Baba, M., Ohsumi, Y. & Novick, P. J. (2000). Geranylgeranylated SNAREs are dominant inhibitors of membrane fusion. *The Journal of cell biology*. Vol. 151, No.2 (October 2000), pp.453-466, 0021-9525

Guo, W., Roth, D., Walch-Solimena, C. & Novick, P. (1999). The exocyst is an effector for Sec4p, targeting secretory vesicles to sites of exocytosis. *The EMBO journal*. Vol. 18, No.4 (February 1999), pp.1071-1080, 0261-4189

Guo, W., Tamanoi, F. & Novick, P. (2001). Spatial regulation of the exocyst complex by Rho1 GTPase. *Nature cell biology*. Vol. 3, No.4 (April 2001), pp.353-360, 1465-7392

Hammer, J. A. & Wu, X. S. (2002). Rabs grab motors: defining the connections between Rab GTPases and motor proteins. *Current opinion in cell biology*. Vol. 14, No.1 (February 2002), pp.69-75, 0955-0674

Harada, A., Furuta, B., Takeuchi, K., Itakura, M., Takahashi, M. & Umeda, M. (2000). Nadrin, a novel neuron-specific GTPase-activating protein involved in regulated exocytosis. *The Journal of biological chemistry*. Vol. 275, No.47 (November 2000), pp.36885-36891, 0021-9258

Hayakawa, M., Kitagawa, H., Miyazawa, K., Kitagawa, M. & Kikugawa, K. (2005). The FWD1/beta-TrCP-mediated degradation pathway establishes a 'turning off switch' of a Cdc42 guanine nucleotide exchange factor, FGD1. *Genes to cells : devoted to molecular & cellular mechanisms*. Vol. 10, No.3 (March 2005), pp.241-251, 1356-9597

Hayakawa, M., Matsushima, M., Hagiwara, H., Oshima, T., Fujino, T., Ando, K., Kikugawa, K., Tanaka, H., Miyazawa, K. & Kitagawa, M. (2008). Novel insights into FGD3, a putative GEF for Cdc42, that undergoes SCF(FWD1/beta-TrCP)-mediated proteasomal degradation analogous to that of its homologue FGD1 but regulates cell morphology and motility differently from FGD1. *Genes to cells : devoted to molecular & cellular mechanisms*. Vol. 13, No.4 (April 2008), pp.329-342, 1356-9597

Hazelett, C. C., Sheff, D. & Yeaman, C. (2011). RalA and RalB differentially regulate development of epithelial tight junctions. *Molecular biology of the cell*. Vol. 22, No.24 (December 2011), pp.4787-4800, 1059-1524

He, B., Xi, F., Zhang, X., Zhang, J. & Guo, W. (2007). Exo70 interacts with phospholipids and mediates the targeting of the exocyst to the plasma membrane. *The EMBO Journal*. Vol. 26, No.18 (September 2007), pp.4053-4065, 0261-4189

He, B. & Guo, W. (2009). The exocyst complex in polarized exocytosis. *Current opinion in cell biology*. Vol. 21, No.4 (August 2009), pp.537-542, 1879-0410

Heger, C. D., Wrann, C. D. & Collins, R. N. (2011). Phosphorylation provides a negative mode of regulation for the yeast Rab GTPase Sec4p. *PloS one*. Vol. 6, No.9 (September 2011), pp.e24332, 1932-6203

Hershko, A. & Ciechanover, A. (1998). The ubiquitin system. *Annual review of biochemistry*. Vol. 67, (July 1998), pp.425-479, 0066-4154

Hershko, A. (2005). The ubiquitin system for protein degradation and some of its roles in the control of the cell division cycle. *Cell death and differentiation*. Vol. 12, No.9 (September 2005), pp.1191-1197, 1350-9047

Hoffman, G. R., Nassar, N. & Cerione, R. A. (2000). Structure of the Rho family GTP-binding protein Cdc42 in complex with the multifunctional regulator RhoGDI. *Cell*. Vol. 100, No.3 (February 2000), pp.345-356, 0092-8674

Hohlfeld, R. (1990). Myasthenia gravis and thymoma: paraneoplastic failure of neuromuscular transmission. *Laboratory investigation; a journal of technical methods and pathology*. Vol. 62, No.3 (March 1990), pp.241-243, 0023-6837

Holz, R. W., Brondyk, W. H., Senter, R. A., Kuizon, L. & Macara, I. G. (1994). Evidence for the involvement of Rab3A in Ca(2+)-dependent exocytosis from adrenal chromaffin cells. *The Journal of biological chemistry*. Vol. 269, No.14 (April 1994), pp.10229-10234, 0021-9258

Hsu, S. C., TerBush, D., Abraham, M. & Guo, W. (2004). The exocyst complex in polarized exocytosis. *International review of cytology*. Vol. 233, (March 2004), pp.243-265, 0074-7696

Hui, E., Johnson, C. P., Yao, J., Dunning, F. M. & Chapman, E. R. (2009). Synaptotagmin-mediated bending of the target membrane is a critical step in Ca(2+)-regulated fusion. *Cell*. Vol. 138, No.4 (August 2009), pp.709-721, 0092-8674

Hume, A. N. & Seabra, M. C. (2011). Melanosomes on the move: a model to understand organelle dynamics. *Biochemical Society transactions*. Vol. 39, No.5 (October 2011), pp.1191-1196, 0300-5127

Hurley, J. H. (2010). The ESCRT complexes. *Critical reviews in biochemistry and molecular biology*. Vol. 45, No.6 (December 2010), pp.463-487, 1040-9238

Hutagalung, A. H. & Novick, P. J. (2011). Role of Rab GTPases in membrane traffic and cell physiology. *Physiological reviews*. Vol. 91, No.1 (January 2011), pp.119-149, 0031-9333

Hyvonen, M., Macias, M. J., Nilges, M., Oschkinat, H., Saraste, M. & Wilmanns, M. (1995). Structure of the binding site for inositol phosphates in a PH domain. *The EMBO journal*. Vol. 14, No.19 (October 1995), pp.4676-4685, 0261-4189

Inoue, M., Chang, L., Hwang, J., Chiang, S. H. & Saltiel, A. R. (2003). The exocyst complex is required for targeting of Glut4 to the plasma membrane by insulin. *Nature*. Vol. 422, No.6932 (April 2003), pp.629-633, 0028-0836

Jafar-Nejad, H., Andrews, H. K., Acar, M., Bayat, V., Wirtz-Peitz, F., Mehta, S. Q., Knoblich, J. A. & Bellen, H. J. (2005). Sec15, a component of the exocyst, promotes notch signaling during the asymmetric division of *Drosophila* sensory organ precursors. *Developmental cell*. Vol. 9, No.3 (September 2005), pp.351-363, 1534-5807

Jin, R., Junutula, J. R., Matern, H. T., Ervin, K. E., Scheller, R. H. & Brunger, A. T. (2005). Exo84 and Sec5 are competitive regulatory Sec6/8 effectors to the RalA GTPase. *The EMBO journal*. Vol. 24, No.12 (June 2005), pp.2064-2074, 0261-4189

Johannes, L., Lledo, P. M., Roa, M., Vincent, J. D., Henry, J. P. & Darchen, F. (1994). The GTPase Rab3a negatively controls calcium-dependent exocytosis in neuroendocrine cells. *The EMBO journal*. Vol. 13, No.9 (May 1994), pp.2029-2037, 0261-4189

Johnson, J. L., Erickson, J. W. & Cerione, R. A. (2009). New insights into how the Rho guanine nucleotide dissociation inhibitor regulates the interaction of Cdc42 with membranes. *The Journal of biological chemistry*. Vol. 284, No.35 (August 2009), pp.23860-23871, 0021-9258

Jolly, C. & Sattentau, Q. J. (2007). Regulated secretion from CD4+ T cells. *Trends in immunology*. Vol. 28, No.11 (November 2007), pp.474-481, 1471-4906

Jordan, P., Brazao, R., Boavida, M. G., Gespach, C. & Chastre, E. (1999). Cloning of a novel human Rac1b splice variant with increased expression in colorectal tumors. *Oncogene*. Vol. 18, No.48 (November 1999), pp.6835-6839, 0950-9232

Jovanovic, J. N., Sihra, T. S., Nairn, A. C., Hemmings, H. C., Jr., Greengard, P. & Czernik, A. J. (2001). Opposing changes in phosphorylation of specific sites in synapsin I during Ca2+-dependent glutamate release in isolated nerve terminals. *The Journal of neuroscience*. Vol. 21, No.20 (October 2001), pp.7944-7953, 0270-6474

Kang, R., Wan, J., Arstikaitis, P., Takahashi, H., Huang, K., Bailey, A. O., Thompson, J. X., Roth, A. F., Drisdel, R. C., Mastro, R., Green, W. N., Yates, J. R., 3rd, Davis, N. G. & El-Husseini, A. (2008). Neural palmitoyl-proteomics reveals dynamic synaptic palmitoylation. *Nature*. Vol. 456, No.7224 (December 2008), pp.904-909, 0028-0836

Karkkainen, S., van der Linden, M. & Renkema, G. H. (2010). POSH2 is a RING finger E3 ligase with Rac1 binding activity through a partial CRIB domain. *FEBS Letters*. Vol. 584, No.18 (September 2010), pp.3867-3872, 0014-5793

Kawato, M., Shirakawa, R., Kondo, H., Higashi, T., Ikeda, T., Okawa, K., Fukai, S., Nureki, O., Kita, T. & Horiuchi, H. (2008). Regulation of platelet dense granule secretion by the Ral GTPase-exocyst pathway. *The Journal of biological chemistry*. Vol. 283, No.1 (January 2007), pp.166-174, 0021-9258

Keep, N. H., Barnes, M., Barsukov, I., Badii, R., Lian, L. Y., Segal, A. W., Moody, P. C. & Roberts, G. C. (1997). A modulator of rho family G proteins, rhoGDI, binds these G proteins via an immunoglobulin-like domain and a flexible N-terminal arm. *Structure*. Vol. 5, No.5 (May 1997), pp.623-633, 0969-2126

Kepner, E. M., Yoder, S. M., Oh, E., Kalwat, M. A., Wang, Z., Quilliam, L. A. & Thurmond, D. C. (2011). Cool-1/betaPIX functions as a guanine nucleotide exchange factor in the cycling of Cdc42 to regulate insulin secretion. *American journal of physiology. Endocrinology and metabolism*. Vol. 301, No.6 (December 2011), pp.E1072-1080, 0193-1849

Khan, O. M., Ibrahim, M. X., Jonsson, I. M., Karlsson, C., Liu, M., Sjogren, A. K., Olofsson, F. J., Brisslert, M., Andersson, S., Ohlsson, C., Hulten, L. M., Bokarewa, M. & Bergo, M. O. (2011). Geranylgeranyltransferase type I (GGTase-I) deficiency hyperactivates macrophages and induces erosive arthritis in mice. *The Journal of clinical investigation*. Vol. 121, No.2 (February 2011), pp.628-639, 0021-9738

Khandelwal, P., Ruiz, W. G., Balestreire-Hawryluk, E., Weisz, O. A., Goldenring, J. R. & Apodaca, G. (2008). Rab11a-dependent exocytosis of discoidal/fusiform vesicles in bladder umbrella cells. *Proceedings of the National Academy of Sciences of the United States of America*. Vol. 105, No.41 (October 2008), pp.15773-15778, 1091-6490

Khvotchev, M. V., Ren, M., Takamori, S., Jahn, R. & Sudhof, T. C. (2003). Divergent functions of neuronal Rab11b in Ca2+-regulated versus constitutive exocytosis. *The Journal of neuroscience : the official journal of the Society for Neuroscience*. Vol. 23, No.33 (November 2003), pp.10531-10539, 0270-6474

Kim, B. H., Kim, H. K. & Lee, S. J. (2011). Experimental analysis of the blood-sucking mechanism of female mosquitoes. *The Journal of experimental biology*. Vol. 214, No.Pt 7 (April 2011), pp.1163-1169, 0022-0949

Kim, H. Y., Choi, H. J., Lim, J. S., Park, E. J., Jung, H. J., Lee, Y. J., Kim, S. Y. & Kwon, T. H. (2011). Emerging role of Akt substrate protein AS160 in the regulation of AQP2 translocation. *American journal of physiology: Renal physiology*. Vol. 301, No.1 (July 2011), pp.F151-161, 1522-1466

Kim, K. S., Park, J. Y., Jou, I. & Park, S. M. (2010). Regulation of Weibel-Palade body exocytosis by alpha-synuclein in endothelial cells. *The Journal of biological chemistry*. Vol. 285, No.28 (July 2010), pp.21416-21425, 0021-9258

Kim, S. J., Zhang, Z., Sarkar, C., Tsai, P. C., Lee, Y. C., Dye, L. & Mukherjee, A. B. (2008). Palmitoyl protein thioesterase-1 deficiency impairs synaptic vesicle recycling at nerve terminals, contributing to neuropathology in humans and mice. *The Journal of clinical investigation*. Vol. 118, No.9 (September 2008), pp.3075-3086, 0021-9738

Kowluru, A. & Veluthakal, R. (2005). Rho guanosine diphosphate-dissociation inhibitor plays a negative modulatory role in glucose-stimulated insulin secretion. *Diabetes*. Vol. 54, No.12 (December 2005), pp.3523-3529, 0012-1797

Kutateladze, T. G. (2007). Mechanistic similarities in docking of the FYVE and PX domains to phosphatidylinositol 3-phosphate containing membranes. *Progress in lipid research*. Vol. 46, No.6 (November 2007), pp.315-327, 0163-7827

Lacy, P. & Stow, J. L. (2011). Cytokine release from innate immune cells: association with diverse membrane trafficking pathways. *Blood*. Vol. 118, No.1 (July 2011), pp.9-18, 0006-4971

Lalli, G. (2009). RalA and the exocyst complex influence neuronal polarity through PAR-3 and aPKC. *Journal of cell science*. Vol. 122, No.10 (May 2009), pp.1499-1506, 0021-9533

Langevin, J., Morgan, M. J., Rossé, C., Racine, V., Sibarita, J.-B., Aresta, S., Murthy, M., Schwarz, T., Camonis, J. & Bellaïche, Y. (2005). *Drosophila* exocyst components Sec5, Sec6, and Sec15 regulate DE-Cadherin trafficking from recycling endosomes to the plasma membrane. *Developmental cell*. Vol. 9, No.3 pp.365-376, 1534-5807

Lapierre, L. A., Kumar, R., Hales, C. M., Navarre, J., Bhartur, S. G., Burnette, J. O., Provance, D. W., Jr., Mercer, J. A., Bahler, M. & Goldenring, J. R. (2001). Myosin vb is associated with plasma membrane recycling systems. *Molecular biology of the cell.* Vol. 12, No.6 (June 2001), pp.1843-1857, 1059-1524

Lebowitz, P. F., Casey, P. J., Prendergast, G. C. & Thissen, J. A. (1997). Farnesyltransferase inhibitors alter the prenylation and growth-stimulating function of RhoB. *The Journal of biological chemistry.* Vol. 272, No.25 (June 1997), pp.15591-15594, 0021-9258

Lee, S., Fan, S., Makarova, O., Straight, S. & Margolis, B. (2002). A novel and conserved protein-protein interaction domain of mammalian Lin-2/CASK binds and recruits SAP97 to the lateral surface of epithelia. *Molecular and cellular biology.* Vol. 22, No.6 (March 2002), pp.1778-1791, 0270-7306

Li, G., Han, L., Chou, T. C., Fujita, Y., Arunachalam, L., Xu, A., Wong, A., Chiew, S. K., Wan, Q., Wang, L. & Sugita, S. (2007). RalA and RalB function as the critical GTP sensors for GTP-dependent exocytosis. *The Journal of neuroscience.* Vol. 27, No.1 (January 2007), pp.190-202, 0270-6474

Li, H., Li, H. F., Felder, R. A., Periasamy, A. & Jose, P. A. (2008). Rab4 and Rab11 coordinately regulate the recycling of angiotensin II type I receptor as demonstrated by fluorescence resonance energy transfer microscopy. *Journal of biomedical optics.* Vol. 13, No.3 (May 2008), pp.031206, 1083-3668

Li, L. & Chin, L. S. (2003). The molecular machinery of synaptic vesicle exocytosis. *Cellular and molecular life sciences: CMLS.* Vol. 60, No.5 (May 2003), pp.942-960, 1420-682X

Liao, J., Shima, F., Araki, M., Ye, M., Muraoka, S., Sugimoto, T., Kawamura, M., Yamamoto, N., Tamura, A. & Kataoka, T. (2008). Two conformational states of Ras GTPase exhibit differential GTP-binding kinetics. *Biochemical and biophysical research communications.* Vol. 369, No.2 (May 2008), pp.327-332, 0006-291X

Lim, K. H., Brady, D. C., Kashatus, D. F., Ancrile, B. B., Der, C. J., Cox, A. D. & Counter, C. M. (2010). Aurora-A phosphorylates, activates, and relocalizes the small GTPase RalA. *Molecular and cellular biology.* Vol. 30, No.2 (January 2009), pp.508-523, 0270-7306

Lin, C. C., Huang, C. C., Lin, K. H., Cheng, K. H., Yang, D. M., Tsai, Y. S., Ong, R. Y., Huang, Y. N. & Kao, L. S. (2007). Visualization of Rab3A dissociation during exocytosis: a study by total internal reflection microscopy. *Journal of cell physiology.* Vol. 211, No.2 (May 2007), pp.316-326, 0021-9541

Lin, M. Y., Lin, Y. M., Kao, T. C., Chuang, H. H. & Chen, R. H. (2011). PDZ-RhoGEF ubiquitination by Cullin3-KLHL20 controls neurotrophin-induced neurite outgrowth. *The Journal of cell biology.* Vol. 193, No.6 (June 2011), pp.985-994, 0021-9525

Liu, J., Zuo, X., Yue, P. & Guo, W. (2007). Phosphatidylinositol 4,5-bisphosphate mediates the targeting of the exocyst to the plasma membrane for exocytosis in mammalian cells. *Molecular biology of the cell.* Vol. 18, No.11 (November 2007), pp.4483-4492, 1059-1524

Liu, J. & Guo, W. (2011). The exocyst complex in exocytosis and cell migration. *Protoplasma.* (October 2011), 0033-183X

Liu, Y., Ding, X., Wang, D., Deng, H., Feng, M., Wang, M., Yu, X., Jiang, K., Ward, T., Aikhionbare, F., Guo, Z., Forte, J. G. & Yao, X. (2007). A mechanism of Munc18b-syntaxin 3-SNAP25 complex assembly in regulated epithelial secretion. *FEBS Letters.* Vol. 581, No.22 (Sepember 2007), pp.4318-4324, 0014-5793

Ljubicic, S., Bezzi, P., Vitale, N. & Regazzi, R. (2009). The GTPase RalA regulates different steps of the secretory process in pancreatic beta-cells. *PloS one*. Vol. 4, No.11 (November 2009), pp.e7770, 1932-6203

Lobell, R. B., Liu, D., Buser, C. A., Davide, J. P., DePuy, E., Hamilton, K., Koblan, K. S., Lee, Y., Mosser, S., Motzel, S. L., Abbruzzese, J. L., Fuchs, C. S., Rowinsky, E. K., Rubin, E. H., Sharma, S., Deutsch, P. J., Mazina, K. E., Morrison, B. W., Wildonger, L., Yao, S. L. & Kohl, N. E. (2002). Preclinical and clinical pharmacodynamic assessment of L-778,123, a dual inhibitor of farnesyl:protein transferase and geranylgeranyl:protein transferase type-I. *Molecular cancer therapeutics*. Vol. 1, No.9 (July 2002), pp.747-758, 1535-7163

Lonart, G. & Sudhof, T. C. (2001). Characterization of rabphilin phosphorylation using phospho-specific antibodies. *Neuropharmacology*. Vol. 41, No.6 (November 2001), pp.643-649, 0028-3908

Longenecker, K., Read, P., Derewenda, U., Dauter, Z., Liu, X., Garrard, S., Walker, L., Somlyo, A. V., Nakamoto, R. K., Somlyo, A. P. & Derewenda, Z. S. (1999). How RhoGDI binds Rho. *Acta crystallographica. Section D, Biological crystallography*. Vol. 55, No.Pt 9 (September 1999), pp.1503-1515, 0907-4449

Lopez, J. A., Kwan, E. P., Xie, L., He, Y., James, D. E. & Gaisano, H. Y. (2008). The RalA GTPase is a central regulator of insulin exocytosis from pancreatic islet beta cells. *The Journal of biological chemistry*. Vol. 283, No.26 (June 2008), pp.17939-17945, 0021-9258

Luo, H. R., Saiardi, A., Nagata, E., Ye, K., Yu, H., Jung, T. S., Luo, X., Jain, S., Sawa, A. & Snyder, S. H. (2001). GRAB: a physiologic guanine nucleotide exchange factor for Rab3A, which interacts with inositol hexakisphosphate kinase. *Neuron*. Vol. 31, No.3 (August 2001), pp.439-451, 0896-6273

Macara, I. G. (1994). Role of the Rab3A GTPase in regulated secretion from neuroendocrine cells. *Trends in endocrinology and metabolism: TEM*. Vol. 5, No.7 (September 1994), pp.267-271, 1043-2760

Malsam, J., Kreye, S. & Sollner, T. H. (2008). Membrane fusion: SNAREs and regulation. *Cellular and molecular life sciences : CMLS*. Vol. 65, No.18 (September 2008), pp.2814-2832, 1420-682X

Manjithaya, R. & Subramani, S. (2011). Autophagy: a broad role in unconventional protein secretion? *Trends in cell biology*. Vol. 21, No.2 (February 2011), pp.67-73, 0962-8924

Marchler-Bauer, A., Lu, S., Anderson, J. B., Chitsaz, F., Derbyshire, M. K., DeWeese-Scott, C., Fong, J. H., Geer, L. Y., Geer, R. C., Gonzales, N. R., Gwadz, M., Hurwitz, D. I., Jackson, J. D., Ke, Z., Lanczycki, C. J., Lu, F., Marchler, G. H., Mullokandov, M., Omelchenko, M. V., Robertson, C. L., Song, J. S., Thanki, N., Yamashita, R. A., Zhang, D., Zhang, N., Zheng, C. & Bryant, S. H. (2011). CDD: a Conserved Domain Database for the functional annotation of proteins. *Nucleic acids research*. Vol. 39, No.Database issue (January 2011), pp.D225-229, 0305-1048

Mark, B. L., Jilkina, O. & Bhullar, R. P. (1996). Association of Ral GTP-binding protein with human platelet dense granules. *Biochemical and biophysical research communications*. Vol. 225, No.1 (August 1996), pp.40-46, 0006-291X

Masedunskas, A., Sramkova, M., Parente, L., Sales, K. U., Amornphimoltham, P., Bugge, T. H. & Weigert, R. (2011). Role for the actomyosin complex in regulated exocytosis revealed by intravital microscopy. *Proceedings of the National Academy of Sciences of the United States of America*. Vol. 108, No.33 (August 2011), pp.13552-13557, 0027-8424

Matsubara, K., Hinoi, T., Koyama, S. & Kikuchi, A. (1997). The post-translational modifications of Ral and Rac1 are important for the action of Ral-binding protein 1, a putative effector protein of Ral. *FEBS Letters*. Vol. 410, No.2-3 (June 1997), pp.169-174, 0014-5793

Mattera, R., Tsai, Y. C., Weissman, A. M. & Bonifacino, J. S. (2006). The Rab5 guanine nucleotide exchange factor Rabex-5 binds ubiquitin (Ub) and functions as a Ub ligase through an atypical Ub-interacting motif and a zinc finger domain. *The Journal of biological chemistry*. Vol. 281, No.10 (March 2006), pp.6874-6883, 0021-9258

Medkova, M., France, Y. E., Coleman, J. & Novick, P. (2006). The rab exchange factor Sec2p reversibly associates with the exocyst. *Molecular biology of the cell*. Vol. 17, No.6 (June 2006), pp.2757-2769, 1059-1524

Menegon, A., Bonanomi, D., Albertinazzi, C., Lotti, F., Ferrari, G., Kao, H. T., Benfenati, F., Baldelli, P. & Valtorta, F. (2006). Protein kinase A-mediated synapsin I phosphorylation is a central modulator of Ca2+-dependent synaptic activity. *The Journal of neuroscience*. Vol. 26, No.45 (November 2006), pp.11670-11681, 0270-6474

Messa, M., Congia, S., Defranchi, E., Valtorta, F., Fassio, A., Onofri, F. & Benfenati, F. (2010). Tyrosine phosphorylation of synapsin I by Src regulates synaptic-vesicle trafficking. *Journal of cell science*. Vol. 123, No.Pt 13 (July 2010), pp.2256-2265, 0021-9533

Mitchison, H. M., Hofmann, S. L., Becerra, C. H., Munroe, P. B., Lake, B. D., Crow, Y. J., Stephenson, J. B., Williams, R. E., Hofman, I. L., Taschner, P. E., Martin, J. J., Philippart, M., Andermann, E., Andermann, F., Mole, S. E., Gardiner, R. M. & O'Rawe, A. M. (1998). Mutations in the palmitoyl-protein thioesterase gene (PPT; CLN1) causing juvenile neuronal ceroid lipofuscinosis with granular osmiophilic deposits. *Human molecular genetics*. Vol. 7, No.2 (February 1998), pp.291-297, 0964-6906

Mitin, N., Roberts, P. J., Chenette, E. J. & Der, C. J. (2012). Posttranslational lipid modification of rho family small GTPases. *Methods in molecular biology*. Vol. 827, (n. d.), pp.87-95, 1064-3745 1064-3745

Mohrmann, K., Leijendekker, R., Gerez, L. & van Der Sluijs, P. (2002). rab4 regulates transport to the apical plasma membrane in Madin-Darby canine kidney cells. *The Journal of biological chemistry*. Vol. 277, No.12 (March 2002), pp.10474-10481, 0021-9258

Moores, S. L., Schaber, M. D., Mosser, S. D., Rands, E., O'Hara, M. B., Garsky, V. M., Marshall, M. S., Pompliano, D. L. & Gibbs, J. B. (1991). Sequence dependence of protein isoprenylation. *The Journal of biological chemistry*. Vol. 266, No.22 (August 1991), pp.14603-14610, 0021-9258

Morgera, F., Sallah, M. R., Dubuke, M. L., Gandhi, P., Brewer, D. N., Carr, C. M. & Munson, M. (2012). Regulation of exocytosis by the exocyst subunit Sec6 and the SM protein Sec1. *Molecular biology of the cell*. Vol. 23, No. 2 (January 2012), pp337-346, 1939-4586

Moskalenko, S., Henry, D. O., Rosse, C., Mirey, G., Camonis, J. H. & White, M. A. (2002). The exocyst is a Ral effector complex. *Nature cell biology*. Vol. 4, No.1 (January 2001), pp.66-72, 1465-7392

Moskalenko, S., Tong, C., Rosse, C., Mirey, G., Formstecher, E., Daviet, L., Camonis, J. & White, M. A. (2003). Ral GTPases regulate exocyst assembly through dual subunit interactions. *The Journal of biological chemistry*. Vol. 278, No.51 (December 2003), pp.51743-51748, 0021-9258

Mott, H. R., Nietlispach, D., Hopkins, L. J., Mirey, G., Camonis, J. H. & Owen, D. (2003). Structure of the GTPase-binding domain of Sec5 and elucidation of its Ral binding site. *The Journal of biological chemistry*. Vol. 278, No.19 (May 2003), pp.17053-17059, 0021-9258

Munson, M. & Novick, P. (2006). The exocyst defrocked, a framework of rods revealed. *Nature structural & molecular biology*. Vol. 13, No.7 (July 2006), pp.577-581, 1545-9993

Nagy, G., Reim, K., Matti, U., Brose, N., Binz, T., Rettig, J., Neher, E. & Sorensen, J. B. (2004). Regulation of releasable vesicle pool sizes by protein kinase A-dependent phosphorylation of SNAP-25. *Neuron*. Vol. 41, No.3 (February 2004), pp.417-429, 0896-6273

Navarro-Lerida, I., Sanchez-Perales, S., Calvo, M., Rentero, C., Zheng, Y., Enrich, C. & Del Pozo, M. A. (2011). A palmitoylation switch mechanism regulates Rac1 function and membrane organization. *The EMBO journal*. Vol. 31, No.3 (December 2011), pp.534-551, 0261-4189

Neco, P., Fernandez-Peruchena, C., Navas, S., Gutierrez, L. M., de Toledo, G. A. & Ales, E. (2008). Myosin II contributes to fusion pore expansion during exocytosis. *The Journal of biological chemistry*. Vol. 283, No.16 (April 2008), pp.10949-10957, 0021-9258

Neel, N. F., Martin, T. D., Stratford, J. K., Zand, T. P., Reiner, D. J. & Der, C. J. (2011). The RalGEF-Ral Effector Signaling Network. *Genes & cancer*. Vol. 2, No.3 (March, 2011), pp.275-287, 1947-6019

Nickel, W. & Seedorf, M. (2008). Unconventional mechanisms of protein transport to the cell surface of eukaryotic cells. *Annual review of cell and developmental biology*. Vol. 24, (November 2008), pp.287-308, 1081-0706

Nickel, W. (2010). Pathways of unconventional protein secretion. *Current opinion in biotechnology*. Vol. 21, No.5 (October 2010), pp.621-626, 0958-1669

Nightingale, T. D., White, I. J., Doyle, E. L., Turmaine, M., Harrison-Lavoie, K. J., Webb, K. F., Cramer, L. P. & Cutler, D. F. (2011). Actomyosin II contractility expels von Willebrand factor from Weibel-Palade bodies during exocytosis. The *Journal of cell biology*. Vol. 194, No.4 (August 2011), pp.613-629, 0021-9525

Nili, U., de Wit, H., Gulyas-Kovacs, A., Toonen, R. F., Sorensen, J. B., Verhage, M. & Ashery, U. (2006). Munc18-1 phosphorylation by protein kinase C potentiates vesicle pool replenishment in bovine chromaffin cells. *Neuroscience*. Vol. 143, No.2 (December 2006), pp.487-500, 0306-4522

Nishimura, N., Nakamura, H., Takai, Y. & Sano, K. (1994). Molecular cloning and characterization of two rab GDI species from rat brain: brain-specific and ubiquitous types. *The Journal of biological chemistry*. Vol. 269, No.19 (May 1994), pp.14191-14198, 0021-9258

Nomanbhoy, T. K. & Cerione, R. (1996). Characterization of the interaction between RhoGDI and Cdc42Hs using fluorescence spectroscopy. *The Journal of biological chemistry*. Vol. 271, No.17 (April 1996), pp.10004-10009, 0021-9258

Novak, E. K., Reddington, M., Zhen, L., Stenberg, P. E., Jackson, C. W., McGarry, M. P. & Swank, R. T. (1995). Inherited thrombocytopenia caused by reduced platelet production in mice with the gunmetal pigment gene mutation. *Blood*. Vol. 85, No.7 (April 1995), pp.1781-1789, 0006-4971

Novick, P. & Guo, W. (2002). Ras family therapy: Rab, Rho and Ral talk to the exocyst. *Trends in cell biology*. Vol. 12, No.6 (June 2002), pp.247-249, 0962-8924

Ohyama, T., Verstreken, P., Ly, C. V., Rosenmund, T., Rajan, A., Tien, A. C., Haueter, C., Schulze, K. L. & Bellen, H. J. (2007). Huntingtin-interacting protein 14, a palmitoyl transferase required for exocytosis and targeting of CSP to synaptic vesicles. *The Journal of cell biology*. Vol. 179, No.7 (December 2007), pp.1481-1496, 0021-9525

Olson, M. F., Pasteris, N. G., Gorski, J. L. & Hall, A. (1996). Faciogenital dysplasia protein (FGD1) and Vav, two related proteins required for normal embryonic development, are upstream regulators of Rho GTPases. *Current biology : CB*. Vol. 6, No.12 (December 1996), pp.1628-1633, 0960-9822

Orlando, K. & Guo, W. (2009). Membrane organization and dynamics in cell polarity. *Cold Spring Harbor perspectives in biology*. Vol. 1, No.5 (November 2010), pp.a001321, 1943-0264

Ortiz, D., Medkova, M., Walch-Solimena, C. & Novick, P. (2002). Ypt32 recruits the Sec4p guanine nucleotide exchange factor, Sec2p, to secretory vesicles; evidence for a Rab cascade in yeast. *The Journal of cell biology*. Vol. 157, No.6 (June 2002), pp.1005-1015, 0021-9525

Ory, S. & Gasman, S. (2011). Rho GTPases and exocytosis: what are the molecular links? *Seminars in cell & developmental biology*. Vol. 22, No.1 (February 2011), pp.27-32, 1084-9521

Ostrowski, M., Carmo, N. B., Krumeich, S., Fanget, I., Raposo, G., Savina, A., Moita, C. F., Schauer, K., Hume, A. N., Freitas, R. P., Goud, B., Benaroch, P., Hacohen, N., Fukuda, M., Desnos, C., Seabra, M. C., Darchen, F., Amigorena, S., Moita, L. F. & Thery, C. (2010). Rab27a and Rab27b control different steps of the exosome secretion pathway. *Nature cell biology*. Vol. 12, No.1 (January 2010), pp.19-30; sup pp 11-13, 1465-7392

Pai, E. F., Krengel, U., Petsko, G. A., Goody, R. S., Kabsch, W. & Wittinghofer, A. (1990). Refined crystal structure of the triphosphate conformation of H-ras p21 at 1.35 A resolution: implications for the mechanism of GTP hydrolysis. *The EMBO journal*. Vol. 9, No.8 (August 1990), pp.2351-2359, 0261-4189

Pasteris, N. G., Nagata, K., Hall, A. & Gorski, J. L. (2000). Isolation, characterization, and mapping of the mouse Fgd3 gene, a new Faciogenital Dysplasia (FGD1; Aarskog Syndrome) gene homologue. *Gene*. Vol. 242, No.1-2 (January 2000), pp.237-247, 0378-1119

Pereira-Leal, J. B. & Seabra, M. C. (2000). The mammalian Rab family of small GTPases: definition of family and subfamily sequence motifs suggests a mechanism for functional specificity in the Ras superfamily. *Journal of molecular biology*. Vol. 301, No.4 (August 2000), pp.1077-1087, 0022-2836

Pereira-Leal, J. B. & Seabra, M. C. (2001). Evolution of the Rab family of small GTP-binding proteins. *Journal of molecular biology*. Vol. 313, No.4 (November 2001), pp.889-901, 0022-2836

Polzin, A., Shipitsin, M., Goi, T., Feig, L. A. & Turner, T. J. (2002). Ral-GTPase influences the regulation of the readily releasable pool of synaptic vesicles. *Molecular and cellular biology*. Vol. 22, No.6 (March 2002), pp.1714-1722, 0270-7306

Prekeris, R., Klumperman, J. & Scheller, R. H. (2000). A Rab11/Rip11 protein complex regulates apical membrane trafficking via recycling endosomes. *Molecular cell*. Vol. 6, No.6 (December 2001), pp.1437-1448, 1097-2765

Radisky, D. C., Levy, D. D., Littlepage, L. E., Liu, H., Nelson, C. M., Fata, J. E., Leake, D., Godden, E. L., Albertson, D. G., Nieto, M. A., Werb, Z. & Bissell, M. J. (2005). Rac1b and reactive oxygen species mediate MMP-3-induced EMT and genomic instability. *Nature*. Vol. 436, No.7047 (July 2005), pp.123-127, 0028-0836

Reid, T., Bathoorn, A., Ahmadian, M. R. & Collard, J. G. (1999). Identification and characterization of hPEM-2, a guanine nucleotide exchange factor specific for Cdc42. *The Journal of biological chemistry*. Vol. 274, No.47 (November 1999), pp.33587-33593, 0021-9258

Resh, M. D. (2004). Membrane targeting of lipid modified signal transduction proteins. *Sub-cellular biochemistry*. Vol. 37, (n. d.), pp.217-232, 0306-0225

Ridley, A. J. (2006). Rho GTPases and actin dynamics in membrane protrusions and vesicle trafficking. *Trends in cell biology*. Vol. 16, No.10 (October 2006), pp.522-529, 0962-8924

Roberts, P. J., Mitin, N., Keller, P. J., Chenette, E. J., Madigan, J. P., Currin, R. O., Cox, A. D., Wilson, O., Kirschmeier, P. & Der, C. J. (2008). Rho Family GTPase modification and dependence on CAAX motif-signaled posttranslational modification. *The Journal of biological chemistry*. Vol. 283, No.37 (September 2008), pp.25150-25163, 0021-9258

Rodriguez, F., Bustos, M. A., Zanetti, M. N., Ruete, M. C., Mayorga, L. S. & Tomes, C. N. (2011). alpha-SNAP prevents docking of the acrosome during sperm exocytosis because it sequesters monomeric syntaxin. *PloS one*. Vol. 6, No.7 (July 2011), pp.e21925, 1932-6203

Rondaij, M. G., Sellink, E., Gijzen, K. A., ten Klooster, J. P., Hordijk, P. L., van Mourik, J. A. & Voorberg, J. (2004). Small GTP-binding protein Ral is involved in cAMP-mediated release of von Willebrand factor from endothelial cells. *Arteriosclerosis, thrombosis, and vascular biology*. Vol. 24, No.7 (July 2004), pp.1315-1320, 1079-5642

Rondaij, M. G., Bierings, R., van Agtmaal, E. L., Gijzen, K. A., Sellink, E., Kragt, A., Ferguson, S. S., Mertens, K., Hannah, M. J., van Mourik, J. A., Fernandez-Borja, M. & Voorberg, J. (2008). Guanine exchange factor RalGDS mediates exocytosis of Weibel-Palade bodies from endothelial cells. *Blood*. Vol. 112, No.1 (Jully 2008), pp.56-63, 0006-4971

Rosse, C., Hatzoglou, A., Parrini, M. C., White, M. A., Chavrier, P. & Camonis, J. (2006). RalB mobilizes the exocyst to drive cell migration. *Molecular and cellular biology*. Vol. 26, No.2 (January 2006), pp.727-734, 0270-7306

Sakane, A., Hatakeyama, S. & Sasaki, T. (2007). Involvement of Rabring7 in EGF receptor degradation as an E3 ligase. *Biochemical and biophysical research communications*. Vol. 357, No.4 (June 2007), pp.1058-1064, 0006-291X

Sakisaka, T., Baba, T., Tanaka, S., Izumi, G., Yasumi, M. & Takai, Y. (2004). Regulation of SNAREs by tomosyn and ROCK: implication in extension and retraction of neurites. *The Journal of cell biology*. Vol. 166, No.1 (July 2004), pp.17-25, 0021-9525

Sano, H., Peck, G. R., Kettenbach, A. N., Gerber, S. A. & Lienhard, G. E. (2011). Insulin-stimulated GLUT4 protein translocation in adipocytes requires the Rab10 guanine nucleotide exchange factor Dennd4C. *The Journal of biological chemistry*. Vol. 286, No.19 (May 2011), pp.16541-16545, 0021-9258

Sato, Y., Fukai, S., Ishitani, R. & Nureki, O. (2007). Crystal structure of the Sec4p.Sec2p complex in the nucleotide exchanging intermediate state. *Proceedings of the National Academy of Sciences of the United States of America*. Vol. 104, No.20 (May 2007), pp.8305-8310, 0027-8424

Scheffzek, K., Stephan, I., Jensen, O. N., Illenberger, D. & Gierschik, P. (2000). The Rac-RhoGDI complex and the structural basis for the regulation of Rho proteins by RhoGDI. *Nature structural & molecular biology*. Vol. 7, No.2 (February 2000), pp.122-126, 1072-8368

Schmidt, A. & Hall, A. (2002). Guanine nucleotide exchange factors for Rho GTPases: turning on the switch. *Genes & development*. Vol. 16, No.13 (Jul I 2002), pp.1587-1609, 0890-9369

Schnelzer, A., Prechtel, D., Knaus, U., Dehne, K., Gerhard, M., Graeff, H., Harbeck, N., Schmitt, M. & Lengyel, E. (2000). Rac1 in human breast cancer: overexpression, mutation analysis, and characterization of a new isoform, Rac1b. *Oncogene*. Vol. 19, No.26 (June 2000), pp.3013-3020, 0950-9232

Schonn, J. S., van Weering, J. R., Mohrmann, R., Schluter, O. M., Sudhof, T. C., de Wit, H., Verhage, M. & Sorensen, J. B. (2010). Rab3 proteins involved in vesicle biogenesis and priming in embryonic mouse chromaffin cells. *Traffic*. Vol. 11, No.11 (November 2010), pp.1415-1428, 1398-9219

Seabra, M. C., Goldstein, J. L., Sudhof, T. C. & Brown, M. S. (1992). Rab geranylgeranyl transferase. A multisubunit enzyme that prenylates GTP-binding proteins terminating in Cys-X-Cys or Cys-Cys. *The Journal of biological chemistry*. Vol. 267, No.20 (July 1992), pp.14497-14503, 0021-9258

Seabra, M. C. (1998). Membrane association and targeting of prenylated Ras-like GTPases. *Cellular signalling*. Vol. 10, No.3 (March 1998), pp.167-172, 0898-6568

Seabra, M. C. & Coudrier, E. (2004). Rab GTPases and myosin motors in organelle motility. *Traffic*. Vol. 5, No.6 (June 2004), pp.393-399, 1398-9219

Sebti, S. M. & Der, C. J. (2003). Opinion: Searching for the elusive targets of farnesyltransferase inhibitors. *Nature reviews. Cancer*. Vol. 3, No.12 (December 2004), pp.945-951, 1474-175X

Sebti, S. M. (2005). Protein farnesylation: implications for normal physiology, malignant transformation, and cancer therapy. *Cancer Cell*. Vol. 7, No.4 (April 2005), pp.297-300, 1535-6108

Segev, N. (2011). GTPases in intracellular trafficking: an overview. *Seminars in cell & developmental biology*. Vol. 22, No.1 (February 2011), pp.1-2, 1084-9521

Shandala, T., Woodcock, J. M., Ng, Y., Biggs, L., Skoulakis, E. M., Brooks, D. A. & Lopez, A. F. (2011). *Drosophila* 14-3-3epsilon has a crucial role in anti-microbial peptide secretion and innate immunity. *Journal of cell science*. Vol. 124, No.Pt 13 (July 2011), pp.2165-2174, 0021-9533

Shipitsin, M. & Feig, L. A. (2004). RalA but not RalB enhances polarized delivery of membrane proteins to the basolateral surface of epithelial cells. *Molecular and cellular biology*. Vol. 24, No.13 (July 2004), pp.5746-5756, 0270-7306

Shmueli, A., Segal, M., Sapir, T., Tsutsumi, R., Noritake, J., Bar, A., Sapoznik, S., Fukata, Y., Orr, I., Fukata, M. & Reiner, O. (2010). Ndel1 palmitoylation: a new mean to regulate cytoplasmic dynein activity. *The EMBO journal*. Vol. 29, No.1 (January 2010), pp.107-119, 0261-4189

Simons, K. & Ikonen, E. (1997). Functional rafts in cell membranes. *Nature*. Vol. 387, No.6633 (June 1997), pp.569-572, 0028-0836

Sivaram, M. V., Saporita, J. A., Furgason, M. L., Boettcher, A. J. & Munson, M. (2005). Dimerization of the exocyst protein Sec6p and its interaction with the t-SNARE Sec9p. *Biochemistry*. Vol. 44, No.16 (April 2005), pp.6302-6311, 0006-2960

Sivaram, M. V., Furgason, M. L., Brewer, D. N. & Munson, M. (2006). The structure of the exocyst subunit Sec6p defines a conserved architecture with diverse roles. *Nature structural & molecular biology*. Vol. 13, No.6 (June 2006), pp.555-556, 11545-9985

Sjogren, A. K., Andersson, K. M., Liu, M., Cutts, B. A., Karlsson, C., Wahlstrom, A. M., Dalin, M., Weinbaum, C., Casey, P. J., Tarkowski, A., Swolin, B., Young, S. G. & Bergo, M. O. (2007). GGTase-I deficiency reduces tumor formation and improves survival in mice with K-RAS-induced lung cancer. *The Journal of clinical investigation*. Vol. 117, No.5 (May 2007), pp.1294-1304, 0021-9738

Sollner, T., Whiteheart, S. W., Brunner, M., Erdjument-Bromage, H., Geromanos, S., Tempst, P. & Rothman, J. E. (1993). SNAP receptors implicated in vesicle targeting and fusion. *Nature*. Vol. 362, No.6418 (March 1993), pp.318-324, 0028-0836

Spaargaren, M. & Bischoff, J. R. (1994). Identification of the guanine nucleotide dissociation stimulator for Ral as a putative effector molecule of R-ras, H-ras, K-ras, and Rap. *Proceedings of the National Academy of Sciences of the United States of America*. Vol. 91, No.26 (December 1994), pp.12609-12613, 0027-8424

Spiczka, K. S. & Yeaman, C. (2008). Ral-regulated interaction between Sec5 and paxillin targets Exocyst to focal complexes during cell migration. *Journal of cell science*. Vol. 121, No.Pt 17 (September 2008), pp.2880-2891, 0021-9533

Stenmark, H. (2009). Rab GTPases as coordinators of vesicle traffic. *Nature reviews. Molecular cell biology*. Vol. 10, No.8 (August 2009), pp.513-525, 1471-0072

Stinchcombe, J. C., Barral, D. C., Mules, E. H., Booth, S., Hume, A. N., Machesky, L. M., Seabra, M. C. & Griffiths, G. M. (2001). Rab27a is required for regulated secretion in cytotoxic T lymphocytes. *The Journal of cell biology*. Vol. 152, No.4 (February 2001), pp.825-834, 0021-9525

Su, L., Lineberry, N., Huh, Y., Soares, L. & Fathman, C. G. (2006). A novel E3 ubiquitin ligase substrate screen identifies Rho guanine dissociation inhibitor as a substrate of gene related to anergy in lymphocytes. *The Journal of immunology : official journal of the American Association of Immunologists*. Vol. 177, No.11 (December 2006), pp.7559-7566, 0022-1767

Sudhof, T. C. & Rothman, J. E. (2009). Membrane fusion: grappling with SNARE and SM proteins. *Science*. Vol. 323, No.5913 (January 2009), pp.474-477, 0036-8075

Sugawara, K., Shibasaki, T., Mizoguchi, A., Saito, T. & Seino, S. (2009). Rab11 and its effector Rip11 participate in regulation of insulin granule exocytosis. *Genes to cells : devoted to molecular & cellular mechanisms*. Vol. 14, No.4 (April 2009), pp.445-456, 1356-9597

Sugihara, K., Asano, S., Tanaka, K., Iwamatsu, A., Okawa, K. & Ohta, Y. (2002). The exocyst complex binds the small GTPase RalA to mediate filopodia formation. *Nature cell biology*. Vol. 4, No.1 (January 2002), pp.73-78, 1465-7392

Sun, L., Bittner, M. A. & Holz, R. W. (2003). Rim, a component of the presynaptic active zone and modulator of exocytosis, binds 14-3-3 through its N terminus. *The Journal of biological chemistry*. Vol. 278, No.40 (October 2003), pp.38301-38309, 0021-9258

Swank, R. T., Jiang, S. Y., Reddington, M., Conway, J., Stephenson, D., McGarry, M. P. & Novak, E. K. (1993). Inherited abnormalities in platelet organelles and platelet formation and associated altered expression of low molecular weight guanosine triphosphate-binding proteins in the mouse pigment mutant gunmetal. *Blood*. Vol. 81, No.10 (May 1993), pp.2626-2635, 0006-4971

Takaya, A., Kamio, T., Masuda, M., Mochizuki, N., Sawa, H., Sato, M., Nagashima, K., Mizutani, A., Matsuno, A., Kiyokawa, E. & Matsuda, M. (2007). R-Ras regulates exocytosis by Rgl2/Rlf-mediated activation of RalA on endosomes. *Molecular biology of the cell*. Vol. 18, No.5 (May 2007), pp.1850-1860, 1059-1524

Taylor, A., Mules, E. H., Seabra, M. C., Helfrich, M. H., Rogers, M. J. & Coxon, F. P. (2011). Impaired prenylation of Rab GTPases in the gunmetal mouse causes defects in bone cell function. *Small GTPases*. Vol. 2, No.3 (May 2011), pp.131-142, 2154-1256

TerBush, D. R., Maurice, T., Roth, D. & Novick, P. (1996). The Exocyst is a multiprotein complex required for exocytosis in Saccharomyces cerevisiae. *The EMBO journal*. Vol. 15, No.23 (December 1996), pp.6483-6494, 0261-4189

Tong, J., Borbat, P. P., Freed, J. H. & Shin, Y. K. (2009). A scissors mechanism for stimulation of SNARE-mediated lipid mixing by cholesterol. *Proceedings of the National Academy of Sciences of the United States of America*. Vol. 106, No.13 (March 2009), pp.5141-5146, 0027-8424

Trahey, M. & McCormick, F. (1987). A cytoplasmic protein stimulates normal N-ras p21 GTPase, but does not affect oncogenic mutants. *Science*. Vol. 238, No.4826 (October 1987), pp.542-545, 0036-8075

Tsuboi, T. & Fukuda, M. (2006). Rab3A and Rab27A cooperatively regulate the docking step of dense-core vesicle exocytosis in PC12 cells. *Journal of cell science*. Vol. 119, No.Pt 11 (June 2006), pp.2196-2203, 0021-9533

Tsuboi, T., Kanno, E. & Fukuda, M. (2007). The polybasic sequence in the C2B domain of rabphilin is required for the vesicle docking step in PC12 cells. *Journal of neurochemistry*. Vol. 100, No.3 (February 2007), pp.770-79, 0022-3042

Tsuboi, T. (2009). Molecular mechanism of attachment process of dense-core vesicles to the plasma membrane in neuroendocrine cells. *Neuroscience Research*. Vol. 63, No.2 (February 2009), pp.83-88, 0168-0102

Uno, T., Moriwaki, T., Isoyama, Y., Uno, Y., Kanamaru, K., Yamagata, H., Nakamura, M. & Takagi, M. (2010). Rab14 from *Bombyx mori* (Lepidoptera: Bombycidae) shows ATPase activity. *Biology letters*. Vol. 6, No.3 (June 2010), pp.379-381, 1744-9561

Urbe, S., Huber, L. A., Zerial, M., Tooze, S. A. & Parton, R. G. (1993). Rab11, a small GTPase associated with both constitutive and regulated secretory pathways in PC12 cells. *FEBS letters*. Vol. 334, No.2 (November 1993), pp.175-182, 0014-5793

van Dam, E. M. & Robinson, P. J. (2006). Ral: mediator of membrane trafficking. *The international journal of biochemistry & cell biology*. Vol. 38, No.11 (n. d.), pp.1841-1847, 1357-2725

van der Meulen, J., Bhullar, R. P. & Chancellor-Maddison, K. A. (1991). Association of a 24-kDa GTP-binding protein, Gn24, with human platelet alpha-granule membranes. *FEBS letters*. Vol. 291, No.1 (October 1991), pp.122-126, 0014-5793 van der Sluijs, P., Hull, M., Webster, P., Male, P., Goud, B. & Mellman, I. (1992). The small GTP-binding protein rab4 controls an early sorting event on the endocytic pathway. *Cell*. Vol. 70, No.5 (Sepember 1992), pp.729-740, 0092-8674

van Rijssel, J., Hoogenboezem, M., Wester, L., Hordijk, P. L. & Van Buul, J. D. (2012). The N-Terminal DH-PH Domain of Trio Induces Cell Spreading and Migration by Regulating Lamellipodia Dynamics in a Rac1-Dependent Fashion. *PloS one*. Vol. 7, No.1 (January 2012), pp.e29912, 1932-6203

Vesa, J., Hellsten, E., Verkruyse, L. A., Camp, L. A., Rapola, J., Santavuori, P., Hofmann, S. L. & Peltonen, L. (1995). Mutations in the palmitoyl protein thioesterase gene causing infantile neuronal ceroid lipofuscinosis. *Nature*. Vol. 376, No.6541 (August 1995), pp.584-587, 0028-0836

Visvikis, O., Lores, P., Boyer, L., Chardin, P., Lemichez, E. & Gacon, G. (2008). Activated Rac1, but not the tumorigenic variant Rac1b, is ubiquitinated on Lys 147 through a JNK-regulated process. *The FEBS journal*. Vol. 275, No.2 (January 2008), pp.386-396, 1742-464X

Wang, H., Owens, C., Chandra, N., Conaway, M. R., Brautigan, D. L. & Theodorescu, D. (2010). Phosphorylation of RalB is important for bladder cancer cell growth and metastasis. *Cancer research*. Vol. 70, No.21 (November 2010), pp.8760-8769, 0008-5472

Wang, H. R., Zhang, Y., Ozdamar, B., Ogunjimi, A. A., Alexandrova, E., Thomsen, G. H. & Wrana, J. L. (2003). Regulation of cell polarity and protrusion formation by targeting RhoA for degradation. *Science*. Vol. 302, No.5651 (December 2003), pp.1775-1779, 0036-8075

Wang, Y., Jerdeva, G., Yarber, F. A., da Costa, S. R., Xie, J., Qian, L., Rose, C. M., Mazurek, C., Kasahara, N., Mircheff, A. K. & Hamm-Alvarez, S. F. (2003). Cytoplasmic dynein participates in apically targeted stimulated secretory traffic in primary rabbit lacrimal acinar epithelial cells. *Journal of cell science*. Vol. 116, No.Pt 10 (May 2003), pp.2051-2065, 0021-9533

Wang, Y. & Dasso, M. (2009). SUMOylation and deSUMOylation at a glance. *Journal of cell science*. Vol. 122, No.Pt 23 (December 2009), pp.4249-4252, 0021-9533 0021-9533

Wang, Z. & Thurmond, D. C. (2010). Differential phosphorylation of RhoGDI mediates the distinct cycling of Cdc42 and Rac1 to regulate second-phase insulin secretion. *The Journal of biological chemistry*. Vol. 285, No.9 (February 2009), pp.6186-6197, 0021-9258 0021-9258

Ward, E. S., Martinez, C., Vaccaro, C., Zhou, J., Tang, Q. & Ober, R. J. (2005). From sorting endosomes to exocytosis: association of Rab4 and Rab11 GTPases with the Fc receptor, FcRn, during recycling. *Molecular biology of the cell*. Vol. 16, No.4 (April 2005), pp.2028-2038, 1059-1524

Weimbs, T., Low, S. H., Chapin, S. J., Mostov, K. E., Bucher, P. & Hofmann, K. (1997). A conserved domain is present in different families of vesicular fusion proteins: a new superfamily. *Proceedings of the National Academy of Sciences of the United States of America*. Vol. 94, No.7 (April 1997), pp.3046-3051, 0027-8424

Wilkinson, K. A. & Henley, J. M. (2010). Mechanisms, regulation and consequences of protein SUMOylation. *The Biochemical journal*. Vol. 428, No.2 (June 2010), pp.133-145, 0264-6021

Williams, A. L., Bielopolski, N., Meroz, D., Lam, A. D., Passmore, D. R., Ben-Tal, N., Ernst, S. A., Ashery, U. & Stuenkel, E. L. (2011). Structural and functional analysis of tomosyn identifies domains important in exocytotic regulation. *The Journal of biological chemistry*. Vol. 286, No.16 (April 2011), pp.14542-14553, 0021-9258

Williams, C. (1991). Polypes degeneres du colon. [Degenerated colonic polyps]. *Acta gastro-enterologica Belgica*. Vol. 54, No.3-4 (May 1991), pp.276-278, 0001-5644

Williams, J. A., Chen, X. & Sabbatini, M. E. (2009). Small G proteins as key regulators of pancreatic digestive enzyme secretion. *American journal of physiology. Endocrinology and metabolism*. Vol. 296, No.3 (March 2009), pp.E405-414, 0193-1849

Winter-Vann, A. M. & Casey, P. J. (2005). Post-prenylation-processing enzymes as new targets in oncogenesis. *Nature reviews. Cancer*. Vol. 5, No.5 (May 2005), pp.405-412, 1474-175X

Wolthuis, R. M., Bauer, B., van 't Veer, L. J., de Vries-Smits, A. M., Cool, R. H., Spaargaren, M., Wittinghofer, A., Burgering, B. M. & Bos, J. L. (1996). RalGDS-like factor (Rlf) is a novel Ras and Rap 1A-associating protein. *Oncogene*. Vol. 13, No.2 (July 1996), pp.353-362, 0950-9232

Wu, H. & Brennwald, P. (2010). The function of two Rho family GTPases is determined by distinct patterns of cell surface localization. *Molecular and cellular biology*. Vol. 30, No.21 (November 2010), pp.5207-5217, 0270-7306

Wu, H., Turner, C., Gardner, J., Temple, B. & Brennwald, P. (2010). The Exo70 subunit of the exocyst is an effector for both Cdc42 and Rho3 function in polarized exocytosis. *Molecular biology of the cell*. Vol. 21, No.3 (February 2009), pp.430-442, 1059-1524

Wu, J. C., Chen, T. Y., Yu, C. T., Tsai, S. J., Hsu, J. M., Tang, M. J., Chou, C. K., Lin, W. J., Yuan, C. J. & Huang, C. Y. (2005). Identification of V23RalA-Ser194 as a critical mediator for Aurora-A-induced cellular motility and transformation by small pool expression screening. *The Journal of biological chemistry*. Vol. 280, No.10 (March 2005), pp.9013-9022, 0021-9258

Wu, S., Mehta, S. Q., Pichaud, F., Bellen, H. J. & Quiocho, F. A. (2005). Sec15 interacts with Rab11 via a novel domain and affects Rab11 localization in vivo. *Nature structural & molecular biology*. Vol. 12, No.10 (October 2005), pp.879-885, 1545-9993

Wu, Z., Owens, C., Chandra, N., Popovic, K., Conaway, M. & Theodorescu, D. (2010). RalBP1 is necessary for metastasis of human cancer cell lines. *Neoplasia : an international journal for oncology research*. Vol. 12, No.12 (December 2010), pp.1003-1012, 1476-5586

Xiong, X., Xu, Q., Huang, Y., Singh, R. D., Anderson, R., Leof, E., Hu, J. & Ling, K. (2012). An association between type Igamma PI4P 5-kinase and Exo70 directs E-cadherin clustering and epithelial polarization. *Molecular biology of the cell*. Vol. 23, No.1 (January 2012), pp.87-98, 1059-1524

Xu, L., Lubkov, V., Taylor, L. J. & Bar-Sagi, D. (2010). Feedback regulation of Ras signaling by Rabex-5-mediated ubiquitination. *Current biology: CB*. Vol. 20, No.15 (August 2010), pp.1372-1377, 0960-9822

Yalovsky, S., Rodr Guez-Concepcion, M. & Gruissem, W. (1999). Lipid modifications of proteins - slipping in and out of membranes. *Trends in plant science*. Vol. 4, No.11 (November 1999), pp.439-445, 1360-1385

Yamagata, Y., Jovanovic, J. N., Czernik, A. J., Greengard, P. & Obata, K. (2002). Bidirectional changes in synapsin I phosphorylation at MAP kinase-dependent sites by acute neuronal excitation in vivo. *Journal of neurochemistry*. Vol. 80, No.5 (March 2002), pp.835-842, 0022-3042

Yamaguchi, K., Ohara, O., Ando, A. & Nagase, T. (2008). Smurf1 directly targets hPEM-2, a GEF for Cdc42, via a novel combination of protein interaction modules in the ubiquitin-proteasome pathway. *Biological chemistry*. Vol. 389, No.4 (April 2008), pp.405-413, 1431-6730

Yan, H., Jahanshahi, M., Horvath, E. A., Liu, H. Y. & Pfleger, C. M. (2010). Rabex-5 ubiquitin ligase activity restricts Ras signaling to establish pathway homeostasis in Drosophila. *Current biology: CB*. Vol. 20, No.15 (August 2010), pp.1378-1382, 0960-9822

Yeaman, C., Grindstaff, K. K. & Nelson, W. J. (2004). Mechanism of recruiting Sec6/8 (exocyst) complex to the apical junctional complex during polarization of epithelial cells. *Journal of cell science*. Vol. 117, No.Pt 4 (February 2004), pp.559-570, 0021-9533

Zajac, A., Sun, X., Zhang, J. & Guo, W. (2005). Cyclical regulation of the exocyst and cell polarity determinants for polarized cell growth. *Molecular biology of the cell*. Vol. 16, No.3 (March 2005), pp.1500-1512, 1059-1524

Zarelli, V. E., Ruete, M. C., Roggero, C. M., Mayorga, L. S. & Tomes, C. N. (2009). PTP1B dephosphorylates N-ethylmaleimide-sensitive factor and elicits SNARE complex disassembly during human sperm exocytosis. *The Journal of biological chemistry*. Vol. 284, No.16 (April 2009), pp.10491-10503, 0021-9258

Zhang, F. L. & Casey, P. J. (1996). Protein prenylation: molecular mechanisms and functional consequences. *Annual review of biochemistry*. Vol. 65, (July 1996), pp.241-269, 0066-4154

Zhang, Q., Zhen, L., Li, W., Novak, E. K., Collinson, L. M., Jang, E. K., Haslam, R. J., Elliott, R. W. & Swank, R. T. (2002). Cell-specific abnormal prenylation of Rab proteins in platelets and melanocytes of the gunmetal mouse. *British journal of haematology*. Vol. 117, No.2 (May 2002), pp.414-423, 0007-1048

Zhang, X., Wang, P., Gangar, A., Zhang, J., Brennwald, P., TerBush, D. & Guo, W. (2005). Lethal giant larvae proteins interact with the exocyst complex and are involved in polarized exocytosis. *The Journal of cell biology*. Vol. 170, No.2 (July 2005), pp.273-283, 0021-9525

Zhang, X. M., Ellis, S., Sriratana, A., Mitchell, C. A. & Rowe, T. (2004). Sec15 is an effector for the Rab11 GTPase in mammalian cells. *The Journal of biological chemistry*. Vol. 279, No.41 (October 2004), pp.43027-43034, 0021-9258

Zhou, C. Z., Li de La Sierra-Gallay, I., Quevillon-Cheruel, S., Collinet, B., Minard, P., Blondeau, K., Henckes, G., Aufrere, R., Leulliot, N., Graille, M., Sorel, I., Savarin, P., de la Torre, F., Poupon, A., Janin, J. & van Tilbeurgh, H. (2003). Crystal structure of the yeast Phox homology (PX) domain protein Grd19p complexed to phosphatidylinositol-3-phosphate. *The Journal of biological chemistry*. Vol. 278, No.50 (December 2003), pp.50371-50376, 0021-9258

5

At the Intersection of the Pathways for Exocytosis and Autophagy

D.A. Brooks, C. Bader, Y.S. Ng,
R.D. Brooks, G.N. Borlace and T. Shandala
Mechanisms in Cell Biology and Diseases Research Group,
School of Pharmacy and Medical Sciences, Sansom Institute for Health Research
University of South Australia
Australia

1. Introduction

Recent studies have suggested that there are molecular links between the two critical biological processes of exocytosis and autophagy. Exocytosis involves the transport of intracellular vesicles to the plasma membrane of the cell, where vesicular fusion results in the delivery of membrane and protein to the cell surface, and secretion of the vesicular contents. Exocytosis is utilized in, for example, hormone or antimicrobial peptide secretion, the delivery of proteoglycans to the cell surface, cell-cell communication and neurotransmission (Brennwald & Rossi, 2007; He & Guo, 2009). Autophagy is a mechanism for the recycling and degradation of cytoplasmic content, which involves surrounding an area of cytoplasm with a double membrane structure, which then interacts with degradative endosome-lysosome compartments (He & Klionsky, 2009). Autophagy has important functions in a range of cell processes including the maintenance of cellular homeostasis, starvation adaption, energy balance, organelle clearance, immunity and cell death. In human diseases, such as cancers, neurodegenerative disorders (e.g. Huntington's disease), and chronic inflammatory diseases (e.g. Crohn's disease), there have been reports of functional disparity in both of these important membrane-related cellular pathways. There is now increasing evidence that exocytosis and autophagy share molecular machinery and there are a number of reasons why this would be beneficial in terms of cellular function.

Exocytosis and autophagy may be competitively, cooperatively or independently regulated, depending upon the nature of the intracellular and/or extracellular environment. In response to conditions of low or high energy demand, there would be an advantage to the cell in reducing the energy consuming process of secretion, where the membrane from exocytic vesicles could be utilized to enable rapid expansion of the autophagic compartment (i.e. competitive regulation). During an immune response there may be concomitant stimulation of autophagy to degrade an intracellular bacterial pathogen, and exocytosis to release second messengers and antimicrobial effectors. Alternatively, it may be necessary to only upregulate an individual process, which is the case for increased autophagy during organelle and cytoplasm turnover under restricted nutrient supply (e.g. in the bone growth plate) (He & Klionsky, 2009), or increased exocytosis during proteoglycan delivery to the cell surface (Franken *et al.*, 2003). Finally, during the removal of dysfunctional exocytic vesicles or the rapid cessation of secretion (e.g. during neurotransmission), the cell requires organelle-specific molecular machinery, for the nucleation of autophagy (Geng *et al.*, 2010).

Despite defects in multiple human syndromes that demonstrate changes in both exocytosis and autophagy, the mutual dependence of these processes on common molecular machinery has only recently been investigated. Evidence indicates that the exocyst complex and its regulator Ral (Ras like GTPase), both of which are known to have a critical function in exocytosis, also appear to be essential for the initiation of autophagy (Bodemann *et al.*, 2011). Similarly, the small GTPase Rab11 has a critical role in exocytosis at the recycling endosome and in exocytic vesicle function (van Ijzendoorn, 2006), although in times of starvation where autophagy is induced, Rab11 containing vesicular compartments have been shown to asscociate with autophagosomes (Rab11 positive amphisomes; (Fader *et al.*, 2008)). Finally, disruption of the exocytic Rab GTPase Sec4 and its guanine nucleotide exchange factor Sec2, can have significant effects on the anterograde movement of the integral autophagosome membrane protein Atg9 (Autophagy related protein 9), thereby influencing the recruitment of Atg8 to the phagophore assembly site (PAS) (Bodemann *et al.*, 2011). The movements of Atg9 and Atg8 are of particular interest as they are both important during the initiation of autophagy (Geng *et al.*, 2010; Wang *et al.*, 2009). Atg9 has, in turn, been reported to reside on exocytic vesicles that can be converted into a phagophore assembly site (Mari *et al.*, 2010; Mari & Reggiori, 2010). The aim of this chapter is to provide an overview of the exocytic and autophagic processes with a focus on the common molecular machinery acting at critical control points. It is this machinery that may facilitate communication between these functionally distinct vesicular compartments and may act as potential sites for regulation.

2. Exocytosis

The exocytic pathway delivers cargo carrying vesicles from either the *trans*-Golgi network (TGN) or recycling endosomes to the plasma membrane, where membrane fusion occurs to release the vesicular content (Figure 1). This vesicular content may be either vesicle membrane proteins directed to the cell surface, or lumenal contents for secretion into the extracellular milieu. This anterograde trafficking route may vary depending upon the cargo and cell type involved (Wurster *et al.*, 1990), such as in melanocytes for melanin exocytosis and in neurons for neurotransmission. Exocytosis is also involved in numerous other cellular functions, including immune responses, cell-cell communication, cell growth, cell polarity and neurotransmission.

There are two main exocytic routes from the *trans*-Golgi network to the plasma membrane: the constitutive and the regulated routes (Stow *et al.*, 2009). The constitutive route continuously delivers membrane and cargo from the *trans*-Golgi network to the cell surface, and is thought to be utilized for housekeeping functions. Although this process can be up-regulated in response to environmental stress, it is generally representative of a basal level of cell activity and secretion. However, a number of pro-inflammatory cytokines, including TNFα and IL-6, are released from macrophages via the constitutive route, in response to pro-inflammatory stimuli (Shurety *et al.*, 2000). The regulated route involves the redirection of newly synthesized cargo to compartments where these molecules are stored until their release is triggered by a specific stimulus (mediated by calcium ion mobilization). In this

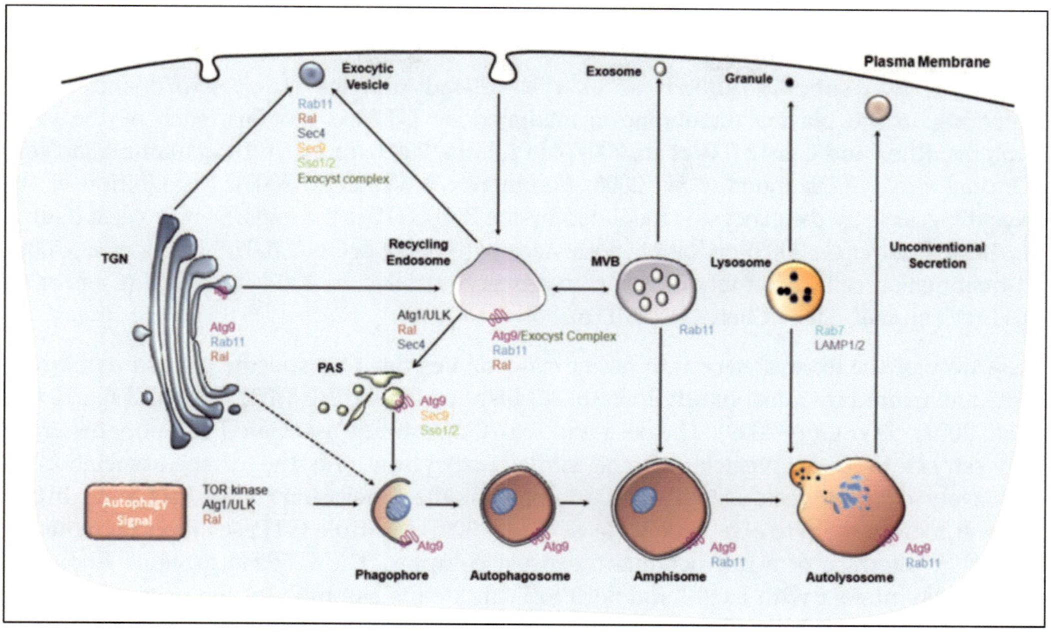

Fig. 1. Proposed model for crosstalk between autophagy and exocytosis.

route, cargo can be trafficked via a number of compartments including recycling endosomes, early endosomes, multivesicular bodies, secretory granules and secretory lysosomes. This pathway is utilized for the immune related secretion of cytokines and antimicrobial peptides, following exposure to pathogens or inflammatory stimuli. Similarly, for neurotransmitter release, exocytosis is stimulated by an increase in the intracellular calcium ion concentration in neurons; which allows the propagation of neuronal function. These different pathways are in dynamic balance with the endocytic pathway, which, apart from facilitating uptake into the cell, allows the recovery of membrane from the plasma membrane following exocytosis, enabling the cell to control its surface area (Khandelwal *et al.*, 2010). The molecular machinery that drives exocytosis therefore operates in conjunction with the endocytic machinery, and in some cases may involve common elements that have dual function.

2.1 The molecular machinery for exocytosis

The molecular machinery that facilitates the process of exocytosis can vary with respect to the specific cell type and specialist cargo being transported, although two key molecular complexes are conserved for most membrane associated exocytic events; the exocyst and the soluble N-ethylmaleimide sensitive factor attachment protein receptor (SNARE) complex (Liu & Guo, 2011; Nair *et al.*, 2011). Through interaction with a number of effector molecules, these complexes mediate the tethering, docking and fusion of vesicles with the plasma membrane.

The exocyst is an octomeric complex that is required for the efficient delivery of exocytic vesicles to the plasma membrane (TerBush *et al.*, 1996). The components of the exocyst complex were first identified for yeast in the early 1990's, with mammalian orthologues subsequently being identified (Lipschutz & Mostov, 2002). In yeast, this complex consists of six secretion related proteins (Sec), Sec3, Sec5, Sec6, Sec8, Sec10 and Sec15, with an additional two subunits known as exocyst related proteins (Exo), Exo70 and Exo84. Tethering to the plasma membrane is mediated by GTPase proteins, such as the yeast proteins, Rho3 and Cdc42 (He *et al.*, 2007; He & Guo, 2009), or by TC10 in mammalian cells (Dupraz *et al.*, 2009; Inoue *et al.*, 2006; Pommereit & Wouters, 2007). Recognition of the exocytic vesicle by the exocyst is mediated by the Rab GTPase proteins, Sec4 in yeast (Guo *et al.*, 1999; Zajac *et al.*, 2005) or Rab11 in metazoans (Novick & Guo, 2002; Novick *et al.*, 2006). In mammalian cells, assembly of this complex is controlled by RalA and RalB (Chen *et al.*, 2011a; Chen *et al.*, 2007; Chen *et al.*, 2011b).

Assembly of the exocyst serves to tether exocytic vesicles to a specific plasma membrane site, demarcated by phosphatidylinositol 4,5-bisphosphate (PI(4,5)P2), Sec3 and Exo70 (He *et al.*, 2007; He & Guo, 2009). The Sec3 and Exo70 membrane associated components of the exocyst act to target vesicles to the site of exocytosis, via the direct association of positively charged residues in the D domain at the C-terminus with PI(4,5)P2, in the plasma membrane (He *et al.*, 2007; He & Guo, 2009). Multiple GTPases then regulate the assembly interface of a full octameric exocyst complex. The GTPase proteins Rho3 and Cdc42 also interact with Exo70 and Sec3 to facilitate the assembly of the exocyst complex at the plasma membrane (He *et al.*, 2007; Moskalenko *et al.*, 2002). In mammalian cells, Exo70 facilitates exocyst association with the plasma membrane through its interaction with TC10 (the orthologue of Cdc42) (Inoue *et al.*, 2003; Liu *et al.*, 2007). In yeast, secretory vesicles acquire the Rab GTPase protein Sec4, which directly interacts with the exocyst, via the Sec15 subunit, thus allowing the secretory vesicle to be recognised by the exocyst plasma membrane complex. Unlike in yeast, the vesicular targeting of the exocyst in metazoans is thought to occur through interactions between Sec15 and Rab11 (Langevin *et al.*, 2005; Wu *et al.*, 2005; Zhang *et al.*, 2004), and the tethering of the secretory vesicles to the plasma membrane is regulated by Sec5 and Ral (Brymora *et al.*, 2001; Chen *et al.*, 2011a; Li *et al.*, 2007). Active RalA (GTP bound form) interacts with Sec5, and upon delivery of the vesicles to the plasma membrane, the interaction between RalA-Sec5 is broken through the phosphorylation of Sec5 by protein kinase C (PKC) (Chen *et al.*, 2011a). Detachment of Sec5 from RalA allows the release of the exocyst complex once the vesicle is delivered to the plasma membrane. The emerging model for the assembly of the exocyst suggests that the components are present as distinct sub-complexes on vesicles and the plasma membrane. In this manner, the assembly of the exocyst may integrate various cellular signalling pathways to ensure that exocytosis is tightly controlled (Sugihara *et al.*, 2002).

Following cell surface membrane tethering by the exocyst complex, SNARE proteins facilitate the final step of exocytosis by bringing together the vesicular and plasma membranes for fusion. There are two groups of SNARE proteins; t-SNAREs, such as syntaxin1 and Sec9, which are found on the inner leaflet of the plasma membrane of cells and denote the target membrane; and v-SNAREs, which are found on a range of membrane compartments and denotes the vesicular membrane (Shorer *et al.*, 2005; Stow *et al.*, 2006). These proteins work by cognate pairing of t-SNAREs with their opposing v-SNAREs to form a four helix bundle, which allows the two membranes to be brought into close proximity, and this facilitates membrane fusion (Stow *et al.*, 2006). A number of studies have provided evidence of interactions between SNARE proteins and components of the exocyst complex (Bao *et al.*, 2008; Hattendorf *et al.*, 2007; Wiederkehr *et al.*, 2004; Zhang *et al.*, 2005). In yeast, this interaction is orchestrated though WD-40 adaptor proteins Sro7p and Sro77p (Zhang *et al.*, 2005), which are homologues of lethal giant larvae (Lgl); first identified as a tumor suppressor in *Drosophila* (Gateff, 1978) and since demonstrated to play a role in cell polarity (Bilder *et al.*, 2000). Sro7p and Sro77p interact with the exocyst components Sec6 and Exo84 as well as t-SNARE Sec9, thus providing a link between these two complexes to mediate the final steps of membrane fusion and exocytosis (Zhang *et al.*, 2005).

3. Autophagy

Autophagy is responsible for a number of routine housekeeping functions, including the elimination of defective proteins, the prevention of abnormal protein aggregate accumulation, the turnover of glycogen, the removal of intracellular pathogens and the recycling of aged or dysfunctional organelles. These functions are likely to be critical for autophagy-mediated protection against aging, cancer, neurodegenerative disease and infection (Levine & Kroemer, 2008). Autophagy involves the engulfing of cytoplasmic content into a double membrane vesicle, which is used to mediate the degradation of the internalised contents following interaction with endosome and lysosome compartments (Figure 1). Autophagy normally occurs at a basal level, but stimuli such as starvation, hormonal and developmental signals, accumulation of unfolded proteins or invasion of microorganisms, can each modulate the rate of autophagic activity (Meijer & Codogno, 2004).

3.1 The induction and sequence of the autophagic process

The process of autophagy is mediated by the recruitment of autophagy related proteins to the limiting membranes of the forming phagosome, where they assemble the so-called pre-autophagosomal structure. This nucleation step is known to occur at sites adjacent to mitochondria in yeast (Mari & Reggiori, 2010), while other eukaryotes are thought to have multiple nucleation sites that may include the endoplasmic reticulum, Golgi, mitochondria and secretory vesicles (Hailey *et al.*, 2010; Hamasaki & Yoshimori, 2010; Militello & Colombo, 2011; Tooze & Yoshimori, 2010; Weidberg *et al.*, 2011). Autophagosomes are formed via the expansion of the isolation membrane to completely surround an area of cytoplasm. Maturation of the autophagosome involves fusion with a multivesicular body to form an amphisome, which subsequently fuses with a lysosome to become a fully functional autolysosome. Through the action of lysosomal enzymes, the degradation process then recycles molecular constituents back into the cytoplasm.

3.2 The molecular machinery involved in autophagosome formation and maturation

The induction and nucleation of autophagy is dependent on the successive assembly of a number of complexes within the cytoplasm, such as the Atg1-Atg13 (or mammalian unc-51-like kinase 1(Ulk1)–Atg13) kinase complex and the Atg5-Atg12 ubiquitin-like conjugation system. Up-stream signalling pathways lead to the activation of the Atg1/Ulk1 complex, which in turn recruits other members of the autophagic machinery to the site of nucleation. The exact mode for this recruitment is as yet unknown, however the individual step-specific complexes are well described for yeast and higher eukaryote systems.

One of the upstream regulators, targets of rapamycin (TOR) acts as an inhibitor of autophagy. Inactivation of TOR leads to the assembly of an active Atg1 complex; Atg1:Atg13:Atg17 in yeast (Kamada *et al.,* 2000; Nakatogawa *et al.,* 2009), and Ulk1:Atg13:FIP200 in higher eukaryotes (Chang & Neufeld, 2010; Mehrpour *et al.*, 2010). In the latter case, activation is thought to occur via a change in Atg13-mediated phosphorylation of Ulk1 (Chang & Neufeld, 2010), although the exact site and the induction signal for this initiation step remains unclear (Chang & Neufeld, 2009, 2010; Mehrpour *et al.*, 2010).

The solution to the mystery of the origin of autophagic compartment may lay in the biology of Atg9, the only trans-membrane autophagic protein that is present throughout autophagosome maturation. In yeast, Atg9 has been observed to form clusters near the mitochondria, suggesting the possibility of a membrane pool for autophagy (Mari & Reggiori, 2010). However, an equivalent structure has yet to be identified in other eukaryotes, and there may be multiple nucleation sites, including the endoplasmic reticulum, Golgi, mitochondria (Hailey *et al.*, 2010; Hamasaki & Yoshimori, 2010; Tooze & Yoshimori, 2010) and plasma membrane (Ravikumar *et al.*, 2010). Recent studies in mammalian cells showed that Atg9 initially resides at the Golgi and is trafficked to recycling endosomes (Wang *et al.*, 2011; Webber *et al.*, 2007; Webber & Tooze, 2010). This suggests the involvement of the Golgi complex in the autophagic pathway. Starvation dependent trafficking of mammalian Atg9 to the pre-autophagosomal structure requires the Atg1/Ulk1 kinase, Atg13, as well as p38 MAPK interaction protein, p38IP (Webber & Tooze, 2010). Following Atg9 recruitment, nucleation of the pre-autophagosomal structure limiting membrane is controlled by a protein complex containing a member of the vacuolar protein sorting family, Vps34, and Atg6/Beclin1. Atg6 is crucial for the recruitment of other autophagic proteins to the pre-autophagosomal structure, while Vps34 kinase phosphorylates phosphatidylinositol (PI_3P) in order to recruit Atg8 and Atg18 (Kundu & Thompson, 2008; Polson *et al.*, 2010).

Two ubiquitin-like conjugation systems are required for the expansion and closure of the autophagosome, Atg5-Atg12 and the Atg8-phosphotidylethanolamine complex (Ichimura *et al.*, 2000; Mizushima *et al.*, 1998). In the first of these systems the conjugated form of Atg5-Atg12 associates with Atg16 dimers to become a multimeric Atg5-Atg12-Atg16 complex. It is believed that this Atg5-Atg12-Atg16 complex is required for the formation of pre-autophagosomal structures, and allows association with the second Atg8 related conjugation system. In the second ubiquitin-like conjugation system, cytosolic Atg8, or LC3 (microtubule-associated protein 1 light chain 3) in mammals, is modified by the attachment of the phospholipid anchor phosphatidylethanolamine, or PE. This step results in the localisation of Atg8/LC3-PE to the isolation membrane of the phagophore and may contribute to the expansion of autophagic membranes (McPhee & Baehrecke, 2009; Nakatogawa *et al.*, 2007).

Once the autophagosome is closed by fusion of the expanding edges of the phagophore, its maturation proceeds through fusion with multi-vesicular bodies, late endosomes and lysosomes (Razi *et al.*, 2009). It has been suggested that fusion of the autophagosome with endocytic compartments is facilitated by endosome membrane fusion machinery (Eskelinen, 2005) including the membrane targeting proteins Rab11 and Rab7 GTPases and membrane fusion protein complexes, such as SNAREs, ESCRT proteins, Vps28, Vps25, Vps32, Deep

Orange (Dor)/Vps18 and Carnation (Car)/Vps33a (Fader *et al.*, 2008; Gutierrez *et al.*, 2004a; Gutierrez *et al.*, 2004b; Simonsen & Tooze, 2009), as well as the lysosomal membrane proteins, Lamp-1 and Lamp-2 (Tanaka *et al.*, 2000). After fusion with the lysosome, degradation of protein, lipid, glycogen, RNA, DNA and other contents is dependent upon the action of lysosomal acid hydrolases (Koike *et al.*, 2005; Tanaka *et al.*, 2000; Tanida *et al.*, 2005). The small molecules (e.g. amino acids and sugars) are then transported back to the cytosol for protein synthesis and the maintenance of other cellular functions (He & Klionsky, 2009).

4. Diseases that show links between exocytosis and autophagy

Altered regulation of exocytosis and autophagy has been shown in a number of debilitating diseases including cancer (Gozuacik & Kimchi, 2006; Levine, 2007; Miracco *et al.*, 2007; Pattingre *et al.*, 2005; Tayeb *et al.*, 2005), neurodegenerative diseases (Gao & Hong, 2008; Keating, 2008; Yu *et al.*, 2005), and chronic inflammatory diseases (Barbier, 2003; Barrett, 2008; Cadwell *et al.*, 2010; Fujita *et al.*, 2008; Homer *et al.*, 2010; Rioux, 2007; Saitoh *et al.*, 2008).

In cancer, the uncontrolled cell proliferation that results in tumor outgrowth is associated with increased secretion of pro-oncogenic proteins and lysosomal enzymes. Thus, lysosomal cathepsins, acid phosphatase and various glycosidases have been used as diagnostic markers and to define metastatic potential in a range of cancers (Tappel, 2005). The underlying reason for this increase in lysosomal enzyme secretion may be linked to the increase in endosome-lysosome membrane recycling that is required to maintain plasma membrane area during rapid cell division (Boucrot & Kirchhausen, 2007). Increased lysosomal enzyme secretion has also been associated with extracellular matrix degradation and this can facilitate metastasis (Tayeb *et al.*, 2005). The migration of metastatic cancer cells also involves upregulated exocytosis, as a means of membrane delivery to the leading edge of the migrating cell. This allows the formation of lamellipodia and filopodia, and thereby cellular movement. Exocytosis and cell division are both high energy demand cellular processes, and so it is not too surprising that autophagy has also been implicated in the carcinogenic process, as a means of energy supply.

There is, however, controversy in the literature regarding the pro-survival and pro-death functions of autophagy (Hippert *et al.*, 2006; Kundu & Thompson, 2008; Levine, 2007; Levine & Kroemer, 2008). The cyto-protective role that autophagy has under conditions of starvation or low energy supply, prevents apoptosis (Boya *et al.*, 2005), and is therefore thought to promote cancer cell growth and survival within solid tumors prior to vascularization. In stark contrast, the suppression of autophagy via a number of regulatory pathways can lead to tumorigenesis (Gozuacik & Kimchi, 2006; Levine, 2007; Miracco *et al.*, 2007; Pattingre *et al.*, 2005). The increased tumorigenesis observed in *beclin1/Atg6* and *Atg5* murine mutants, and the high number of mono-allelic deletion mutations in these genes observed in different types of human cancer, indicate a direct tumor suppressor role for autophagy (Aita *et al.*, 1999; Hippert *et al.*, 2006; Kundu & Thompson, 2008; Levine, 2007). In addition, p53 and PTEN, which are frequently mutated in cancer patients, can stimulate autophagy (Bae *et al.*, 2007; Lindmo *et al.*, 2006; Shin *et al.*, 2011; Wang *et al.*, 2011); while PI3K, p38 MAPK and Akt, which are often activated in cancer, can suppress autophagy (Webber & Tooze, 2010). The apparent disparate roles of autophagy in cancer make it difficult to ascertain its exact function, and it also remains unclear whether exocytosis and autophagy are acting independently or as inter-linked processes in this disease. However, in one study, the trafficking of lysosomes in cancer cells was found to be linked to autophagosome formation through the common molecular machinery of the microtubule -dependent motor protein KIF5B (kinesin heavy chain protein 5B; Cardoso *et al.*, 2009) a protein previously demonstrated to be involved in exocytosis (Varadi *et al.*, 2002).

Neurodegenerative disorders including Parkinson's, Huntington's and Alzheimer's disease are progressive disorders, which have in common the loss of function of neurons in discrete areas of the central nervous system. This loss of function is thought to be a result of aggregation of misfolded proteins. Autophagy has a role in the degradation of misfolded protein (Yu *et al.*, 2005), and the functional loss of *Atg5* or *Atg7* results in the accumulation of ubiquitinated protein aggregates, and a neurodegenerative phenotype (Hara, 2006; Komatsu *et al.*, 2006). Furthermore, altered autophagy has been shown to be linked with altered exocytosis in a number of neurodegenerative disorders, leading to impaired release of neurotransmitters and increased inflammation (Gao & Hong, 2008; Keating, 2008). This highlights a direct link between autophagy and the recycling of the specialist secretory vesicles that control neurotransmission at the synaptic terminals of neurons.

Some inflammatory diseases, such as Crohn's disease, are thought to be caused by a breakdown in the regulation of exocytosis leading to increased secretion of pro-inflammatory factors (Barbier, 2003; Cadwell *et al.*, 2010). In addition, a number of genetic screens of patients suffering from Crohn's disease have identified mutations in autophagy related genes (Barrett, 2008; Rioux, 2007). The autophagy related protein Atg16L is thought to function as a scaffold for LC3 lipidation, by dynamically localizing to the source of membrane involved in autophagosome formation (Fujita *et al.*, 2008). A genetic defect in *Atg16L* may decrease the efficiency by which pathogens can be cleared from cells via autophagy, evoking an increased inflammatory response (Fujita *et al.*, 2008; Homer *et al.*, 2010; Saitoh *et al.*, 2008). In addition, there is mounting evidence that defects in this autophagy gene can also lead to defects in exocytosis, causing a build-up of secretory granules in specific cell types (Cadwell *et al.*, 2009). These concurrent defects in both exocytosis and autophagy may be one more piece of evidence for co-regulation and a shared molecular link between these two cellular processes, and raises the important question: is there common molecular machinery for exocytosis and autophagy?

5. Exocytosis and autophagy: Common cellular functions and molecular machinery

Exocytosis and autophagy are essential for a number of common biological processes, including; the immune response (Govind, 2008; Minty *et al.*, 1983; Murray *et al.*, 1998; Ostenson *et al.*, 2006), cell growth (Brennwald & Rossi, 2007; Orlando & Guo, 2009; Wei & Zheng, 2011; Zhang *et al.*, 2005), cell proliferation and apoptosis (Kundu, 2011; Shin *et al.*, 2011; Zeng *et al.*, 2012), and multicellular organism development (Gutnick *et al.*, 2011; Hu *et al.*, 2011; Sato & Sato, 2011; Tra *et al.*, 2011). Autophagy and exocytosis both involve membrane trafficking and fusion events and so similar groups of molecular machinery may be required for both processes: such as GTPase proteins, that facilitate membrane tethering and SNARE proteins which are involved in membrane fusion. There is increasing evidence indicating shared molecular machinery between these processes, which provokes questions concerning possible dual regulation as a means of balance for these pathways.

Autophagy and exocytosis can have opposing or synergistic roles in cell function. For example, during times of reduced nutrient availability autophagy is stimulated, allowing cells to recycle cytoplasmic components, and exocytosis is reduced to conserve cellular constituents and energy (Shorer *et al.*, 2005). In response to stimulation, specialised secretory cells (e.g. chromaffin neuroendicrine cells) divert energy utilization towards the exocytic pathway (Malacombe *et al.*, 2006). Conversely, when a cell is presented with an immune challenge, both exocytosis and autophagy can be upregulated; exocytosis for the release of immune response factors and autophagy to clear invading pathogens from cells (Stow *et al.*, 2009). Given these findings, it would appear to be of advantage to cells to have a mechanism coordinating the activity of these two processes.

5.1 The endosomal network is involved in both exocytosis and autophagy

The dynamic flow of membrane and membrane proteins within a cell is mediated through the endosomal network (Figure 1). For example, lipids and proteins from the plasma membrane are recovered by the cell for cytosolic recycling via compensatory endocytosis, which also allows for the maintenance of membrane homeostasis at the sites of active exocytosis; directing endocytosed membrane back into the endosomal network or Golgi for degradation or recycling (Sramkova *et al.*, 2009). This type of endocytosis is of particular importance in specialized secretory cells, such as, bladder umbrella cells (Khandelwal *et al.*, 2008; Khandelwal *et al.*, 2010), neurons (Kim & von Gersdorff, 2009; Llobet *et al.*, 2011; Logiudice *et al.*, 2009) and neuroendocrine cells (Engisch & Nowycky, 1998; Barg & Machado, 2008). This allows for the rapid recycling of secretory vesicles back into the reserve pool. Endosomes are at the nexus of the exocytic and autophagic pathways allowing for the sorting and directing of membrane. Thus, in yeast, Atg9 clusters are connected with both the endocytic and exocytic systems, and delivered to the phagophore assembly site via recycling endosomes (Geng *et al.*, 2010; Mari *et al.*, 2010). The recycling endosome's exocytic function is involved in the maintenance of cell polarity through the sorting of membrane proteins such as clathrin and cadherin (Farr *et al.*, 2009). The recycling endosome machinery also plays a role in the fusion of multivesicular bodies with autophagosomes, which is an essential step in phagosome maturation (Fader & Colombo, 2009; Razi *et al.*, 2009; Tooze & Razi, 2009). Recent studies have suggested a significant overlap of the molecular machinery used in these two biological processes (Bodemann *et al.*, 2011; Geng *et al.*, 2010). This involves the exocyst complex and its regulators (e.g. small GTPases), as well as membrane fusion machinery (e.g. SNAREs; Table 1).

5.2 Small GTPases at the cross road of exocytosis and autophagy

5.2.1 Ral small GTPase

Ras-like proteins (Ral) are small GTPases that function as an essential component of the cellular machinery regulating the post-Golgi targeting of exocytic vesicles to the plasma membrane (Balasubramanian *et al.*, 2010; Chen *et al.*, 2007; Kawato *et al.*, 2008; Kim *et al.*, 2010; Ljubicic *et al.*, 2009; Lopez *et al.*, 2008; Rondaij *et al.*, 2008; Rosse *et al.*, 2006; Shipitsin & Feig, 2004; Spiczka & Yeaman, 2008). Ral function is directly mediated by its interaction with the exocyst complex (Feig, 2003; Kawato *et al.*, 2008; Mark *et al.*, 1996; Mott *et al.*, 2003), in particular Sec5 which has been shown to be essential for Ral-exocyst dependent exocytosis (Fukai *et al.*, 2003; Moskalenko *et al.*, 2002).

Protein	Role in Exocytosis	References	Role in Autophagy	References
Ral	Interacts with exocyst via Sec5 to facilitate the tethering of vesicles to the plasma membrane.	(Balasubramanian *et al.*, 2010; Brymora *et al.*, 2001; Chen *et al.*, 2011a; Chen *et al.*, 2007; Fukai *et al.*, 2003; Kawato *et al.*, 2008; Li *et al.*, 2007; Ljubicic *et al.*, 2009; Lopez *et al.*, 2008; Mark *et al.*, 1996; Moskalenko *et al.*, 2002; Mott *et al.*, 2003; Shipitsin & Feig, 2004)	RalB but not RalA involved in initation of autophagy in mammalian cell lines. Over expression of active RalB enhances autophagy while depletion decreases autophagy.	(Bodemann *et al.*, 2011)
Rab11	Bound to exocytic vesicles and is involved in the anterograde trafficking of vesicles from recycling endosomes to the plasma membrane. Interacts with the exocyst component Sec15 to assist tethering of vesicles to the plasma membrane.	(Langevin *et al.*, 2005; Oztan *et al.*, 2007; Shandala *et al.*, 2011; Ward *et al.*, 2005; Wu *et al.*, 2005; Zhang *et al.*, 2004)	Facilitates fusion of the autophagosome with endocytic compartments.	(Fader *et al.*, 2008)
Sec4	Allows the interaction of the secretory vesicle with the exocyst complex via Sec15 to facilitate tethering to the plasma membrane.	(Guo *et al.*, 1999)	Involved in the recruitment of Atg9 to the PAS.	(Geng *et al.*, 2010)
Exocyst Complex	Octomeric complex required for tethering of exocytic vesicles to the plasma membrane in a site specific manner	(He *et al.*, 2007; He & Guo, 2009; Jin *et al.*, 2011; Langevin *et al.*, 2005; Morgera *et al.*, 2012)	Proposed as a scaffold for the initiation of autophagy complexes.	(Bodemann *et al.*, 2011; Farré & Subramani, 2011)
Sso1/2-Sec9	t-SNARE that denotes the site of exocytosis on the plasma membrane, possibly through interactions with the exocyst complex and its effectors	(Aalto *et al.*, 1993; Brennwald *et al.*, 1994)	Involved in the formation of Atg9 associated tubule-vesicular clusters emanating from the PAS	(Nair *et al.*, 2011)

Protein	Role in Exocytosis	References	Role in Autophagy	References
VAMP7	Involved in constitutive exocytosis in a number of cell types	(Galli *et al.*, 1998; Oishi *et al.*, 2006)	Involved in lysosome fusion during autophagosome maturation	(Fader *et al.*, 2009)
Atg9	Unknown role but has been found on secretory vesicles. May have a role in unconventional secretion	(Bruns *et al.*, 2011; Mari *et al.*, 2010)	Transmembrane protein required for the transport and assembly of membrane during autophagosome formation	(He *et al.*, 2009)
Atg16L	Involved in secretion from secretory granules in intestinal Paneth cells	(Cadwell *et al.*, 2008; Cadwell *et al.*, 2009)	Functions as a scaffold for LC3 lipidation, required during autophagosome formation	(Fujita *et al.*, 2008)

Table 1. Proteins involved in both autophagy and exocytosis

In addition to its well documented role in exocytosis, recent evidence from mammalian cell cultures indicates that RalB is involved in the formation of autophagosomes (Bodemann *et al.*, 2011). The crucial role for RalB as an upstream activator of autophagy is illustrated by the fact that the over-expression of its active GTP-bound form was sufficient to induce autophagy, even in the absence of autophagy-specific stimuli (Bodemann *et al.*, 2011). RalB is present on sites of nascent autophagosome formation, together with Beclin1 and Atg5, and its depletion, similar to the depletion of Atg5 and Beclin1, significantly impaired the formation of starvation-induced LC3/Atg8 punctae and the turnover of LC3/Agt8. Interestingly, depletion of RalB also impaired the digestion of autophagocytosed *Salmonella typhimurium*. The autophagy-related function of RalB appears to be mediated by its effector Exo84, a component of the exocyst complex (Bodemann *et al.*, 2011). Activated by starvation, RalB triggers Exo84 interaction with the autophagy initiation component Beclin1. Intriguingly, the alternative RalB roles in exocytosis and autophagy appear to be driven by environmental signal/s, as nutrient availability determines the RalB coupling preferences to a down-stream effector; endogenous RalB preferentially associates with Exo84 in nutrient poor conditions and with Sec5 under nutrient rich conditions (Bodemann *et al.*, 2011). This model has not been investigated in higher eukaryotes, but these findings in yeast suggest a role for the exocyst complex as a scaffold for the assembly of a number of important autophagy initiators.

5.2.2 Yeast Rab small GTPase Sec4

The yeast Rab GTPase Sec4 and its activator Sec2 have well-established roles in the tethering of secretory vesicles to sites of active exocytosis, in a process mediated by interaction with the exocyst complex component Sec15 (Geng *et al.*, 2010) . Recent studies indicate that Sec2 and Sec4 also have a role in anterograde trafficking of the autophagic membrane protein Atg9, as silencing of Sec4 blocked the delivery of Atg9 to the pre-autophagosomal structure

(Geng *et al.*, 2010). Furthermore, when the domain of Sec4 that is known to interact with Sec15 was altered, the effect on autophagy was equivalent to the effect of Sec4 silencing. Taking into account that there is no apparent role for Sec15 in autophagy, this suggests that autophagy-specific proteins may compete for this Sec15-binding domain in order to switch the function of activated GTP-bound Sec4 between exocytosis and autophagy.

5.2.3 Metazoan Rab11 small GTPase

Rab11 is a small GTPase, which is most often referred to as a recycling endosome marker. However, it has also been observed on vesicles bound for exocytosis (Shandala *et al.*, 2011; Ward *et al.*, 2005), and amphisomes; an intermediate compartment that is formed during autophagosome maturation, prior to lysosomal fusion (Fader & Colombo, 2009)

The exocytic role for Rab11 is mediated by its association with the Sec15 exocyst component. This has been shown in MSCK cells (Oztan *et al.*, 2007; Zhang *et al.*, 2004), and in *Drosophila* photoreceptor and sensory neuron cells (Wu *et al.*, 2005). Rab11 is important for the anterograde trafficking of; numerous membrane receptors (Chernyshova *et al.*, 2011), the epithelial sodium channel complex of the cortical collecting duct of the kidneys (Butterworth *et al.*, 2012), and DE-Cadherin in polarised cells (Langevin *et al.*, 2005; Wu *et al.*, 2005; Zhang *et al.*, 2004), as well as the calcium dependent exocytosis of growth hormones (Ren *et al.*, 1998; Takaya *et al.*, 2007). A number of intracellular pathogens, such as *Porphyromonas gingivalis*, influenza A and HIV, have been reported to hi-jack Rab11 dependent anterograde trafficking as a means of escape from host cells (Kadiu & Gendelman, 2011; Momose *et al.*, 2011; Takeuchi *et al.*, 2011).

An example of coordinated exocytosis and autophagy comes from the biology of multivesicular bodies (MVBs). MVBs are specialised late endosomes, a crucial intermediate in the internalization of nutrients, ligands and receptors into small intraluminal vesicles, also known as exosomes (Fader & Colombo, 2009). Rab11 decorates MVBs and is involved in both the biogenesis of MVBs and exosome release (Fader *et al.*, 2008). During the maturation of hematopoietic progenitors into reticulocytes and erythrocytes, proteins that are not required at the mature stage are sequestered into exosomes of MVBs. In this scenario Rab11 is involved in the targeting of MVBs to the plasma membrane, where exosomes are released into the extracellular milieu (Fader & Colombo, 2006). Active Rab11 is also required for the interaction of MVBs with autophagosomes, where the resulting calcium-stimulated fusion of these organelles promotes efficient degradation of autophagic contents (Fader *et al.*, 2008; Savina *et al.*, 2005). Thus, Rab11 may represent a critical regulator of membrane flow between recycling endosomes (as a source of exocytic vesicles) and multivesicular bodies, where it can be engaged in both autophagic maturation and secretion.

5.3 The exocyst and the initiation of autophagy

Components of the exocyst complex involved in regulated and polarized exocytosis have also been shown to associate with a number of essential autophagy proteins (Bodemann *et al.*, 2011). Exocyst components Sec3 and Sec8 interact *in vitro* with positive (FIP200, ATG14L) and negative (RUBICON) regulators of autophagy, as well as with the phagophore expansion complex Atg5-Atg12 (Bodemann *et al.*, 2011). The functionality of these physical interactions is confirmed by the fact that LC3/Atg8 autophagosome formation was impaired in cells depleted for core exocyst components. For example, the depletion of Sec8 rendered cells insensitive to starvation stimulation, and impaired autophagy to the same extent as seen for the depletion of Atg5 and Beclin1 (Bodemann *et al.*, 2011). Further interrogation of this system showed that the localization of exocyst components, with the autophagy initiator Atg1/Ulk1 and other proteins involved in isolation membrane formation, was altered following the induction of autophagy (Bodemann *et al.*, 2011). As these processes are under the control of Ral or Rab, the differential recruitment of the exocyst to the target membrane might depend on the signals upstream of these small GTPases.

5.4 SNARE proteins and membrane fusion during exocytosis and autophagy

Both autophagosome maturation and anterograde vesicle trafficking via the exocytic route involve a series of membrane fusion steps, the execution of which is controlled by SNAREs. Recent studies in yeast have indicated that some exocytic t-SNAREs may also play a role in membrane dynamics during autophagy (Geng *et al.*, 2010; Nair *et al.*, 2011). The anterograde trafficking of the key autophagic membrane determinant, Atg9, depends on interaction with exocytic Sso1-Sec9, as well as on the endosomal t-SNARE Tlg2 and the v-SNAREs Sec22 and Ykt6. Sso1/2 and Sec9 SNAREs are also responsible for the formation of the Atg9 associated tubular-vesicular clusters emanating from the pre-autophagosomal structure, and their depletion results in Atg9 localization to small vesicular structures, possibly *trans*-Golgi network derived secretory vesicles, that fail to be delivered to the pre-autophagosomal structure (Nair *et al.*, 2011). This failure of Atg9 delivery to pre-autophagosomal structure abolishes Atg8 recruitment, and thereby abrogates autophagosome biogenesis (Geng *et al.*, 2010; Nair *et al.*, 2011) .

Another group of SNAREs, the vesicle-associated membrane proteins (VAMPs), appear to be involved in membrane fusion events in both autophagy and exocytosis. One the one hand, during autophagosome maturation, it has been shown that VAMP3 and VAMP7 are required for sequential fusion with multivesicular bodies and lysosomes respectively (Fader *et al.*, 2009). In HeLa cells, VAMP7 has been shown to be involved in homotypic fusion of Atg16L1 positive vesicles to allow autophagosome biogenesis (Moreau *et al.*, 2011). On the other hand, it has been recently demonstrated that VAMP7 is involved in constitutive exocytosis in HSY cells (Oishi *et al.*, 2006), and in apical trafficking of exocytic vesicles in polarized epithelial cells, such as MDCK cells and CaCo-2 cells (Galli *et al.*, 1998). VAMP3 has been postulated to be a v-SNARE for early and recycling endosomes, with a role in constitutive exocytosis, but its role might be redundant as mice with a null mutation for this gene were normal in most endocytic and exocytic pathways, including constitutive exocytosis (Wang *et al.*, 2004; Yang *et al.*, 2001). The question remains whether these functions of SNAREs are restricted to specific tissues, or universal for all tissues, and if so, what are the upstream signals that direct these SNAREs to either the exocytic or autophagic membranes?

5.5 The role of autophagy genes in secretion

There is emerging evidence of involvement of autophagy proteins in polarized secretion involving lysosomes. In bone resorptive osteoclasts, Atg5, Atg7, Atg4B, and LC3/Atg8 participate in directing lysosomes to fuse with the plasma membrane and in the release of the lytic enzyme cathepsin K into the extracellular space (DeSelm *et al.*, 2011). This type of lysosome-regulated exocytosis is not restricted to osteoclasts involved in bone remodelling, and has been described for lysosome related organelles in many other specialist cell types, such as melanosomes in melanocytes and lytic granules in neutrophils (Blott & Griffiths, 2002; Chen *et al.*, 2012; Luzio *et al.*, 2007). In yeast grown under conditions of nitrogen starvation, autophagy genes are required for the secretion of an acyl coenzyme A binding protein (Acb1) (Bruns *et al.*, 2011; Duran *et al.*, 2010; Manjithaya *et al.*, 2010; Skinner, 2010). This unconventional route of secretion is initiated at sites that are positive for the Golgi assembly and stacking protein (GRASP65) homologue 1 (Grh1), which attracts core autophagy-related proteins Atg9 and Atg8 to a novel compartment (Bruns *et al.*, 2011). These Acb1-containing autophagosomes then evade fusion with the lytic vacuole, fusing instead with recycling endosomes to form multivesicular body carriers, which then fuse with the plasma membrane in a t-SNARE Sso1 dependent fashion, to release Acb1. It is still not clear how beneficial or economical it is for cells to use these unconventional routes of secretion.

6. Summary

Despite the similarity in requirements for membrane dynamics in the processes of exocytosis and autophagy, correlations between the molecular machinery used for both of these processes are only beginning to be elucidated. The exocytic and autophagic functions of cells are critical for the maintenance of cell homeostasis and the exchange of membrane between intracellular compartments and the cell surface. In addition, the fusion and fission events that remodel the exocytic vesicle and the autophagosome are likely to require much of the same molecular machinery. Therefore, it is likely that there is co-ordinated control of these two processes to ensure that they can be regulated with respect to each other. Members of the exocyst complex, and some autophagy related proteins, have already been shown to have functions in their opposite processes, and the involvement of the Ral small GTPases in the global control of exocytosis and autophagy mirrors the role of Rab small GTPases in the control of endosome trafficking. There are many intriguing questions brought about by recent findings. What is the decision making signal that diverges the components of the shared machinery from one pathway to another? Is there a common upstream signal for both pathways, be it through the insulin receptor/mTOR (Webber & Tooze, 2010), MAPK (Webber & Tooze, 2010), redox (Lee *et al.*, 2012), or are there combinations of these signals? Or is there a yet to be defined intrinsic factor of the autophagic or exocytic membrane, with a changing affinity for vesicular compartments? This is a very interesting time to be exploring the intersection of the exocytosis and autophagy pathways, particularly while we are is still looking for the key controllers of cellular homeostasis in cancers, neurodegenerative and immune disorders.

7. References

Aalto, M. K., Ronne, H. & Keranen, S. (1993). Yeast syntaxins Sso1p and Sso2p belong to a family of related membrane proteins that function in vesicular transport. *The EMBO journal*, Vol. 12, No.11 (November 1993), pp.4095-4104, 0261-4189

Aita, V. M., Liang, X. H., Murty, V. V., Pincus, D. L., Yu, W., Cayanis, E., Kalachikov, S., Gilliam, T. C. & Levine, B. (1999). Cloning and genomic organization of beclin 1, a candidate tumor suppressor gene on chromosome 17q21. *Genomics,* Vol. 59, No.1 (July 1999), pp.59-65, 0888-7543

Bae, Y. J., Kang, S. J. & Park, K. S. (2007). *Drosophila melanogaster* Parkin ubiquitinates peanut and septin1 as an E3 ubiquitin-protein ligase. *Insect biochemistry and molecular biology,* Vol. 37, No.5 (May 2007), pp.430-439, 0965-1748

Balasubramanian, N., Meier, J. A., Scott, D. W., Norambuena, A., White, M. A. & Schwartz, M. A. (2010). RalA-exocyst complex regulates integrin-dependent membrane raft exocytosis and growth signaling. *Current biology : CB,* Vol. 20, No.1 (January 2010), pp.75-79, 0960-9822

Bao, Y., Lopez, J. A., James, D. E. & Hunziker, W. (2008). Snapin interacts with the Exo70 subunit of the exocyst and modulates GLUT4 trafficking. *The Journal of biological chemistry,* Vol. 283, No.1 (January 2008), pp.324-331, 0021-9258

Barbier, M. (2003). Overexpression of leptin mRNA in mesenteric adipose tissue in inflammatory bowel diseases. *Gastroentérologie clinique et biologique,* Vol. 27, No.11 (November 2003), pp.987-991, 0399-8320

Barg, S. & Machado, J. D. (2008). Compensatory endocytosis in chromaffin cells. *Acta physiologica,* Vol. 192, No.2 (February 2008), pp.195-201, 1748-1708

Barrett, J. C. (2008). Genome-wide association defines more than 30 distinct susceptibility loci for Crohn's disease. *Nature genetics,* Vol. 40, No.8 (August 2008), pp.955-962, 1061-4036

Bilder, D., Li, M. & Perrimon, N. (2000). Cooperative regulation of cell polarity and growth by *Drosophila* tumor suppressors. *Science,* Vol. 289, No.5476 (July 2000), pp.113-116, 0036-8075

Blott, E. J. & Griffiths, G. M. (2002). Secretory lysosomes. *Nature reviews. Molecular cell biology,* Vol. 3, No.2 (February 2002), pp.122-131, 1471-0072

Bodemann, B. O., Orvedahl, A., Cheng, T., Ram, R. R., Ou, Y. H., Formstecher, E., Maiti, M., Hazelett, C. C., Wauson, E. M., Balakireva, M., Camonis, J. H., Yeaman, C., Levine, B. & White, M. A. (2011). RalB and the exocyst mediate the cellular starvation response by direct activation of autophagosome assembly. *Cell,* Vol. 144, No.2 (January 2011), pp.253-267, 0092-8674

Boucrot, E. & Kirchhausen, T. (2007). Endosomal recycling controls plasma membrane area during mitosis. *Proceedings of the National Academy of Sciences of the United States of America,* Vol. 104, No.19 (May 2007), pp.7939-7944, 0027-8424

Boya, P., Gonzalez-Polo, R. A., Casares, N., Perfettini, J. L., Dessen, P., Larochette, N., Metivier, D., Meley, D., Souquere, S., Yoshimori, T., Pierron, G., Codogno, P. & Kroemer, G. (2005). Inhibition of macroautophagy triggers apoptosis. *Molecular and cellular biology,* Vol. 25, No.3 (February 2005), pp.1025-1040, 0270-7306

Brennwald, P., Kearns, B., Champion, K., Keranen, S., Bankaitis, V. & Novick, P. (1994). Sec9 is a SNAP-25-like component of a yeast SNARE complex that may be the effector of Sec4 function in exocytosis. *Cell,* Vol. 79, No.2 (October 1994), pp.245-258, 0092-8674

Brennwald, P. & Rossi, G. (2007). Spatial regulation of exocytosis and cell polarity: yeast as a model for animal cells. *FEBS letters,* Vol. 581, No.11 (May 2007), pp.2119-2124, 0014-5793

Bruns, C., McCaffery, J. M., Curwin, A. J., Duran, J. M. & Malhotra, V. (2011). Biogenesis of a novel compartment for autophagosome-mediated unconventional protein secretion. *The Journal of cell biology,* Vol. 195, No.6 (December 2011), pp.979-992, 0021-9525

Brymora, A., Valova, V. A., Larsen, M. R., Roufogalis, B. D. & Robinson, P. J. (2001). The brain exocyst complex interacts with RalA in a GTP-dependent manner: identification of a novel mammalian Sec3 gene and a second Sec15 gene. *The Journal of biological chemistry,* Vol. 276, No.32 (August 2001), pp.29792-29797, 0021-9258

Butterworth, M. B., Edinger, R. S., Silvis, M. R., Gallo, L. I., Liang, X., Apodaca, G., Frizzell, R. A. & Johnson, J. P. (2012). Rab11b regulates the trafficking and recycling of the epithelial sodium channel (ENaC). *American journal of physiology. Renal physiology,* Vol. 302, No.5 (March 2012), pp.F581-F590, 1522-1466

Cadwell, K., Liu, J. Y., Brown, S. L., Miyoshi, H., Loh, J., Lennerz, J. K., Kishi, C., Kc, W., Carrero, J. A., Hunt, S., Stone, C. D., Brunt, E. M., Xavier, R. J., Sleckman, B. P., Li, E., Mizushima, N., Stappenbeck, T. S. & Virgin Iv, H. W. (2008). A key role for autophagy and the autophagy gene Atg16l1 in mouse and human intestinal Paneth cells. *Nature,* Vol. 456, No.7219 (November 2008), pp.259-263, 0028-0836

Cadwell, K., Patel, K. K., Komatsu, M., Virgin, H. W. t. & Stappenbeck, T. S. (2009). A common role for Atg16L1, Atg5 and Atg7 in small intestinal Paneth cells and Crohn disease. *Autophagy,* Vol. 5, No.2 (February 2009), pp.250-252, 1554-8627

Cadwell, K., Patel, K. K., Maloney, N. S., Liu, T.-C., Ng, A. C. Y., Storer, C. E., Head, R. D., Xavier, R., Stappenbeck, T. S. & Virgin, H. W. (2010). Virus-plus-susceptibility gene interaction determines Crohn's disease gene Atg16L1 phenotypes in intestine. *Cell,* Vol. 141, No.7 (June 2010), pp.1135-1145, 0092-8674

Cardoso, C. M., Groth-Pedersen, L., Hoyer-Hansen, M., Kirkegaard, T., Corcelle, E., Andersen, J. S., Jaattela, M. & Nylandsted, J. (2009). Depletion of kinesin 5B affects lysosomal distribution and stability and induces peri-nuclear accumulation of autophagosomes in cancer cells. *PLoS one,* Vol. 4, No.2 (February 2009), pp.e4424, 1932-6203

Chang, Y. Y. & Neufeld, T. P. (2009). An Atg1/Atg13 complex with multiple roles in TOR-mediated autophagy regulation. *Molecular biology of the cell,* Vol. 20, No.7 (April 2009), pp.2004-2014, 1059-1524

Chang, Y. Y. & Neufeld, T. P. (2010). Autophagy takes flight in *Drosophila. FEBS letters,* Vol. 584, No.7 (April 2010), pp.1342-1349, 0014-5793

Chen, G., Zhang, Z., Wei, Z., Cheng, Q., Li, X., Li, W., Duan, S. & Gu, X. (2012). Lysosomal exocytosis in Schwann cells contributes to axon remyelination. *Glia,* Vol. 60, No.2 (February 2012), pp.295-305, 0894-1491

Chen, X. W., Leto, D., Chiang, S. H., Wang, Q. & Saltiel, A. R. (2007). Activation of RalA is required for insulin-stimulated Glut4 trafficking to the plasma membrane via the exocyst and the motor protein Myo1c. *Developmental cell,* Vol. 13, No.3 (September 2007), pp.391-404, 1534-5807

Chen, X. W., Leto, D., Xiao, J., Goss, J., Wang, Q., Shavit, J. A., Xiong, T., Yu, G., Ginsburg, D., Toomre, D., Xu, Z. & Saltiel, A. R. (2011a). Exocyst function is regulated by effector phosphorylation. *Nature cell biology,* Vol. 13, No.5 (May 2011), pp.580-588, 1465-7392

Chen, X. W., Leto, D., Xiong, T., Yu, G., Cheng, A., Decker, S. & Saltiel, A. R. (2011b). A Ral GAP complex links PI 3-kinase/Akt signaling to RalA activation in insulin action. *Molecular biology of the cell,* Vol. 22, No.1 (January 2011), pp.141-152, 1939-4586

Chernyshova, Y., Leshchyns'ka, I., Hsu, S. C., Schachner, M. & Sytnyk, V. (2011). The neural cell adhesion molecule promotes FGFR-dependent phosphorylation and membrane targeting of the exocyst complex to induce exocytosis in growth cones. *The Journal of neuroscience,* Vol. 31, No.10 (March 2011), pp.3522-3535, 0270-6474

DeSelm, C. J., Miller, B. C., Zou, W., Beatty, W. L., van Meel, E., Takahata, Y., Klumperman, J., Tooze, S. A., Teitelbaum, S. L. & Virgin, H. W. (2011). Autophagy proteins regulate the secretory component of osteoclastic bone resorption. *Developmental cell,* Vol. 21, No.5 (November 2011), pp.966-974, 1534-5807

Dupraz, S., Grassi, D., Bernis, M. E., Sosa, L., Bisbal, M., Gastaldi, L., Jausoro, I., Caceres, A., Pfenninger, K. H. & Quiroga, S. (2009). The TC10-Exo70 complex is essential for membrane expansion and axonal specification in developing neurons. *The Journal of neuroscience,* Vol. 29, No.42 (October 2009), pp.13292-13301, 0270-6474

Duran, J. M., Anjard, C., Stefan, C., Loomis, W. F. & Malhotra, V. (2010). Unconventional secretion of Acb1 is mediated by autophagosomes. *The Journal of cell biology,* Vol. 188, No.4 (February 2010), pp.527-536, 0021-9525

Engisch, K. L. & Nowycky, M. C. (1998). Compensatory and excess retrieval: two types of endocytosis following single step depolarizations in bovine adrenal chromaffin cells. *The Journal of physiology,* Vol. 506, No.3 (February 1998), pp.591-608, 0022-3751

Eskelinen, E. L. (2005). Maturation of autophagic vacuoles in mammalian cells. *Autophagy,* Vol. 1, No.1 (April 2006), pp.1-10, 1554-8627

Fader, C. M. & Colombo, M. I. (2006). Multivesicular bodies and autophagy in erythrocyte maturation. *Autophagy,* Vol. 2, No.2 (April 2006), pp.122-125, 1554-8627

Fader, C. M., Sanchez, D., Furlan, M. & Colombo, M. I. (2008). Induction of autophagy promotes fusion of multivesicular bodies with autophagic vacuoles in k562 cells. *Traffic,* Vol. 9, No.2 (February 2008), pp.230-250, 1398-9219

Fader, C. M. & Colombo, M. I. (2009). Autophagy and multivesicular bodies: two closely related partners. *Cell death and differentiation,* Vol. 16, No.1 (January 2009), pp.70-78, 1350-9047

Fader, C. M., Sanchez, D. G., Mestre, M. B. & Colombo, M. I. (2009). TI-VAMP/VAMP7 and VAMP3/cellubrevin: two v-SNARE proteins involved in specific steps of the autophagy/multivesicular body pathways. *Biochimica et biophysica acta,* Vol. 1793, No.12 (December 2009), pp.1901-1916, 0006-3002

Farr, G. A., Hull, M., Mellman, I. & Caplan, M. J. (2009). Membrane proteins follow multiple pathways to the basolateral cell surface in polarized epithelial cells. *The Journal of cell biology,* Vol. 186, No.2 (July 2009), pp.269-282, 0021-9525

Farré, J.-C. & Subramani, S. (2011). Rallying the exocyst as an autophagy scaffold. *Cell,* Vol. 144, No.2 (January 2011), pp.172-174, 0092-8674

Feig, L. A. (2003). Ral-GTPases: approaching their 15 minutes of fame. *Trends in cell biology,* Vol. 13, No.8 (August 2003), pp.419-425, 0962-8924

Franken, S., Junghans, U., Rosslenbroich, V., Baader, S. L., Hoffmann, R., Gieselmann, V., Viebahn, C. & Kappler, J. (2003). Collapsin response mediator proteins of neonatal rat brain interact with chondroitin sulfate. *The Journal of biological chemistry,* Vol. 278, No.5 (January 2003), pp.3241-3250, 0021-9258

Fujita, N., Itoh, T., Fukuda, M., Noda, T. & Yoshimori, T. (2008). The Atg16L complex specifies the site of LC3 lipidation for membrane biogenesis in autophagy. *Molecular biology of the cell,* Vol. 19, No.5 (May 2008), pp.2092-2100, 1059-1524

Fukai, S., Matern, H. T., Jagath, J. R., Scheller, R. H. & Brunger, A. T. (2003). Structural basis of the interaction between RalA and Sec5, a subunit of the sec6/8 complex. *The EMBO journal,* Vol. 22, No.13 (July 2003), pp.3267-3278, 0261-4189

Galli, T., Zahraoui, A., Vaidyanathan, V. V., Raposo, G., Tian, J. M., Karin, M., Niemann, H. & Louvard, D. (1998). A novel tetanus neurotoxin-insensitive vesicle-associated membrane protein in SNARE complexes of the apical plasma membrane of epithelial cells. *Molecular biology of the cell,* Vol. 9, No.6 (June 1998), pp.1437-1448, 1059-1524

Gao, H.-M. & Hong, J.-S. (2008). Why neurodegenerative diseases are progressive: uncontrolled inflammation drives disease progression. *Trends in immunology,* Vol. 29, No.8 (August 2008), pp.357-365, 1471-4906

Gateff, E. (1978). Malignant neoplasms of genetic origin in *Drosophila melanogaster. Science,* Vol. 200, No.4349 (June 1978), pp.1448-1459, 0036-8075

Geng, J., Nair, U., Yasumura-Yorimitsu, K. & Klionsky, D. J. (2010). Post-Golgi Sec proteins are required for autophagy in *Saccharomyces cerevisiae. Molecular biology of the cell,* Vol. 21, No.13 (July 2010), pp.2257-2269, 1059-1524

Govind, S. (2008). Innate immunity in *Drosophila*: Pathogens and pathways. *Insect science,* Vol. 15, No.1 (February 2008), pp.29-43, 1672-9609

Gozuacik, D. & Kimchi, A. (2006). DAPk protein family and cancer. *Autophagy,* Vol. 2, No.2 (April 2006), pp.74-79, 1554-8627

Guo, W., Roth, D., Walch-Solimena, C. & Novick, P. (1999). The exocyst is an effector for Sec4p, targeting secretory vesicles to sites of exocytosis. *The EMBO journal,* Vol. 18, No.4 (February 1999), pp.1071-1080, 0261-4189

Gutierrez, M. G., Master, S. S., Singh, S. B., Taylor, G. A., Colombo, M. I. & Deretic, V. (2004a). Autophagy is a defense mechanism inhibiting BCG and *Mycobacterium tuberculosis* survival in infected macrophages. *Cell,* Vol. 119, No.6 (December 2004), pp.753-766, 0092-8674

Gutierrez, M. G., Munafo, D. B., Beron, W. & Colombo, M. I. (2004b). Rab7 is required for the normal progression of the autophagic pathway in mammalian cells. *Journal of cell science,* Vol. 117, No.13 (June 2004), pp.2687-2697, 0021-9533

Gutnick, A., Blechman, J., Kaslin, J., Herwig, L., Belting, H. G., Affolter, M., Bonkowsky, J. L. & Levkowitz, G. (2011). The hypothalamic neuropeptide oxytocin is required for formation of the neurovascular interface of the pituitary. *Developmental cell,* Vol. 21, No.4 (October 2011), pp.642-654, 1534-5807

Hailey, D.W., Rambold, A.S., Satpute-Krishnan, P., Mitra, K., Sougrat, R., Kim, P.K. & Lippincott-Schwartz, J. (2010). Mitochondria supply membranes for autophagosome biogenesis during starvation. *Cell,* Vol. 141, No.4 (May 2010), pp.656-667, 0092-8674

Hamasaki, M. & Yoshimori, T. (2010). Where do they come from? Insights into autophagosome formation. *FEBS letters,* Vol. 584, No.7 (April 2010), pp.1296-1301, 0014-5793

Hara, T. (2006). Suppression of basal autophagy in neural cells causes neurodegenerative disease in mice. *Nature,* Vol. 441, No.7095 (June 2006), pp.885-889, 0028-0836

Hattendorf, D. A., Andreeva, A., Gangar, A., Brennwald, P. J. & Weis, W. I. (2007). Structure of the yeast polarity protein Sro7 reveals a SNARE regulatory mechanism. *Nature,* Vol. 446, No.7135 (March 2007), pp.567-571, 0028-0836

He, B., Xi, F., Zhang, X., Zhang, J. & Guo, W. (2007). Exo70 interacts with phospholipids and mediates the targeting of the exocyst to the plasma membrane. *The EMBO journal,* Vol. 26, No.18 (September 2007), pp.4053-4065, 0261-4189

He, B. & Guo, W. (2009). The exocyst complex in polarized exocytosis. *Current opinion in cell biology,* Vol. 21, No.4 (August 2009), pp.537-542, 0955-0674

He, C., Baba, M. & Klionsky, D. J. (2009). Double duty of Atg9 self-association in autophagosome biogenesis. *Autophagy,* Vol. 5, No.3 (April 2009), pp.385-387, 1554-8627

He, C. & Klionsky, D. J. (2009). Regulation mechanisms and signaling pathways of autophagy. *Annual review of genetics,* Vol. 43, (December 2009), pp.67-93, 0066-4197

Hippert, M. M., O'Toole, P. S. & Thorburn, A. (2006). Autophagy in cancer: good, bad, or both? *Cancer research,* Vol. 66, No.19 (October 2006), pp.9349-9351, 0008-5472

Homer, C. R., Richmond, A. L., Rebert, N. A., Achkar, J. P. & McDonald, C. (2010). ATG16L1 and NOD2 interact in an autophagy-dependent antibacterial pathway implicated in Crohn's disease pathogenesis. *Gastroenterology,* Vol. 139, No.5 (November 2010), pp.1630-1641, 0016-5085

Hu, Z., Zhang, J. & Zhang, Q. (2011). Expression pattern and functions of autophagy-related gene atg5 in zebrafish organogenesis. *Autophagy,* Vol. 7, No.12 (December 2011), pp.1514-1527, 1554-8627

Ichimura, Y., Kirisako, T., Takao, T., Satomi, Y., Shimonishi, Y., Ishihara, N., Mizushima, N., Tanida, I., Kominami, E., Ohsumi, M., Noda, T. & Ohsumi, Y. (2000). A ubiquitin-like system mediates protein lipidation. *Nature,* Vol. 408, No.6811 (November 2000), pp.488-492, 0028-0836

Inoue, M., Chang, L., Hwang, J., Chiang, S. H. & Saltiel, A. R. (2003). The exocyst complex is required for targeting of Glut4 to the plasma membrane by insulin. *Nature,* Vol. 422, No.6932 (April 2003), pp.629-633, 0028-0836

Inoue, M., Chiang, S. H., Chang, L., Chen, X. W. & Saltiel, A. R. (2006). Compartmentalization of the exocyst complex in lipid rafts controls Glut4 vesicle tethering. *Molecular biology of the cell,* Vol. 17, No.5 (May 2006), pp.2303-2311, 1059-1524

Jin, Y., Sultana, A., Gandhi, P., Franklin, E., Hamamoto, S., Khan, A. R., Munson, M., Schekman, R. & Weisman, L. S. (2011). Myosin V transports secretory vesicles via a Rab GTPase cascade and interaction with the exocyst complex. *Developmental cell,* Vol. 21, No.6 (December 2011), pp.1156-1170, 1534-5807

Kadiu, I. & Gendelman, H. E. (2011). Human Immunodeficiency Virus type 1 endocytic trafficking through macrophage bridging conduits facilitates spread of infection. *Journal of neuroimmune pharmacology,* Vol. 6, No.4 (December 2011), pp.658-675, 1557-1890

Kamada, Y., Funakoshi, T., Shintani, T., Nagano, K., Ohsumi, M. & Ohsumi, Y. (2000). Tor-mediated induction of autophagy via an Apg1 protein kinase complex. *The Journal of cell biology,* Vol. 150, No.6 (Sepember 2000), pp.1507-1513, 0021-9525

Kawato, M., Shirakawa, R., Kondo, H., Higashi, T., Ikeda, T., Okawa, K., Fukai, S., Nureki, O., Kita, T. & Horiuchi, H. (2008). Regulation of platelet dense granule secretion by the Ral GTPase-exocyst pathway. *The Journal of biological chemistry,* Vol. 283, No.1 (January 2007), pp.166-174, 0021-9258

Keating, D. J. (2008). Mitochondrial dysfunction, oxidative stress, regulation of exocytosis and their relevance to neurodegenerative diseases. *Journal of neurochemistry,* Vol. 104, No.2 (January 2008), pp.298-305, 0022-3042

Khandelwal, P., Ruiz, W. G., Balestreire-Hawryluk, E., Weisz, O. A., Goldenring, J. R. & Apodaca, G. (2008). Rab11a-dependent exocytosis of discoidal/fusiform vesicles in bladder umbrella cells. *Proceedings of the National Academy of Sciences of the United States of America,* Vol. 105, No.41 (October 2008), pp.15773-15778, 0027-8424

Khandelwal, P., Ruiz, W. G. & Apodaca, G. (2010). Compensatory endocytosis in bladder umbrella cells occurs through an integrin-regulated and RhoA- and dynamin-dependent pathway. *The EMBO journal,* Vol. 29, No.12 (June 2010), pp.1961-1975, 0261-4189

Kim, J. H. & von Gersdorff, H. (2009). Traffic jams during vesicle cycling lead to synaptic depression. *Neuron,* Vol. 63, No.2 (July 2009), pp.143-145, 0896-6273

Kim, K. S., Park, J. Y., Jou, I. & Park, S. M. (2010). Regulation of Weibel-Palade body exocytosis by alpha-synuclein in endothelial cells. *The Journal of biological chemistry,* Vol. 285, No.28 (July 2010), pp.21416-21425, 0021-9258

Koike, M., Shibata, M., Waguri, S., Yoshimura, K., Tanida, I., Kominami, E., Gotow, T., Peters, C., von Figura, K., Mizushima, N., Saftig, P. & Uchiyama, Y. (2005). Participation of autophagy in storage of lysosomes in neurons from mouse models of neuronal ceroid-lipofuscinoses (Batten disease). *American journal of pathology,* Vol. 167, No.6 (December 2005), pp.1713-1728, 0002-9440

Komatsu, M., Waguri, S., Chiba, T., Murata, S., Iwata, J., Tanida, I., Ueno, T., Koike, M., Uchiyama, Y., Kominami, E. & Tanaka, K. (2006). Loss of autophagy in the central nervous system causes neurodegeneration in mice. *Nature,* Vol. 441, No.7095 (June 2006), pp.880-884, 0028-0836

Kundu, M. & Thompson, C. B. (2008). Autophagy: basic principles and relevance to disease. *Annual review of pathology,* Vol. 3, (February 2008), pp.427-455, 1553-4006

Kundu, M. (2011). ULK1, mammalian target of rapamycin, and mitochondria: linking nutrient availability and autophagy. *Antioxidants & redox signaling,* Vol. 14, No.10 (May 2011), pp.1953-1958, 1523-0864

Langevin, J., Morgan, M. J., Sibarita, J. B., Aresta, S., Murthy, M., Schwarz, T., Camonis, J. & Bellaiche, Y. (2005). *Drosophila* exocyst components Sec5, Sec6, and Sec15 regulate

DE-Cadherin trafficking from recycling endosomes to the plasma membrane. *Developmental cell,* Vol. 9, No.3 (September 2005), pp.365-376, 1534-5807

Lee, J., Giordano, S. & Zhang, J. (2012). Autophagy, mitochondria and oxidative stress: cross-talk and redox signalling. *The biochemical journal,* Vol. 441, No.2 (January 2012), pp.523-540, 0264-6021

Levine, B. (2007). Cell biology: autophagy and cancer. *Nature,* Vol. 446, No.7137 (April 2007), pp.745-747, 0028-0836

Levine, B. & Kroemer, G. (2008). Autophagy in the pathogenesis of disease. *Cell,* Vol. 132, No.1 (January 2008), pp.27-42, 0092-8674

Li, G., Han, L., Chou, T. C., Fujita, Y., Arunachalam, L., Xu, A., Wong, A., Chiew, S. K., Wan, Q., Wang, L. & Sugita, S. (2007). RalA and RalB function as the critical GTP sensors for GTP-dependent exocytosis. *The Journal of neuroscience,* Vol. 27, No.1 (January 2007), pp.190-202, 0270-6474

Lindmo, K., Simonsen, A., Brech, A., Finley, K., Rusten, T. E. & Stenmark, H. (2006). A dual function for Deep orange in programmed autophagy in the *Drosophila melanogaster* fat body. *Experimental cell research,* Vol. 312, No.11 (July 2006), pp.2018-2027, 0014-4827

Lipschutz, J. H. & Mostov, K. E. (2002). Exocytosis: the many masters of the exocyst. *Current biology: CB,* Vol. 12, No.6 (March 2002), pp.R212-214, 0960-9822

Liu, J., Zuo, X., Yue, P. & Guo, W. (2007). Phosphatidylinositol 4,5-bisphosphate mediates the targeting of the exocyst to the plasma membrane for exocytosis in mammalian cells. *Molecular biology of the cell,* Vol. 18, No.11 (November 2007), pp.4483-4492, 1059-1524

Liu, J. & Guo, W. (2011). The exocyst complex in exocytosis and cell migration. *Protoplasma,* DOI: 10.1007/s00709-011-0330-1 (October 2011), 0033-183X

Ljubicic, S., Bezzi, P., Vitale, N. & Regazzi, R. (2009). The GTPase RalA regulates different steps of the secretory process in pancreatic beta-cells. *PloS one,* Vol. 4, No.11 (November 2009), pp.e7770, 1932-6203

Llobet, A., Gallop, J. L., Burden, J. J., Camdere, G., Chandra, P., Vallis, Y., Hopkins, C. R., Lagnado, L. & McMahon, H. T. (2011). Endophilin drives the fast mode of vesicle retrieval in a ribbon synapse. *The Journal of neuroscience,* Vol. 31, No.23 (June 2011), pp.8512-8519, 0270-6474

Logiudice, L., Sterling, P. & Matthews, G. (2009). Vesicle recycling at ribbon synapses in the finely branched axon terminals of mouse retinal bipolar neurons. *Neuroscience,* Vol. 164, No.4 (December 2009), pp.1546-1556, 0306-4522

Lopez, J. A., Kwan, E. P., Xie, L., He, Y., James, D. E. & Gaisano, H. Y. (2008). The RalA GTPase is a central regulator of insulin exocytosis from pancreatic islet beta cells. *The Journal of biological chemistry,* Vol. 283, No.26 (June 2008), pp.17939-17945, 0021-9258

Luzio, J. P., Pryor, P. R. & Bright, N. A. (2007). Lysosomes: fusion and function. *Nature reviews. Molecular cell biology,* Vol. 8, No.8 (August 2007), pp.622-632, 1471-0072

Malacombe, M., Ceridono, M., Calco, V., Chasserot-Golaz, S., McPherson, P. S., Bader, M. F. & Gasman, S. (2006). Intersectin-1L nucleotide exchange factor regulates secretory granule exocytosis by activating Cdc42. *The EMBO journal*, Vol. 25, No.15 (August 2006), pp.3494-3503, 0261-4189

Manjithaya, R., Anjard, C., Loomis, W. F. & Subramani, S. (2010). Unconventional secretion of *Pichia pastoris* Acb1 is dependent on GRASP protein, peroxisomal functions, and autophagosome formation. *The Journal of cell biology*, Vol. 188, No.4 (February 2010), pp.537-546, 0021-9525

Mari, M., Griffith, J., Rieter, E., Krishnappa, L., Klionsky, D. J. & Reggiori, F. (2010). An Atg9-containing compartment that functions in the early steps of autophagosome biogenesis. *The Journal of cell biology*, Vol. 190, No.6 (September 2010), pp.1005-1022, 0021-9525

Mari, M. & Reggiori, F. (2010). Atg9 reservoirs, a new organelle of the yeast endomembrane system? *Autophagy*, Vol. 6, No.8 (November 2010), pp.1221-1223, 1554-8627

Mark, B. L., Jilkina, O. & Bhullar, R. P. (1996). Association of Ral GTP-binding protein with human platelet dense granules. *Biochemical and biophysical research communications*, Vol. 225, No.1 (August 1996), pp.40-46, 0006-291X

McPhee, C. K. & Baehrecke, E. H. (2009). Autophagy in *Drosophila melanogaster*. *Biochimica et biophysica acta*, Vol. 1793, No.9 (September 2009), pp.1452-1460, 0006-3002

Mehrpour, M., Esclatine, A., Beau, I. & Codogno, P. (2010). Overview of macroautophagy regulation in mammalian cells. *Cell research*, Vol. 20, No.7 (July 2010), pp.748-762, 1001-0602

Meijer, A. J. & Codogno, P. (2004). Regulation and role of autophagy in mammalian cells. *The international journal of biochemistry & cell biology*, Vol. 36, No.12 (December 2004), pp.2445-2462, 1357-2725

Militello, R. D. & Colombo, M. I. (2011). A membrane is born: origin of the autophagosomal compartment. *Current molecular medicine*, Vol. 11, No.3 (April 2011), pp.197-203, 1566-5240

Minty, C. A., Hall, N. D. & Bacon, P. A. (1983). Depressed exocytosis by rheumatoid neutrophils *in vitro*. *Rheumatology international*, Vol. 3, No.3 (September 1983), pp.139-142, 0172-8172

Miracco, C., Cosci, E., Oliveri, G., Luzi, P., Pacenti, L., Monciatti, I., Mannucci, S., De Nisi, M. C., Toscano, M., Malagnino, V., Falzarano, S. M., Pirtoli, L. & Tosi, P. (2007). Protein and mRNA expression of autophagy gene Beclin 1 in human brain tumours. *International journal of oncology*, Vol. 30, No.2 (February 2007), pp.429-436, 1019-6439

Mizushima, N., Noda, T., Yoshimori, T., Tanaka, Y., Ishii, T., George, M. D., Klionsky, D. J., Ohsumi, M. & Ohsumi, Y. (1998). A protein conjugation system essential for autophagy. *Nature*, Vol. 395, No.6700 (September 1998), pp.395-398, 0028-0836

Momose, F., Sekimoto, T., Ohkura, T., Jo, S., Kawaguchi, A., Nagata, K. & Morikawa, Y. (2011). Apical transport of influenza A virus ribonucleoprotein requires Rab11-positive recycling endosome. *PloS one*, Vol. 6, No.6 (June 2011), pp.e21123, 1932-6203

Moreau, K., Ravikumar, B., Renna, M., Puri, C. & Rubinsztein, D. C. (2011). Autophagosome precursor maturation requires homotypic fusion. *Cell*, Vol. 146, No.2 (July 2011), pp.303-317, 0092-8674

Morgera, F., Sallah, M. R., Dubuke, M. L., Gandhi, P., Brewer, D. N., Carr, C. M. & Munson, M. (2012). Regulation of exocytosis by the exocyst subunit Sec6 and the SM protein Sec1. *Molecular biology of the cell,* Vol. 23, No.2 (January 2012), pp. 337-346, 1059-1524

Moskalenko, S., Henry, D. O., Rosse, C., Mirey, G., Camonis, J. H. & White, M. A. (2002). The exocyst is a Ral effector complex. *Nature cell biology,* Vol. 4, No.1 (January 2002), pp.66-72, 1465-7392

Mott, H. R., Nietlispach, D., Hopkins, L. J., Mirey, G., Camonis, J. H. & Owen, D. (2003). Structure of the GTPase-binding domain of Sec5 and elucidation of its Ral binding site. *The Journal of biological chemistry,* Vol. 278, No.19 (May 2003), pp.17053-17059, 0021-9258

Murray, P. D., McGavern, D. B., Lin, X., Njenga, M. K., Leibowitz, J., Pease, L. R. & Rodriguez, M. (1998). Perforin-dependent neurologic injury in a viral model of multiple sclerosis. *The Journal of neuroscience,* Vol. 18, No.18 (September 1998), pp.7306-7314, 0270-6474

Nair, U., Jotwani, A., Geng, J., Gammoh, N., Richerson, D., Yen, W. L., Griffith, J., Nag, S., Wang, K., Moss, T., Baba, M., McNew, J. A., Jiang, X., Reggiori, F., Melia, T. J. & Klionsky, D. J. (2011). SNARE proteins are required for macroautophagy. *Cell,* Vol. 146, No.2 (July 2011), pp.290-302, 0092-8674

Nakatogawa, H., Ichimura, Y. & Ohsumi, Y. (2007). Atg8, a ubiquitin-like protein required for autophagosome formation, mediates membrane tethering and hemifusion. *Cell,* Vol. 130, No.1 (July 2007), pp.165-178, 0092-8674

Nakatogawa, H., Suzuki, K., Kamada, Y. & Ohsumi, Y. (2009). Dynamics and diversity in autophagy mechanisms: lessons from yeast. *Nature reviews. Molecular cell biology,* Vol. 10, No.7 (July 2009), pp.458-467, 1471-0072

Novick, P. & Guo, W. (2002). Ras family therapy: Rab, Rho and Ral talk to the exocyst. *Trends in cell biology,* Vol. 12, No.6 (June 2002), pp.247-249, 0962-8924

Novick, P., Medkova, M., Dong, G., Hutagalung, A., Reinisch, K. & Grosshans, B. (2006). Interactions between Rabs, tethers, SNAREs and their regulators in exocytosis. *Biochemical Society transactions,* Vol. 34, No.5 (November 2006), pp.683-686, 0300-5127

Oishi, Y., Arakawa, T., Tanimura, A., Itakura, M., Takahashi, M., Tajima, Y., Mizoguchi, I. & Takuma, T. (2006). Role of VAMP-2, VAMP-7, and VAMP-8 in constitutive exocytosis from HSY cells. *Histochemistry and cell biology,* Vol. 125, No.3 (March 2006), pp.273-281, 0948-6143

Orlando, K. & Guo, W. (2009). Membrane organization and dynamics in cell polarity. *Cold Spring Harbor perspectives in biology,* Vol. 1, No.5 (November 2009), pp.a001321, 1943-0264

Ostenson, C.-G., Gaisano, H., Sheu, L., Tibell, A. & Bartfai, T. (2006). Impaired gene and protein expression of exocytotic soluble N-ethylmaleimide attachment protein receptor complex proteins in pancreatic islets of type 2 diabetic patients. *Diabetes,* Vol. 55, No.2 (February 2006), pp.435-440, 0012-1797

Oztan, A., Silvis, M., Weisz, O. A., Bradbury, N. A., Hsu, S. C., Goldenring, J. R., Yeaman, C. & Apodaca, G. (2007). Exocyst requirement for endocytic traffic directed toward the apical and basolateral poles of polarized MDCK cells. *Molecular biology of the cell,* Vol. 18, No.10 (October 2007), pp.3978-3992, 1059-1524

Pattingre, S., Tassa, A., Qu, X., Garuti, R., Liang, X. H., Mizushima, N., Packer, M., Schneider, M. D. & Levine, B. (2005). Bcl-2 antiapoptotic proteins inhibit Beclin 1-dependent autophagy. *Cell,* Vol. 122, No.6 (September 2005), pp.927-939, 0092-8674

Polson, H. E., de Lartigue, J., Rigden, D. J., Reedijk, M., Urbe, S., Clague, M. J. & Tooze, S. A. (2010). Mammalian Atg18 (WIPI2) localizes to omegasome-anchored phagophores and positively regulates LC3 lipidation. *Autophagy,* Vol. 6, No.4 (May 2010), pp. 506-522, 1554-8627

Pommereit, D. & Wouters, F. S. (2007). An NGF-induced Exo70-TC10 complex locally antagonises Cdc42-mediated activation of N-WASP to modulate neurite outgrowth. *Journal of cell science,* Vol. 120, No.15 (August 2007), pp.2694-2705, 0021-9533

Ravikumar, B., Moreau, K., Jahreiss, L., Puri, C. & Rubinsztein, D. C. (2010). Plasma membrane contributes to the formation of pre-autophagosomal structures. *Nature cell biology,* Vol. 12, No.8 (August 2010), pp.747-757, 1465-7392

Razi, M., Chan, E. Y. & Tooze, S. A. (2009). Early endosomes and endosomal coatomer are required for autophagy. *The Journal of cell biology,* Vol. 185, No.2 (April 2009), pp.305-321, 0021-9525

Ren, M., Xu, G., Zeng, J., De Lemos-Chiarandini, C., Adesnik, M. & Sabatini, D. D. (1998). Hydrolysis of GTP on Rab11 is required for the direct delivery of transferrin from the pericentriolar recycling compartment to the cell surface but not from sorting endosomes. *Proceedings of the National Academy of Sciences of the United States of America,* Vol. 95, No.11 (May 1998), pp.6187-6192, 0027-8424

Rioux, J. D. (2007). Genome-wide association study identifies new susceptibility loci for Crohn disease and implicates autophagy in disease pathogenesis. *Nature genetics,* Vol. 39, No.5 (May 2007), pp.596-604, 1061-4036

Rondaij, M. G., Bierings, R., van Agtmaal, E. L., Gijzen, K. A., Sellink, E., Kragt, A., Ferguson, S. S., Mertens, K., Hannah, M. J., van Mourik, J. A., Fernandez-Borja, M. & Voorberg, J. (2008). Guanine exchange factor RalGDS mediates exocytosis of Weibel-Palade bodies from endothelial cells. *Blood,* Vol. 112, No.1 (July 2008), pp.56-63, 0006-4971

Rosse, C., Hatzoglou, A., Parrini, M. C., White, M. A., Chavrier, P. & Camonis, J. (2006). RalB mobilizes the exocyst to drive cell migration. *Molecular and cellular biology,* Vol. 26, No.2 (January 2006), pp.727-734, 0270-7306

Saitoh, T., Fujita, N., Jang, M. H., Uematsu, S., Yang, B. G., Satoh, T., Omori, H., Noda, T., Yamamoto, N. & Komatsu, M. (2008). Loss of the autophagy protein Atg16L1 enhances endotoxin-induced IL-1beta production. *Nature,* Vol. 456, No.7219 (November 2008), pp.264-268, 0028-0836

Sato, M. & Sato, K. (2011). Degradation of paternal mitochondria by fertilization-triggered autophagy in *C. elegans* embryos. *Science,* Vol. 334, No.6059 (November 2011), pp.1141-1144, 0036-8075

Savina, A., Fader, C. M., Damiani, M. T. & Colombo, M. I. (2005). Rab11 promotes docking and fusion of multivesicular bodies in a calcium-dependent manner. *Traffic*, Vol. 6, No.2 (February 2005), pp.131-143, 1398-9219

Shandala, T., Woodcock, J. M., Ng, Y., Biggs, L., Skoulakis, E. M., Brooks, D. A. & Lopez, A. F. (2011). *Drosophila* 14-3-3epsilon has a crucial role in anti-microbial peptide secretion and innate immunity. *Journal of cell science*, Vol. 124, No.13 (July 2011), pp.2165-2174, 0021-9533

Shin, S. W., Kim, S. Y. & Park, J. W. (2011). Autophagy inhibition enhances ursolic acid-induced apoptosis in PC3 cells. *Biochimica et biophysica acta*, Vol. 1823, No.2 (December 2011), pp.451-457, 0006-3002

Shipitsin, M. & Feig, L. A. (2004). RalA but not RalB enhances polarized delivery of membrane proteins to the basolateral surface of epithelial cells. *Molecular and cellular biology*, Vol. 24, No.13 (July 2004), pp.5746-5756, 0270-7306

Shorer, H., Amar, N., Meerson, A. & Elazar, Z. (2005). Modulation of N-ethylmaleimide-sensitive factor activity upon amino acid deprivation. *The Journal of biological chemistry*, Vol. 280, No.16 (April 2005), pp.16219-16226, 0021-9258

Shurety, W., Merino-Trigo, A., Brown, D., Hume, D. A. & Stow, J. L. (2000). Localization and post-Golgi trafficking of tumor necrosis factor-alpha in macrophages. *Journal of interferon & cytokine research*, Vol. 20, No.4 (April 2000), pp.427-438, 1079-9907

Simonsen, A. & Tooze, S. A. (2009). Coordination of membrane events during autophagy by multiple class III PI3-kinase complexes. *Journal of cell biology*, Vol. 186, No.6 (September 2009), pp.773-782, 0021-9525

Skinner, M. A. (2010). Membrane trafficking: Mapping a new route. *Nature reviews. Molecular cell biology*, Vol. 11, No.4 (April 2010), pp.234-235, 1471-0072

Spiczka, K. S. & Yeaman, C. (2008). Ral-regulated interaction between Sec5 and paxillin targets Exocyst to focal complexes during cell migration. *Journal of cell science*, Vol. 121, No.17 (September 2008), pp.2880-2891, 0021-9533

Sramkova, M., Masedunskas, A., Parente, L., Molinolo, A. & Weigert, R. (2009). Expression of plasmid DNA in the salivary gland epithelium: novel approaches to study dynamic cellular processes in live animals. *American journal of physiology. Cell physiology*, Vol. 297, No.6 (December 2009), pp.C1347-1357, 0363-6143

Stow, J. L., Manderson, A. P. & Murray, R. Z. (2006). SNAREing immunity: the role of SNAREs in the immune system. *Nature reviews. Immunology*, Vol. 6, No.12 (December 2006), pp.919-929, 1474-1733

Stow, J. L., Low, P. C., Offenhauser, C. & Sangermani, D. (2009). Cytokine secretion in macrophages and other cells: pathways and mediators. *Immunobiology*, Vol. 214, No.7 (July 2009), pp.601-612, 0171-2985

Sugihara, K., Asano, S., Tanaka, K., Iwamatsu, A., Okawa, K. & Ohta, Y. (2002). The exocyst complex binds the small GTPase RalA to mediate filopodia formation. *Nature cell biology*, Vol. 4, No.1 (January 2002), pp.73-78, 1465-7392

Takaya, A., Kamio, T., Masuda, M., Mochizuki, N., Sawa, H., Sato, M., Nagashima, K., Mizutani, A., Matsuno, A., Kiyokawa, E. & Matsuda, M. (2007). R-Ras regulates exocytosis by Rgl2/Rlf-mediated activation of RalA on endosomes. *Molecular biology of the cell*, Vol. 18, No.5 (May 2007), pp.1850-1860, 1059-1524

Takeuchi, H., Furuta, N., Morisaki, I. & Amano, A. (2011). Exit of intracellular *Porphyromonas gingivalis* from gingival epithelial cells is mediated by endocytic recycling pathway. *Cellular microbiology,* Vol. 13, No.5 (May 2011), pp.677-691, 1462-5814

Tanaka, Y., Guhde, G., Suter, A., Eskelinen, E. L., Hartmann, D., Lullmann-Rauch, R., Janssen, P. M., Blanz, J., von Figura, K. & Saftig, P. (2000). Accumulation of autophagic vacuoles and cardiomyopathy in LAMP-2-deficient mice. *Nature,* Vol. 406, No.6798 (August 2000), pp.902-906, 0028-0836

Tanida, I., Minematsu-Ikeguchi, N., Ueno, T. & Kominami, E. (2005). Lysosomal turnover, but not a cellular level, of endogenous LC3 is a marker for autophagy. *Autophagy,* Vol. 1, No.2 (July 2005), pp.84-91, 1554-8627

Tappel, A. (2005). Lysosomal and prostasomal hydrolytic enzymes and redox processes and initiation of prostate cancer. *Medical hypotheses,* Vol. 64, No.6 (February 2005), pp.1170-1172, 0306-9877

Tayeb, M. A., Skalski, M., Cha, M. C., Kean, M. J., Scaife, M. & Coppolino, M. G. (2005). Inhibition of SNARE-mediated membrane traffic impairs cell migration. *Experimental cell research,* Vol. 305, No.1 (April 2005), pp.63-73, 0014-4827

TerBush, D. R., Maurice, T., Roth, D. & Novick, P. (1996). The Exocyst is a multiprotein complex required for exocytosis in *Saccharomyces cerevisiae. The EMBO journal,* Vol. 15, No.23 (December 1996), pp.6483-6494, 0261-4189

Tooze, S. A. & Razi, M. (2009). The essential role of early endosomes in autophagy is revealed by loss of COPI function. *Autophagy,* Vol. 5, No.6 (August 2009), pp.874-875, 1554-8627

Tooze, S. A. & Yoshimori, T. (2010). The origin of the autophagosomal membrane. *Nature cell biology,* Vol. 12, No.9 (September 2010), pp.831-835, 1465-7392

Tra, T., Gong, L., Kao, L. P., Li, X. L., Grandela, C., Devenish, R. J., Wolvetang, E. & Prescott, M. (2011). Autophagy in human embryonic stem cells. *PloS one,* Vol. 6, No.11 (November 2011), pp.e27485, 1932-6203

van Ijzendoorn, S. C. (2006). Recycling endosomes. *Journal of cell science,* Vol. 119, No.9 (May 2006), pp.1679-1681, 0021-9533

Varadi, A., Ainscow, E. K., Allan, V. J. & Rutter, G. A. (2002). Involvement of conventional kinesin in glucose-stimulated secretory granule movements and exocytosis in clonal pancreatic β-cells. *Journal of cell science,* Vol. 115, No.21 (Nov 2002), pp.4177-4189, 0021-9533

Wang, A. L., Lukas, T. J., Yuan, M., Du, N., Tso, M. O. & Neufeld, A. H. (2009). Autophagy and exosomes in the aged retinal pigment epithelium: possible relevance to drusen formation and age-related macular degeneration. *PloS one,* Vol. 4, No.1 (January 2009) pp.e4160, 1932-6203

Wang, C., Wang, Y., McNutt, M. A. & Zhu, W. G. (2011). Autophagy process is associated with anti-neoplastic function. *Acta biochimica et biophysica Sinica,* Vol. 43, No.6 (June 2011), pp.425-432, 1672-9145

Wang, C. C., Ng, C. P., Lu, L., Atlashkin, V., Zhang, W., Seet, L. F. & Hong, W. (2004). A role of VAMP8/endobrevin in regulated exocytosis of pancreatic acinar cells. *Developmental cell,* Vol. 7, No.3 (September 2004), pp.359-371, 1534-5807

Wang, T., Ming, Z., Xiaochun, W. & Hong, W. (2011). Rab7: role of its protein interaction cascades in endo-lysosomal traffic. *Cellular signalling,* Vol. 23, No.3 (March 2011), pp.516-521, 0898-6568

Ward, E. S., Martinez, C., Vaccaro, C., Zhou, J., Tang, Q. & Ober, R. J. (2005). From sorting endosomes to exocytosis: association of Rab4 and Rab11 GTPases with the Fc receptor, FcRn, during recycling. *Molecular biology of the cell,* Vol. 16, No.4 (April 2005), pp.2028-2038, 1059-1524

Webber, J. L., Young, A. R. & Tooze, S. A. (2007). Atg9 trafficking in mammalian cells. *Autophagy,* Vol. 3, No.1 (January 2007), pp.54-56, 1554-8627

Webber, J. L. & Tooze, S. A. (2010). New insights into the function of Atg9. *FEBS letters,* Vol. 584, No.7 (April 2010), pp.1319-1326, 0014-5793

Wei, Y. & Zheng, X. F. (2011). Nutritional control of cell growth via TOR signaling in budding yeast. *Methods in molecular biology,* Vol. 759, No.2 (January 2011), pp.307-319, 1064-3745

Weidberg, H., Shvets, E. & Elazar, Z. (2011). Biogenesis and cargo selectivity of autophagosomes. *Annual review of biochemistry,* Vol. 80, (June 2011), pp.125-156, 0066-4154

Wiederkehr, A., De Craene, J. O., Ferro-Novick, S. & Novick, P. (2004). Functional specialization within a vesicle tethering complex. *The Journal of cell biology,* Vol. 167, No.5 (December 2004), pp.875-887, 0021-9525

Wu, S., Mehta, S. Q., Pichaud, F., Bellen, H. J. & Quiocho, F. A. (2005). Sec15 interacts with Rab11 via a novel domain and affects Rab11 localization *in vivo. Nature structural & molecular biology,* Vol. 12, No.10 (October 2005), pp.879-885, 1545-9985

Wurster, S., Nakov, R., Allgaier, C. & Hertting, G. (1990). Involvement of N-ethylmaleimide-sensitive G proteins in the modulation of evoked [(3)H]noradrenaline release from rabbit hippocampus synaptosomes. *Neurochemistry international,* Vol. 17, No.2 (January 1990), pp.149-155, 0197-0186

Yang, C., Mora, S., Ryder, J. W., Coker, K. J., Hansen, P., Allen, L. A. & Pessin, J. E. (2001). VAMP3 null mice display normal constitutive, insulin- and exercise-regulated vesicle trafficking. *Molecular and cellular biology,* Vol. 21, No.5 (March 2001), pp.1573-1580, 0270-7306

Yu, W. H., Cuervo, A. M., Kumar, A., Peterhoff, C. M., Schmidt, S. D., Lee, J. H., Mohan, P. S., Mercken, M., Farmery, M. R., Tjernberg, L. O., Jiang, Y., Duff, K., Uchiyama, Y., Naslund, J., Mathews, P. M., Cataldo, A. M. & Nixon, R. A. (2005). Macroautophagy--a novel Beta-amyloid peptide-generating pathway activated in Alzheimer's disease. *The Journal of cell biology,* Vol. 171, No.1 (October 2005), pp.87-98, 0021-9525

Zajac, A., Sun, X., Zhang, J. & Guo, W. (2005). Cyclical regulation of the exocyst and cell polarity determinants for polarized cell growth. *Molecular biology of the cell,* Vol. 16, No.3 (March 2005), pp.1500-1512, 1059-1524

Zeng, R., Chen, Y., Zhao, S. & Cui, G. H. (2012). Autophagy counteracts apoptosis in human multiple myeloma cells exposed to oridonin *in vitro* via regulating intracellular ROS and SIRT1. *Acta pharmacologica Sinica,* Vol. 33, No.1 (January 2012), pp.91-100, 1671-4083

Zhang, X., Wang, P., Gangar, A., Zhang, J., Brennwald, P., TerBush, D. & Guo, W. (2005). Lethal giant larvae proteins interact with the exocyst complex and are involved in polarized exocytosis. *The Journal of cell biology,* Vol. 170, No.2 (July 2005), pp.273-283, 0021-9525

Zhang, X. M., Ellis, S., Sriratana, A., Mitchell, C. A. & Rowe, T. (2004). Sec15 is an effector for the Rab11 GTPase in mammalian cells. *The Journal of biological chemistry,* Vol. 279, No.41 (October 2004), pp.43027-43034, 0021-9258

6

Peroxicretion, a Novel Tool for Engineering Membrane Trafficking

Cees M.J. Sagt
DSM Biotechnology Center, Delft
The Netherlands

1. Introduction

The production of proteins by recombinant micro-organisms has been possible since the late 70's. Since then enormous steps have been made to improve protein titers and to expand the range of possible proteins to be produced. The development of genetic modification tools and improved ease of use of cloning strategies have played an important role in this. The time to construct a strain and develop a cost-effective bioprocess has decreased significantly. Concurrently, sequencing data, which serve as a library for donor genes to be overexpressed, have increased exponentially. Therefore, the speed and flexibility to engineer custom-made protein factories has improved tremendously during the last decades. As in any developed technology, a certain degree of standardization is desired. In order to make this ambition more tangible the concept of cell factories is often used. This concept is based on the principle that the cell factory should be able to produce any protein to a certain desired level. Based on the concept of standardized expression vectors and expression hosts the basics of this cell factory concept were built. However, in reality this concept proved to be far too simple, as a standardized input does not result in a standardized output. This is caused by the immense complexity and dynamics of cellular systems. Even when all components are described, by sequencing the genome, the interaction, compartmentalization and dynamics of these components are largely unknown. A striking example is the difference between homologous protein expression which are produced to levels up to 50 g/l in filamentous fungi whereas heterologous proteins are produced at levels which are usually 100-1000 fold less [1]. So even when identical tools are used in a standardized approach the final result can vary over several orders of magnitude.

The general consensus is that this large difference is linked to cellular events and responses which are caused by the overexpression of the heterologous proteins [2], therefore the scientific community, together with biotech industries, have been studying these effects for several decades. The cellular stress responses and intracellular events which are linked to the use of cells as protein production factories are very well described [3-9]. This profound insight in the molecular mechanism of the cellular stress reactions has facilitated the increase in expression and secretion levels of some heterologous proteins. However, levels comparable to homologous proteins have not been reached [1, 10]. It is evident that, when even more demanding sources of biodiversity are tapped, like intracellular proteins, the engineering of a robust and versatile cell factory which is able to produce this wide range of proteins becomes a major challenge.

The production of intracellular proteins in a secreted form, to avoid complex and costly downstream processing, has hardly been touched upon. Several strategies have been developed to increase protein titers, however, these efforts have all been focused on secreted proteins. The functional diversity of intracellular proteins is very broad, while extracellular enzymes are mostly hydrolases, as shown in Figure 1. If the cell would be able to secrete these intracellular enzymes in an active form a wide range of applications could be within reach. Current strategies do not suffice but when these are combined with membrane engineering to redirect vesicle trafficking this could become a very promising approach.

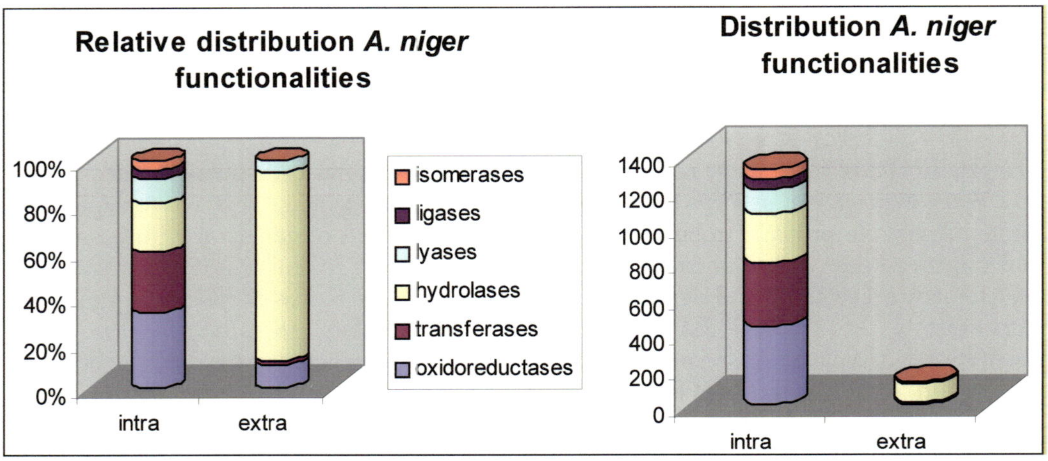

Fig. 1. Distribution of enzyme functionalities in *Aspergillus niger*. The genome sequence of *Aspergillus niger* was mined for annotated functionalities based on EC classifications. The extracellular (extra) or intracellular (intra) protein distribution is based on this annotated genome. The left panel indicates relative function distribution, the right panel shows absolute numbers of enzyme function distribution.

This chapter will discuss the different strategies to improve the productivity of the cellular protein factory. Especially the possibility of engineering membrane trafficking will be addressed. This technology is not only useful for developing a versatile and robust cell factory but can be very valuable to study intracellular membrane trafficking on an academic level as well.

2. Conventional engineering strategies for improved protein production

The conventional strategy to improve protein production is to increase the expression levels of the gene of interest by using multiple gene copies and strong promotors [11, 12]. As a consequence, the cell is faced with very high levels of mRNA which should rapidly be translated into active and correctly folded proteins. To handle this increased demand for folding capacity the cell often reacts with the so called Unfolded Protein Response (UPR). This UPR is a well studied mechanism [13-17] which is also applied to increase folding capacity. By upregulating the transcription factors needed, or the resulting chaperones and folding enzymes, the folding capacity of the cell is enhanced [18-21].

For heterologous proteins the fusion peptide concept has been proven to be successful [22]. This concept is used for heterologous proteins which are secreted by their native host but are overexpressed in an alternative system to increase production titers. By fusing the protein of interest to a homologous protein which is naturally secreted by the host, the desired protein is also transported outside the cell, for example, *Acremonium murorum* phenol oxidase expressed in *Aspergillus awamori* [23]. This carrier protein is often an N - terminal part of a very well secreted homologous protein which is fused to the protein of interest. Often a cleavage site is engineered, like Kex2 with the aim to produce only the full length protein. In reality often a mixture of unprocessed protein, partially processed protein and fully processed protein is produced, which can have consequences for downstream processing in order to meet the required product specifications. On the other hand, there are also reports which describe secretion of heterologous proteins by using only signal sequences without any additional fusion proteins like the production of llama variable heavy-chain antibody fragments (VHHs) [24] and the production of *Arthromyces ramosus* peroxidase [25] both in *Aspergillus awamori*. However, the titers obtained by these systems are in the mg/l range which is generally not sufficient for an economically feasible process.

The next generation of tools to increase protein titers are based on technology developments within bio-IT combined with the possibility of designing custom-made genes. Recently, the concept of codon adaptation was demonstrated. This concept is based on the use of favorable codons so that translation of the mRNA encoding the protein of interest is not a bottleneck. An excellent *in silico* study has recently been published [26] describing that translational speed and, concomitantly, ribosome density are determined by the combination of coding sequences and the tRNA pool. The first dozens of codons are translated at lower rates and create an area where ribosomes are very densely packed, especially on transcripts with a high mRNA level. This strategy could prevent ribosome jamming in later stages of the translational process and therefore be of physiological advantage of the cell. This elevation of translational speed towards the end of transcripts is often seen in fungi (which are biotechnological work-horses) and provides the cell with an effective tool against late abortions of protein synthesis which is an energy-consuming process. As suggested by the authors the fine-tuning of this ramping principle could be an additional tool to increase the efficiency of the protein production factory.

As optimizing codon usage is aimed at improving the translation of mRNA into protein, the folding capacity of the cell can become limiting, especially when high copy numbers and strong promoters are combined with codon optimization. The folding and refolding of proteins is tightly linked to membrane trafficking, a very stringent quality control system existing in the endoplasmic reticulum acts as a gatekeeper for proteins to be transported further into the secretory pathway. This quality control system has been reviewed recently [2]. Also the removal of proteolytic activity has been applied to improve protein production [27-33] this approach has shown to be very effective in specific cases. The success of the different approaches is hard to predict and highly dependent on the protein of interest being overproduced. In addition, all of the strategies above are focused to improve secretion of natural secreted proteins, most of them being hydrolases.

An overview on conventional strategies to improve the productivity of the cellular protein factory is shown in Figure 2. As shown, the modification of DNA, which includes cloning strong promotors in front of the gene of interest and constructing multiple copies of this expression cassette is a generic strategy which can lead to improvement of protein production. Also the optimization of codon(pair) usage and mRNA stabilizing elements can be applied in a generic way. The adaptation of the folding capacity of the host cell is not a generic approach. The folding and modification of proteins is intertwined and very protein specific. Improvement of regular transport of secretory vesicles is a strategy which is less well-known, but some possible routes have been described [34]. In addition the overexpression of SEC4, a Rab protein associated with vesicles, doubled protein production in *Pichia pastoris* [35], which is a direct proof that modification of vesicle flow can result in enhanced protein production. However all these approaches are focused on enhancing the efficiency of existing protein production and secretory processes in the host cell.

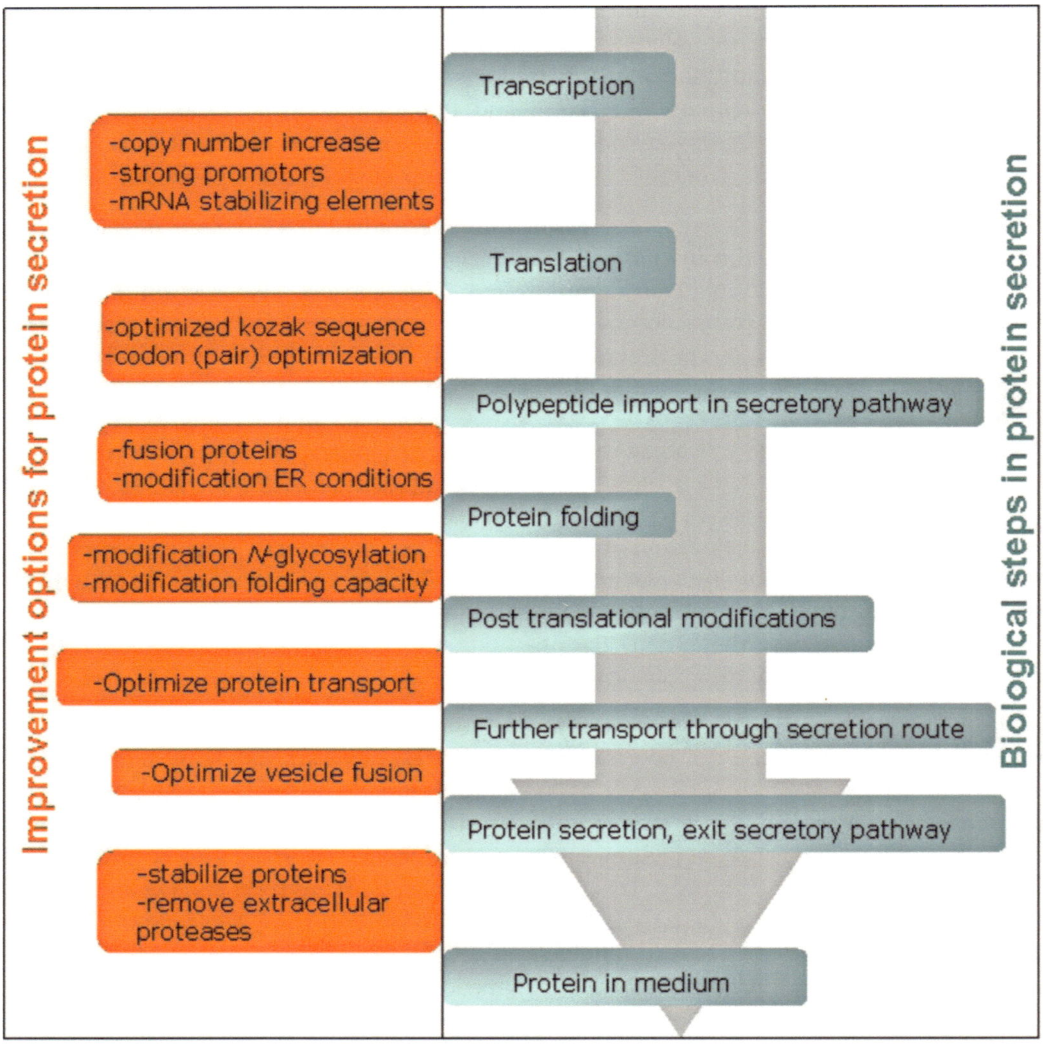

Fig. 2. Conventional improvement options for protein secretion in biological processes. Indicated in blue are the major steps in biological (eukaryotic) systems which are crucial for protein secretion, indicated in orange are improvement strategies which can be applied at the different stages of the protein secretion process.

3. Limitations of conventional strategies

The conventional strategies described above are focused on incremental improvement, overproducing proteins which are naturally secreted products. Therefore, the large diversity of intracellular proteins remains untapped. In addition, these strategies can be seen as generic, at least in part, but very large differences in secretion efficiency exist between different proteins; changes of a few amino acids can often have a detrimental effect on protein secretion [7, 36]. This is probably linked to different interactions with chaperones and folding enzymes. In the case of cutinase secretion in yeast, the bottleneck was circumvented by engineering an N-glycosylation site at the amino terminus of cutinase [8]. The restrictions of conventional strategies are not only illustrated by the order of magnitude of differences in protein secretion but also by the limited class of proteins which are secreted. As shown in Figure 1, in *A. niger* most of the secreted proteins are hydrolases whereas the enzyme variation of intracellular enzymes is much larger. This enormous potential of enzymes is hardly being touched upon. There are strategies to overproduce intracellular proteins inside the cell but this is always followed by a complex downstream processing step to liberate the proteins and to purify them to an acceptable level [37, 38]. In addition, toxic proteins or compounds cannot be produced by these methods at economical levels because they will very likely damage the host cell [39-41]. In order to get access to the large variety of intracellular functionalities and to be able to produce toxic compounds into a cell compartment which does not impact the viability of the host cell, an additional strategy based on membrane engineering would be very valuable.

4. Engineering membrane trafficking as a novel concept

Whereas the conventional approaches are focused on reaching high levels of active proteins, the engineering of membrane trafficking is aimed at redesigning the membrane trafficking in the host cell enabling a custom made flow of vesicles which can expand the possible use of the microbial cell factory. This designed vesicle flow will give access to the large diversity of intracellular enzyme activities and will be important to produce toxic compounds without damaging the host cell.

Recently a novel strategy was described to engineer membrane trafficking, this concept is called peroxicretion [42]. The peroxicretion concept of engineering membrane trafficking is based on two features: 1) the use of cytosolic domains of SNARE proteins (soluble NSF (N-ethylmaleimide-sensitive factor) attachment receptor) which are key for specific membrane trafficking and 2) the use of transmembrane domains, which can reposition the SNARE molecules.

Every cellular compartment of the secretory route contains a specific set of SNARE molecules. They are called v-SNARE or t-SNARE molecules (vesicle or target). SNARE molecules are transmembrane molecules which direct membrane trafficking in eukaryotic cells [43]. More recently, SNAREs are classified into Q and R SNAREs based on structural features. This nomenclature is more precise since certain SNARE molecules act as both v- and t-SNAREs depending on the direction of vesicle flow. Besides transmembrane domains which serve as membrane anchors also lipid modifications like palmitoylation or farnesylation [44, 45] occur. Combinations of transmembrane domains and palmitoylation, preventing degradation of the SNARE, are also reported [46]. SNARE molecules are characterized by the so-called SNARE motif which is a conserved region of approximately 65 amino acids, this region also determines the specificity of the SNARE molecule [47]. This important feature enables modification of the specificity of the SNARE molecules by adapting the SNARE motifs. By fusing these SNARE motifs to transmembrane domains of other proteins it becomes possible to relocalize the SNARE motifs and modify membrane trafficking in eukaryotic systems [42]. Hu *et al.* showed that, by flipping SNARE molecules to the outside of cell membranes, it is possible to fuse these cells, again indicating that SNARE molecules and especially the SNARE motifs are determining the specificity of fusion of membranes [48].The N-terminal domains of SNAREs are thought to act like a zipper which acts from the N-terminus towards the transmembrane domain thereby bringing the membranes in close proximity followed by the actual fusion. This model has been reviewed in an excellent paper [43]. Most research on SNARE molecules has been performed in the field of neuronal cells [49, 50], in relation to synaptic vesicle transport. In *Saccharomyces cerevisiae* a clear set of orthologous proteins can be found as well as in *Aspergillus niger* (see Figure 5).

The concept of SNARE-pin formation has been reviewed by Jahn [51, 43]. In this concept, four helices are interacting to form a so called SNARE pin. The helices are named Q- and R-helices, determined by the glutamine and arginine residues which are participating. The glutamine residues are specified further as Qa, Qb, and Qc. These SNARE complexes are very efficient in mediating membrane fusion. Recently, it was shown very elegantly, by FRET studies, that one SNARE complex is sufficient to enforce membrane fusion [52]. Again, this indicates that modification of these SNARE complexes can have a direct effect on vesicle transport.

It has been described that N-terminal domains and SNARE motifs determine the functionality of SNARE proteins [53]. The functions of the transmembrane domains are not very well known, but one obvious function would be to target and anchor the SNARE in the correct membrane. By using the cytosolic domain of the SNARE and fusing this part to a transmembrane domain of a protein which is located in a different membrane it is possible to engineer membrane trafficking *in vivo* [42].

To reposition the SNARE molecules a transmembrane domain is needed of a protein which is located on the vesicles which need to be modified in order to engineer membrane trafficking. The protein of which the transmembrane part is used should have the N-terminus at the cytosolic side (i.e. surface) of the membrane. The concept of this approach is depicted in Figure 3.

The transmembrane domains of proteins as tools for relocalization of SNARE molecules should be selected carefully. For the proof of principle which was aimed at transforming peroxisomes into secretory vesicles in order to secrete proteins which were located in the peroxisome, the *A. niger* ortholog of a peroxisomal membrane protein described in *Arabidopsis* named PMP22 [54, 55] was used. The final topology of the fusion protein SNARE-membrane anchor is important. The N-terminal domains and SNARE motifs should be positioned at the cytosolic side of the vesicle in order to interact with their SNARE partner. In order to enable peroxisomal fusion with the plasma membrane, it was crucial to identify a peroxisomal membrane anchor, which could be used to decorate the peroxisome with proteins involved in membrane fusion. In order to determine whether peroxisomal membrane proteins have the correct topology we performed topology predictions at the CBS prediction server, as shown in Figure 4 for the PMP22 ortholog of *A. niger* (An04g09130).

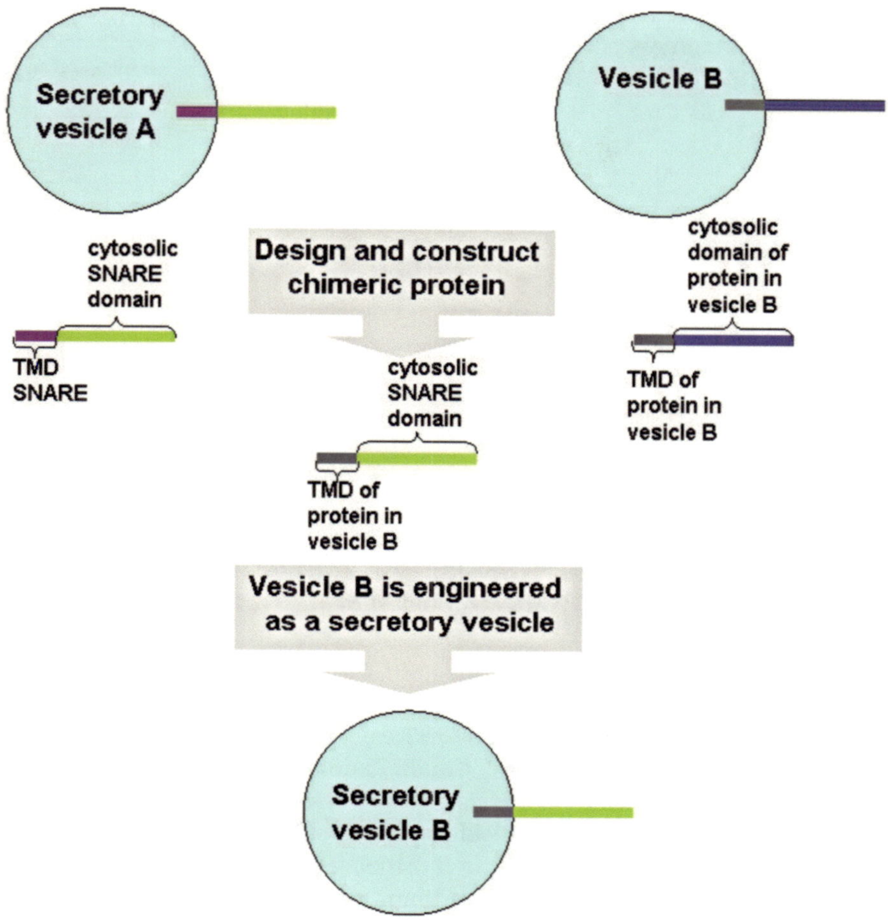

Fig. 3. The membrane trafficking engineering concept. The concept is based on using a cytosolic SNARE domain (green) which is located on a secretory vesicle A. This SNARE domain is fused to a transmembrane domain (grey) of a transmembrane protein originally located on vesicle B. The fusion protein (grey-green) should then be localized at vesicle B, and transforms this vesicle into a secretory vesicle B.

The peroxisomal membrane protein PMP22 has been studied in CHO cells [56], in *Arabidopsis* [55], and in COS cells [57]. All these studies give a description of the membrane topology of PMP22, PMP22 contains 4 TMDs and has the N- and C-termini placed at the cytosolic side of the peroxisome. The precise role of PMP22 is not known. PMP22 contains two peroxisomal targeting regions with almost identical basic clusters which interact with PEX19 [57]. The N-terminus of PMP22 is placed towards the cytosolic side and this topology makes PMP22 useful for v-SNARE fusions at the N-terminus. To determine if PMP22 localizes to the peroxisomes a GFP-PMP22 fusion was constructed and expressed in *A. niger* [42]. Using fluorescence microscopy the localization of GFP-PMP22 was determined. The pattern of the GFP-PMP22 fusion is similar to the GFP-SKL punctuated pattern, which indicates that the PMP22 can be used as peroxisomal membrane anchor [42].

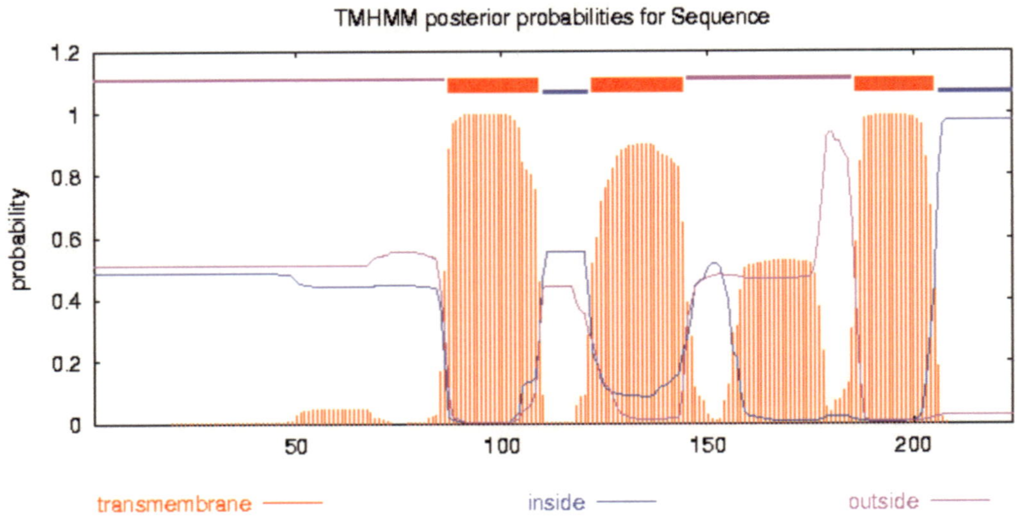

MSAKFQDEAVTSIREDTKELVHKVGNRLTGDGYLALYLRQLQSNPLRTKMLTSGVLSSLQEILASWIA
HDVSKHGHYFSARVPKMALYGMFISAPLGHFLIGILQRVFAGRTSIKAKILQILASNLLVSPIQNAVYL
CCMAVIAGARTFHQVRATVRAGFMPVMKVSWVTSPIALAFAQKFLPEHTWVPFFNIVGFVIGTYVNT
HTKKKRLEALRKCGLPDMVG

Fig. 4. Predicted topology and sequence of the PMP22 ortholog in *A. niger* (An04g09130). The upper panel displays the predicted topology of the PMP22 ortholog by the CBS prediction servers (http://www.cbs.dtu.dk/services/). Four transmembrane domains are predicted, although the TMD between aa 150 and 175 does not exceed the threshold and therefore, 3 TMD are possible as well. It is clear that in both scenarios the N terminus of An04g09130 is positioned towards the outside (cytosolic side) of the vesicle. The lower panel displays the aminoacid sequence of the *A. niger* ortholog of PMP22 (An04g09130) used as input for the CBS prediction server TMD algorithm.

Crucial in this engineering approach is the final orientation of the novel chimeric SNARE molecule, the SNARE domain should be positioned on the outer side (i.e. surface) of the vesicle as shown in Figure 3. The resulting vesicles are decorated with a set of alternative SNARE molecules which can enable fusion of these vesicles with other cellular compartments which contain the appropriate target SNAREs.

Peroxisomes decorated with SNARE proteins have been used as alternative vesicles for transport of intracellular proteins [42]. The use of peroxisomes as alternative vesicles for secretion of proteins which are normally located in the cytosol can be justified by several observations. Peroxisomes are organelles with a single membrane [59] and a very elegant publication shows convincing evidence that peroxisomes originate at the ER [60]. In addition, the identification of the small GTPase Rho1p as being localized on peroxisomes by interaction of the peroxisomal membrane protein Pex25p again indicates a link of peroxisomes with the secretory machinery [61], since Rho1p is known to play a role in actin reorganization and membrane dynamics [62, 63]. In yeast, polarized growth is regulated by Rho1p [64], and also in fungal systems the link between Rho1 and polarized growth has been confirmed [65] in the filamentous fungus *A. niger*, polarized growth is linked to secretion of proteins [66].

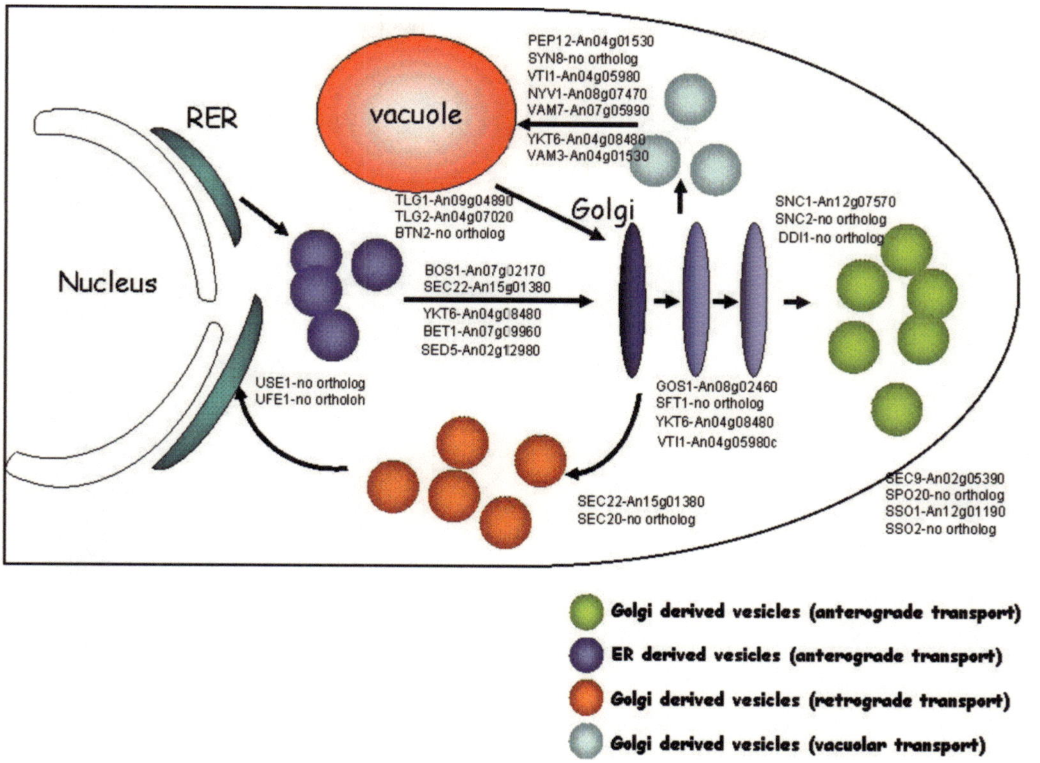

Fig. 5. Overview on SNARE molecules in *Aspergillus niger* and their orthologs in *Saccharomyces cerevisiae*. This overview is based on the *A. niger* sequencing work and the supplementary material of the resulting publication [58]. In green, the vesicle flow form Golgi towards cell membrane is shown, dark blue indicates vesicle flow from ER towards Golgi and the orange vesicles are representing retrograde transport form Golgi towards ER. Finally, the light blue vesicles are representing vacuolar vesicle flow derived from the Golgi apparatus.

More recent papers also very clearly showed that peroxisomes originate from the endoplasmic reticulum [60, 67, 68-70]. The recent finding that insertion of some peroxisomal membrane proteins occurs already at the ER [71] clearly underpins that peroxisomes are linked to the secretory pathway in the eukaryotic cell. However, the same study mentioned that no conventional signal sequences can be found in a set of Peroxisomal Membrane Proteins (PMP) in *S. cerevisiae*. A possible mechanism could be that Pex19p, which is described as a chaperone for proper insertion of PMP into peroxisomes [72-74] also acts a receptor for peroxisomal membrane proteins which are inserted at the ER membrane and become part of peroxisomes at a later stage. An alternative model could be that Pex19p is involved in de budding process of (pre)peroxisomes from the ER where PMP have been already inserted. This yields an interesting model which strongly suggest that peroxisomes are derivatives of the secretory pathway. It has been reported that the transportation of peroxisomes in *Arabidopsis* by actin filaments is dependent on MYA2 (Myosin XI isoform) [75], similar to the transportation of secretory vesicles. This common ground makes peroxisomes an attractive vehicle for transporting intracellular proteins towards the plasma membrane.

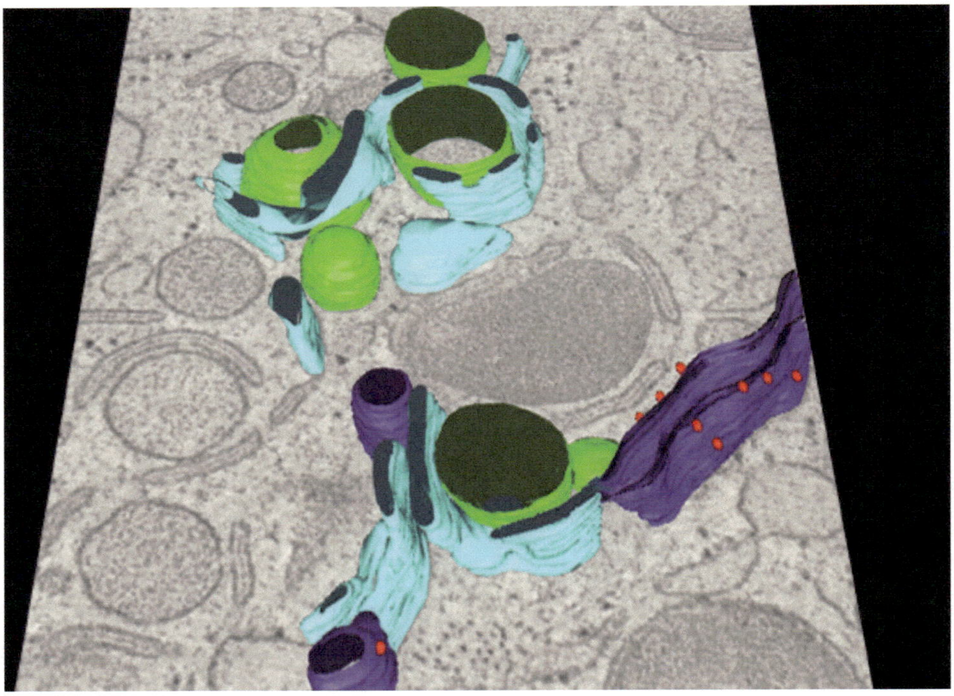

Fig. 6. Peroxisomes originate from the endoplasmic reticulum, taken from Tabak et al. 2003 [67]. In this 3D- reconstituted SEM picture [67] which was processed in silico, the peroxisomes (green) are attached to the smooth ER (light blue), as a continuous membrane structure. Other peroxisomes have already been released from the smooth ER. Rough ER is indicated by dark blue and the red dots represent ribosomes on the RER.

In addition, these organelles contain an import machinery capable of importing completely folded proteins [76-79], and have a controlled way of proliferation [80-82]. So, peroxisomes have the same origin as secretory vesicles and are thus candidates for alternative trafficking of proteins. The first application of such an engineered secretory pathway would be to produce intracellular proteins which show a much broader spectrum of enzymatic activities compared to secreted proteins which have predominantly hydrolase activity (as depicted in Figure 1). Normally, the intracellular proteins are folded in the cytosol, are not N-glycosylated, and disulphide bridges are not formed in this relatively reducing environment [2]. The cytosol does not contain a major machinery for the formation of disulphide bridges or for N-glycosylation as the ER does [83,84]. These basic differences suggest that the conventional route to secrete these intracellular proteins is not compatible with the folding environment, kinetics and modifications of these proteins.

The demonstration of engineering membrane trafficking is based on the ability of peroxisomes to import completely folded proteins. The peroxisomes containing the proteins of interest are equipped with a v-SNARE, which is normally localized at the Golgi apparatus (the study of Marelli *et al.* 2004 did not identify SNARE-like proteins on peroxisomal membranes). This ensures that peroxisomes are able to bind to t-SNAREs which are localized at the plasmamembrane. After the formation of a so called SNARE-pin the

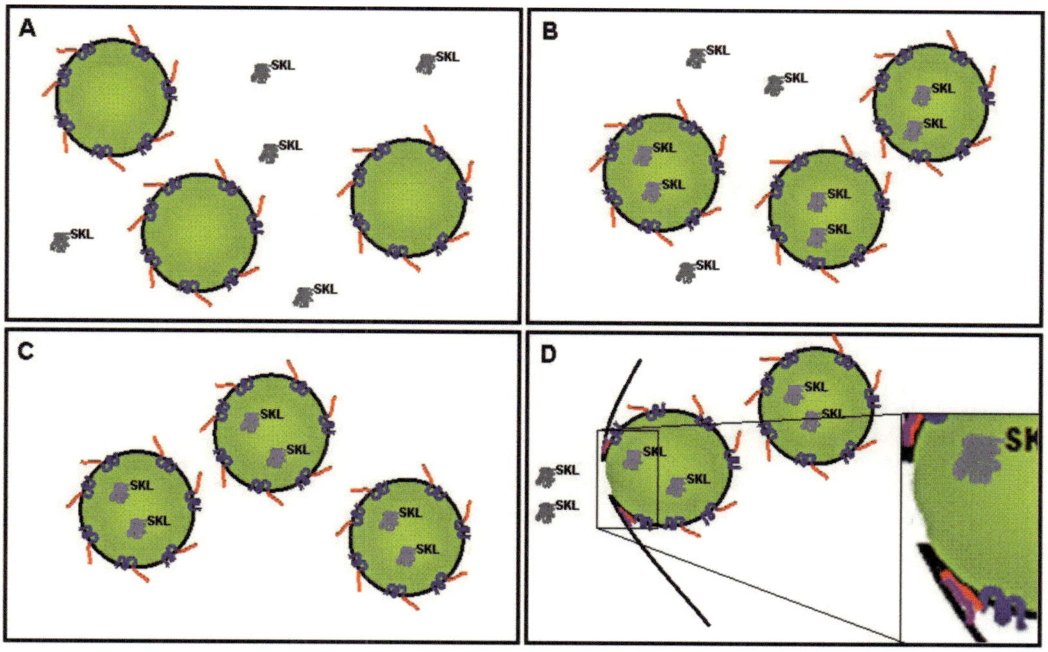

Fig. 7. Overview of the peroxicretion concept. Panel A shows the expression of SKL tagged proteins of interest together with the expression of SNC1-PMP22 fusion protein which are localized on the peroxisomal membrane. The transmembrane anchor encoded by the PMP22 is depicted in blue, the SNC1 SNARE module is encoded by red, together they ensure the decoration of peroxisomes with the *A. niger* ortholog of the v-SNARE SNC1 at the cytosolic side. Panel B shows the translocation of the SKL tagged proteins into the decorated peroxisomes. Panel C shows the complete translocation of the SKL tagged proteins in peroxisomes, the formation of peroxisomes can be increased by overexpression of PEX11 orthologs [86]. Panel D shows the fusion of the peroxisomes with the plasmamembrane, using the ortholog of the t-SNARE SSO1 (purple). The complete SNAREpin structure is more complex, and is not described here for the sake of simplicity.

peroxisomes fuse with the plasma membrane thereby releasing their luminal content. The *A. niger* ortholog of v-SNARE SNC1 (An12g07570) was placed on the peroxisome using the transmembrane domain of the PMP22 ortholog (An04g09130) as a membrane anchor. The transmembrane domain of SNC1 was omitted, the fusion protein is shown in Figure 8. It has been reported that this transmembrane domain of SNC1 is important for the function of SNC1 [85], but the replacement of this TMD by the PMP22 TMD did not abolish the function of SNC1 (An12g07570) in *A. niger* [42]. The replacement of the SNC1 TMD by the PMP22 protein does not diminish the potential of SNC1 to enforce membrane fusion because the peroxisomal content is released extracellularly. The paper of Grote *et al.* 2000 [85] described inhibition of SNC1 function when the TMD was replaced by geranylgeranyl anchors. Probably membrane fusion is only possible when the TMD domains of the SNARE pairs bring the membrane bilayers in close contact so that spontaneous membrane fusion occurs.

Proteins which are produced by using the peroxicretion technology are C-terminally tagged with a PTS1 signal. The most commonly used PTS1 is SKL [87], however variations on this sequence have been described as well [87]. Since the Pex5p (PTS1 receptor) is highly conserved between yeasts and *A. niger*, it has been shown that SKL also functions as a PTS1 signal in *A. niger* [42]. This raises possibilities for purifying the PTS1 tagged products on an affinity column which specifically binds PTS1 sequences. In principal nature has provided us already with a very efficient PTS1 binding protein Pex5p, which is the natural receptor for PTS1.

MSEQPYDPYIPSGANGAGAGASAAQNGDPRTREIDKKIQETVDTMRSNIFKVSERGERLD
SLQDKTDNLATSAQGFRRGANRVRKQMWWKDMKMRSAKFQDEAVTSIREDTKELVHK
VGNRLTGDGYLALYLRQLQSNPLRTKMLTSGVLSSLQEILASWIAHDVSKHGHYFSARVP
KMALYGMFISAPLGHFLIGILQRVFAGRTSIKAKILQILASNLLVSPIQNAVYLCCMAVIAG
ARTFHQVRATVRAGFMPVMKVSWVTSPIALAFAQKFLPEHTWVPFFNIVGFVIGTYVNT
HTKKKRLEALRKCGLPDMVG

Fig. 8. The SNC1-PMP22 fusion protein which enables peroxicretion. SNC1 ortholog sequence (no TMD)is depicted in green (based on An12g07570) the PMP22 ortholog sequence (based on An04g09130) is depicted in blue.

Mutational studies to determine whether shortening of the region between SNC1 and PMP22, thereby placing SNC1 closer to the peroxisomal membrane, could enhance peroxicretion were performed by our group. This approach was partially successful, the most efficient peroxicretion occurred when SNC1 was linked to the full length PMP22 (318 aa) or a truncated version of PMP22 (300 aa). When PMP22 was truncated further the peroxicretion decreased dramatically (< 50%). This is in contrast with a previous study [88], which predicts an increase of vesicle fusion when the hinge region of the v-SNARE and the TMD domain is shortened. This discrepancy is most likely caused by the unnatural TMD of the SNC1-PMP22 fusion protein. The TMD of PMP22 was not evolutionarily selected for efficient membrane fusion. Moreover, the localization information of PMP22 is most likely positioned in this hinge region. GFP fusions with the different truncated PMP22 constructs showed a localization, which is not unambiguously peroxisomal. This indicated (partial) mis-localization of SNC1-PMP22 fusion products when PMP22 is truncated to less than 300 aa, explaining the dramatic decrease of peroxicretion.

Additional modifications to fine-tune the peroxicretion concept can be foreseen, like rebuilding the peroxisome in such a way that all components of the secretory vesicles (proteins and lipids) are present on the peroxisome, making them efficient cell organelles for peroxicretion. Subcellular proteomics could supply us with these leads. One of the most promising options is to use Sec4p a protein involved in SNARE-pin formation. Sec4p is a Rab-GTPase which plays an important role in polarized secretion by tethering secretory vesicles to the plasmamembrane [89, 90]. Sec4p is localized at Golgi-derived vesicles with a geranylgeranyl anchor. Sec4p GEF (Sec2p) and Sec4p GAP (Msb3p and Msb4p) are described and are controlling the activity of Sec4p [91]. This makes Sec4p and its partners potential leads for improving the peroxicretion technology.

In the quest for alternative secretory routes the peroxicretion concept was developed [42], which shows that it is possible to redirect peroxisomes to the cell membrane where they are able to fuse and release their cargo in the extracellular medium, as shown in Figure 7. The peroxisomes can be loaded with intracellular enzymes by adding a C terminal peroxisomal signal sequence like SKL. A combination of the peroxicretion approach with more conventional approaches to improve protein production like removal of proteases and increasing folding capacity of, in this case, the cytosol may further improve protein production. This proof of principle shows that by relocalizing the SNARE molecules to alternative compartments, it is possible to redirect vesicle trafficking in cells, which could also enable production of toxic compounds in micro organisms. This exciting possibility of constructing chimeric proteins of a specific membrane anchor with a specific SNARE functionality is an important tool to engineer membrane trafficking.

Moreover, it can also be hypothesized that due to the modified peroxisomes the complete biomass could now be used for production of enzymes, and not only the hyphal tips as is being suggested for normal secretory proteins [92]. A speculative idea would be to determine the components that allow the peroxisomes to utilize all of the biomass to release their content into the medium and adjust the conventional secretory vesicles in a similar way which would enable secretion of secretory proteins using the complete biomass instead of just the hyphal tips. Using this approach we could even increase fermentation yields of secretory proteins.

5. Designing a robust and versatile cell factory, an outlook

The cell factory of the future should be able to produce a wide range of proteins and metabolites with high productivities and yields in a robust process. It is clear that current strategies contribute in an incremental way, which is important to make the bioprocess economically feasible. In order to expand the product range and design a truly versatile cell factory more innovative approaches are needed. Until now, membrane engineering has hardly been used as an additional tool for designing cell factories. Recent developments have shown that membrane engineering can be a key for designing a truly versatile cell factory. The potential of this approach is enormous, especially in combination with conventional strategies.

In this era of synthetic biology were bio-bricks are used to engineer microbes for new functionalities the concept of membrane trafficking engineering is very well positioned. It enables a true engineering methodology for intracellular trafficking thereby unlocking a complementary approach to already applied strategies like genetic devices based on bio-bricks. In addition, the availability of -omics technologies will function as a starting platform for further fine-tuning of the membrane engineering concept. Based on this holistic approach more key components will be identified which can be used to increase the efficiency of membrane engineering and the application of engineered membranes in microbial cell factories.

This quest for further fine-tuning will be exciting and will have academic impact as well. The membrane trafficking engineering concept can serve the academic society with a new way of studying the complex vesicle flow in biological systems. By combining different components of the secretory pathway into novel chimeric proteins with new functionalities a detailed overview is within reach which can describe the molecular details of the protein secretion mechanism.

This is a major challenge and as with any technology the combined effort of industry and academia can result in significant progress when strong interaction and mutual creativity are applied and recognized.

6. Acknowledgments

Dr. Bert Koekman is acknowledged for editing this chapter.

7. References

[1] Punt PJ, Van Biezen N, Conesa A, Albers A, Mangnus J, Van Den Hondel C. Filamentous fungi as cell factories for heterologous protein production. Trends Biotechnol 2002;20(5):200-6.

[2] Brodsky JL, Skach WR. Protein folding and quality control in the endoplasmic reticulum: Recent lessons from yeast and mammalian cell systems. Curr Opin Cell Biol.

[3] Al-Sheikh H, Watson AJ, Lacey GA, Punt PJ, MacKenzie DA, Jeenes DJ, Pakula T, Penttilä M, Alcocer MJC, Archer DB. Endoplasmic reticulum stress leads to the selective transcriptional downregulation of the glucoamylase gene in *Aspergillus niger*. Mol Microbiol 2004;53(6):1731-42.

[4] Carvalho NDSP, Arentshorst M, Kooistra R, Stam H, Sagt CM, Van Den Hondel CAMJJ, Ram AFJ. Effects of a defective ERAD pathway on growth and heterologous protein production in *Aspergillus niger*. Appl Microbiol Biotechnol 2011;89(2):357-73.

[5] Lombraña M, Moralejo FJ, Pinto R, Martín JF. Modulation of *Aspergillus awamori* thaumatin secretion by modification of bipA gene expression. Appl Environ Microbiol 2004;70(9):5145-52.

[6] Punt PJ, Van Gemeren IA, Drint-Kuijvenhoven J, Hessing JGM, Van Muijlwijk-Harteveld GM, Beijersbergen A, Verrips CT, Van Den Hondel CAMJJ. Analysis of the role of the gene bipA, encoding the major endoplasmic reticulum chaperone protein in the secretion of homologous and heterologous proteins in black aspergilli. Appl Microbiol Biotechnol 1998;50(4):447-54.

[7] Sagt CMJ, Müller WH, Van der Heide L, Boonstra J, Verkleij AJ, Verrips CT. Impaired cutinase secretion in *Saccharomyces cerevisiae* induces irregular endoplasmic reticulum (ER) membrane proliferation, oxidative stress, and ER-associated degradation. Appl Environ Microbiol 2002;68(5):2155-60.

[8] Sagt CMJ, Kleizen B, Verwaal R, De Jong MDM, Muller WH, Smits A, Visser C, Boonstra J, Verkleij AJ, Verrips CT. Introduction of an N-glycosylation site increases secretion of heterologous proteins in yeasts. Appl Environ Microbiol 2000;66(11):4940-4.

[9] Van Gemeren IA, Punt PJ, Drint-Kuyvenhoven A, Broekhuijsen MP, Van 't Hoog A, Beijersbergen A, Verrips CT, Van Den Hondel CAMJJ. The ER chaperone encoding bip A gene of black aspergilli is induced by heat shock and unfolded proteins. Gene 1997;198(2 JAN):43-52.

[10] Meyer V, Wu B, Ram AFJ. Aspergillus as a multi-purpose cell factory: Current status and perspectives. Biotechnol Lett 2011;33(3):469-76.

[11] Verdoes JC, Punt PJ, Schrickx JM, van Verseveld HW, Stouthamer AH, van den Hondel CAMJJ. Glucoamylase overexpression in *Aspergillus niger*: Molecular genetic analysis of strains containing multiple copies of the glaA gene. Transgenic Res 1993;2(2):84-92.

[12] Archer DB, Jeenes DJ, Mackenzie DA. Strategies for improving heterologous protein production from filamentous fungi. Antonie van Leeuwenhoek Int J Gen Mol Microbiol 1994;65(3):245-50.

[13] Saloheimo M, Valkonen M, Penttilä M. Activation mechanisms of the HAC1-mediated unfolded protein response in filamentous fungi. Mol Microbiol 2003;47(4):1149-61.

[14] Mulder HJ, Nikolaev I. HacA-dependent transcriptional switch releases hacA mRNA from a translational block upon endoplasmic reticulum stress. Eukaryotic Cell 2009;8(4):665-75.

[15] Mulder HJ, Nikolaev I, Madrid SM. HacA, the transcriptional activator of the unfolded protein response (UPR) in aspergillus niger, binds to partly palindromic UPR elements of the consensus sequence 5'-CAN(G/A)NTGT/GCCT-3'. Fungal Genet Biol 2006;43(8):560-72.

[16] Davé A, Jeenes DJ, Mackenzie DA, Archer DB. HacA-independent induction of chaperone-encoding gene bipA in *Aspergillus niger* strains overproducing membrane proteins. Appl Environ Microbiol 2006;72(1):953-5.

[17] Mulder HJ, Saloheimo M, Penttilä M, Madrid SM. The transcription factor HacA mediates the unfolded protein response in *Aspergillus niger*, and up-regulates its own transcription. Mol Genet Genomics 2004;271(2):130-40.

[18] Valkonen M, Ward M, Wang H, Penttilä M, Saloheimo M. Improvement of foreign-protein production in *Aspergillus niger var. awamori* by constitutive induction of the unfolded-protein response. Appl Environ Microbiol 2003;69(12):6979-86.

[19] Inan M, Aryasomayajula D, Sinha J, Meagher MM. Enhancement of protein secretion in *Pichia pastoris* by overexpression of protein disulfide isomerase. Biotechnol Bioeng 2006;93(4):771-8.

[20] Conesa A, Jeenes D, Archer DB, Van den Hondel CAMJJ, Punt PJ. Calnexin overexpression increases manganese peroxidase prouction in *Aspergillus niger*. Appl Environ Microbiol 2002;68(2):846-51.

[21] Ngiam C, Jeenes DJ, Punt PJ, Van Den Hondel CAMJJ, Archer DB. Characterization of a foldase, protein disulfide isomerase a, in the protein secretory pathway of *Aspergillus niger*. Appl Environ Microbiol 2000;66(2):775-82.

[22] Gouka RJ, Punt PJ, Van Den Hondel CAMJJ. Glucoamylase gene fusions alleviate limitations for protein production in *Aspergillus awamori* at the transcriptional and (post)translational levels. Appl Environ Microbiol 1997;63(2):488-97.

[23] Gouka RJ, Van Der Heiden M, Swarthoff T, Verrips CT. Cloning of a phenol oxidase gene from *Acremonium murorum* and its expression in *Aspergillus awamori*. Appl Environ Microbiol 2001;67(6):2610-6.

[24] Joosten V, Gouka RJ, Van Den Hondel CAMJJ, Verrips CT, Lokman BC. Expression and production of llama variable heavy-chain antibody fragments (VHHs) by *Aspergillus awamori*. Appl Microbiol Biotechnol 2005;66(4):384-92.

[25] Lokman BC, Joosten V, Hovenkamp J, Gouka RJ, Verrips CT, Van Den Hondel CAMJJ. Efficient production of *Arthromyces ramosus* peroxidase by *Aspergillus awamori*. J Biotechnol 2003;103(2):183-90.

[26] Tuller T, Carmi A, Vestsigian K, Navon S, Dorfan Y, Zaborske J, Pan T, Dahan O, Furman I, Pilpel Y. An evolutionarily conserved mechanism for controlling the efficiency of protein translation. Cell 2010;141(2):344-54.

[27] Nemoto T, Watanabe T, Mizogami Y, Maruyama J-, Kitamoto K. Isolation of *Aspergillus oryzae* mutants for heterologous protein production from a double proteinase gene disruptant. Appl Microbiol Biotechnol 2009;82(6):1105-14.

[28] Yoon J, Kimura S, Maruyama J-, Kitamoto K. Construction of quintuple protease gene disruptant for heterologous protein production in *Aspergillus oryzae.* Appl Microbiol Biotechnol 2009;82(4):691-701.

[29] Van Den Hombergh JPTW, Sollewijn Gelpke MD, Van De Vondervoort PJI, Buxton FP, Visser J. Disruption of three acid proteases in *Aspergillus niger*: Effects on protease spectrum, intracellular proteolysis, and degradation of target proteins. Eur J Biochem 1997;247(2):605-13.

[30] Yoon J, Maruyama J-, Kitamoto K. Disruption of ten protease genes in the filamentous fungus *Aspergillus oryzae* highly improves production of heterologous proteins. Appl Microbiol Biotechnol 2011;89(3):747-59.

[31] Braaksma M, Smilde AK, van der Werf MJ, Punt PJ. The effect of environmental conditions on extracellular protease activity in controlled fermentations of *Aspergillus niger*. Microbiology 2009;155(10):3430-9.

[32] Yoon J, Kimura S, Maruyama J-, Kitamoto K. Construction of quintuple protease gene disruptant for heterologous protein production in *Aspergillus oryzae.* Appl Microbiol Biotechnol 2009;82(4):691-701.

[33] Moralejo FJ, Cardoza RE, Gutierrez S, Lombraña M, Fierro F, Martíni JF. Silencing of the aspergillopepsin B (pepB) gene of *Aspergillus awamori* by antisense RNA expression or protease removal by gene disruption results in a large increase in thaumatin production. Appl Environ Microbiol 2002;68(7):3550-9.

[34] Bankaitis VA, Morris AJ. Lipids and the exocytotic machinery of eukaryotic cells. Curr Opin Cell Biol 2003;15(4):389-95.

[35] Liu S-, Chou W-, Sheu C-, Chang MD-. Improved secretory production of glucoamylase in *Pichia pastoris* by combination of genetic manipulations. Biochem Biophys Res Commun 2005;326(4):817-24.

[36] Sagt CMJ, Müller WH, Boonstra J, Verkleij AJ, Verrips CT. Impaired secretion of a hydrophobic cutinase by *Saccharomyces cerevisiae* correlates with an increased association with immunoglobulin heavy-chain binding protein (BiP). Appl Environ Microbiol 1998;64(1):316-24.

[37] Liu L, Prokopakis GJ, Asenjo JA. Optimization of enzymatic lysis of yeast. Biotechnol Bioeng 1988;32(9):1113-27.

[38] Hopkins TR. Physical and chemical cell disruption for the recovery of intracellular proteins. Bioprocess Technol 1991;12:57-83.

[39] Huang B, Guo J, Yi B, Yu X, Sun L, Chen W. Heterologous production of secondary metabolites as pharmaceuticals in *Saccharomyces cerevisiae.* Biotechnol Lett 2008;30(7):1121-37.

[40] Engels B, Dahm P, Jennewein S. Metabolic engineering of taxadiene biosynthesis in yeast as a first step towards taxol (paclitaxel) production. Metab Eng 2008;10(3-4):201-6.

[41] DeJong JM, Liu Y, Bollon AP, Long RM, Jennewein S, Williams D, Croteau RB. Genetic engineering of taxol biosynthetic genes in *Saccharomyces cerevisiae*. Biotechnol Bioeng 2006;93(2):212-24.

[42] Sagt CMJ, ten Haaft PJ, Minneboo IM, Hartog MP, Damveld RA, van der Laan JM, Akeroyd M, Wenzel TJ, Luesken FA, Veenhuis M, van der Klei I, de Winde JH. Peroxicretion: A novel secretion pathway in the eukaryotic cell. BMC Biotechnol 2009;9.

[43] Jahn R, Scheller RH. SNAREs - engines for membrane fusion. Nat Rev Mol Cell Biol 2006;7(9):631-43.

[44] Sorek N, Bloch D, Yalovsky S. Protein lipid modifications in signaling and subcellular targeting. Curr Opin Plant Biol 2009;12(6):714-20.

[45] Young SG. A thematic review series: Lipid modifications of proteins. J Lipid Res 2005;46(12):2529-30.

[46] Valdez-Taubas J, Pelham H. Swf1-dependent palmitoylation of the SNARE Tlg1 prevents its ubiquitination and degradation. EMBO J 2005;24(14):2524-32.

[47] Paumet F, Rahimian V, Rothman JE. The specificity of SNARE-dependent fusion encoded in the SNARE motif. Proc Natl Acad Sci U S A 2004;101(10):3376-80.

[48] Hu C, Ahmed M, Melia TJ, Söllner TH, Mayer T, Rothman JE. Fusion of cells by flipped SNAREs. Science 2003;300(5626):1745-9.

[49] Jahn R. A neuronal receptor for botulinum toxin. Science 2006;312(5773):540-1.

[50] Hanson PI, Heuser JE, Jahn R. Neurotransmitter release - four years of SNARE complexes. Curr Opin Neurobiol 1997;7(3):310-5.

[51] Jahn R. Principles of exocytosis and membrane fusion. Ann New York Acad Sci 2004;1014:170-8.

[52] Van Den Bogaart G, Holt MG, Bunt G, Riedel D, Wouters FS, Jahn R. One SNARE complex is sufficient for membrane fusion. Nat Struct Mol Biol 2010;17(3):358-64.

[53] Paumet F, Rahimian V, Rothman JE. The specificity of SNARE-dependent fusion encoded in the SNARE motif. Proc Natl Acad Sci U S A 2004;101(10):3376-80.

[54] Murphy MA, Phillipson BA, Baker A, Mullen RT. Characterization of the targeting signal of the arabidopsis 22-kD integral peroxisomal membrane protein. Plant Physiol 2003;133(2):813-28.

[55] Tugal HB, Pool M, Baker A. Arabidopsis 22-kilodalton peroxisomal membrane protein. nucleotide sequence analysis and biochemical characterization. Plant Physiol 1999;120(1):309-20.

[56] Pause B, Saffrich R, Hunziker A, Ansorge W, Just WW. Targeting of the 22 kDa integral peroxisomal membrane protein. FEBS Lett 2000;471(1):23-8.

[57] Brosius U, Dehmel T, Gärtner J. Two different targeting signals direct human peroxisomal membrane protein 22 to peroxisomes. J Biol Chem 2002;277(1):774-84.

[58] Pel HJ, De Winde JH, Archer DB, Dyer PS, Hofmann G, Schaap PJ, Turner G, De Vries RP, Albang R, Albermann K, Andersen MR, Bendtsen JD, Benen JAE, Van Den Berg M, Breestraat S, Caddick MX, Contreras R, Cornell M, Coutinho PM, Danchin EGJ, Debets AJM, Dekker P, Van Dijck PWM, Van Dijk A, Dijkhuizen L, Driessen AJM, D'Enfert C, Geysens S, Goosen C, Groot GSP, De Groot PWJ, Guillemette T, Henrissat B, Herweijer M, Van Den Hombergh JPTW, Van Den Hondel CAMJJ, Van Der Heijden RTJM, Van Der Kaaij RM, Klis FM, Kools HJ, Kubicek CP, Van Kuyk PA, Lauber J, Lu X, Van Der Maarel MJEC, Meulenberg R, Menke H, Mortimer MA, Nielsen J, Oliver SG, Olsthoorn M, Pal K, Van Peij NNME, Ram AFJ, Rinas U, Roubos JA, Sagt CMJ, Schmoll M, Sun J, Ussery D, Varga J, Vervecken W,

Van De Vondervoort PJJ, Wedler H, Wösten HAB, Zeng A-, Van Ooyen AJJ, Visser J, Stam H. Genome sequencing and analysis of the versatile cell factory *Aspergillus niger* CBS 513.88. Nat Biotechnol 2007;25(2):221-31.

[59] Walton PA, Hill PE, Subramani S. Import of stably folded proteins into peroxisomes. Mol Biol Cell 1995;6(6):675-83.

[60] Hoepfner D, Schildknegt D, Braakman I, Philippsen P, Tabak HF. Contribution of the endoplasmic reticulum to peroxisome formation. Cell 2005;122(1):85-95.

[61] Marelli M, Smith JJ, Jung S, Yi E, Nesvizhskii AI, Christmas RH, Saleem RA, Tam YYC, Fagarasanu A, Goodlett DR, Aebersold R, Rachubinski RA, Aitchison JD. Quantitative mass spectrometry reveals a role for the GTPase Rho1p in actin organization on the peroxisome membrane. J Cell Biol 2004;167(6):1099-112.

[62] Roumanie O, Wu H, Molk JN, Rossi G, Bloom K, Brennwald P. Rho GTPase regulation of exocytosis in yeast is independent of GTP hydrolysis and polarization of the exocyst complex. J Cell Biol 2005;170(4):583-94.

[63] MacÍas-Sánchez K, García-Soto J, López-Ramírez A, Martínez-Cadena G. Rho1 and other GTP-binding proteins are associated with vesicles carrying glucose oxidase activity from *Fusarium oxysporum f. sp. lycopersici*. Antonie Van Leeuwenhoek Int J Gen Mol Microbiol 2011;99(3):671-80.

[64] Kozminski KG, Alfaro G, Dighe S, Beh CT. Homologues of oxysterol-binding proteins affect Cdc42p- and Rho1p-mediated cell polarization in *Saccharomyces cerevisiae*. Traffic 2006;7(9):1224-42.

[65] Martínez-Rocha AL, Roncero MIG, López-Ramirez A, Mariné M, Guarro J, Martínez-Cadena G, Di Pietro A. Rho1 has distinct functions in morphogenesis, cell wall biosynthesis and virulence of *Fusarium oxysporum*. Cell Microbiol 2008;10(6):1339-51.

[66] Araujo-Bazán L, Peñalva MA, Espeso EA. Preferential localization of the endocytic internalization machinery to hyphal tips underlies polarization of the actin cytoskeleton in *Aspergillus nidulans*. Mol Microbiol 2008;67(4):891-905.

[67] Tabak HF, Murk JL, Braakman I, Geuze HJ. Peroxisomes start their life in the endoplasmic reticulum. Traffic 2003;4(8):512-8.

[68] Tabak HF, van der Zand A, Braakman I. Peroxisomes: Minted by the ER. Curr Opin Cell Biol 2008;20(4):393-400.

[69] Tabak HF, Hoepfner D, Zand Avd, Geuze HJ, Braakman I, Huynen MA. Formation of peroxisomes: Present and past. Biochim Biophys Acta Mol Cell Res 2006;1763(12):1647-54.

[70] Geuze HJ, Murk JL, Stroobants AK, Griffith JM, Kleijmeer MJ, Koster AJ, Verkleij AJ, Distel B, Tabak HF. Involvement of the endoplasmic reticulum in peroxisome formation. Mol Biol Cell 2003;14(7):2900-7.

[71] Van Der Zand A, Braakman I, Tabak HF. Peroxisomal membrane proteins insert into the endoplasmic reticulum. Mol Biol Cell 2010;21(12):2057-65.

[72] Girzalsky W, Hoffmann LS, Schemenewitz A, Nolte A, Kunau W-, Erdmann R. Pex19p-dependent targeting of Pex17p, a peripheral component of the peroxisomal protein import machinery. J Biol Chem 2006;281(28):19417-25.

[73] Halbach A, Landgraf C, Lorenzen S, Rosenkranz K, Volkmer-Engert R, Erdmann R, Rottensteiner H. Targeting of the tail-anchored peroxisomal membrane proteins

PEX26 and PEX15 occurs through C-terminal PEX19-binding sites. J Cell Sci 2006;119(12):2508-17.
[74] Matsuzono Y, Fujiki Y. In vitro transport of membrane proteins to peroxisomes by shuttling receptor Pex19p. J Biol Chem 2006;281(1):36-42.
[75] Hashimoto K, Igarashi H, Mano S, Nishimura M, Shimmen T, Yokota E. Peroxisomal localization of a myosin XI isoform in arabidopsis thaliana. Plant Cell Physiol 2005;46(5):782-9.
[76] Baker A, Sparkes IA. Peroxisome protein import: Some answers, more questions. Curr Opin Plant Biol 2005;8(6):640-7.
[77] Van Der Klei IJ, Veenhuis M. Peroxisomes: Flexible and dynamic organelles. Curr Opin Cell Biol 2002;14(4):500-5.
[78] Rucktäschel R, Girzalsky W, Erdmann R. Protein import machineries of peroxisomes. Biochim Biophys Acta Biomembr 2011;1808(3):892-900.
[79] de Jonge W, Tabak HF, Braakman I. Chaperone proteins and peroxisomal protein import. Top Curr Genet 2006;16:149-83.
[80] Nagotu S, Veenhuis M, Van der Klei IJ. Divide et impera: The dictum of peroxisomes. Traffic 2010;11(2):175-84.
[81] Orth T, Reumann S, Zhang X, Fan J, Wenzel D, Quan S, Hu J. The PEROXIN11 protein family controls peroxisome proliferation in arabidopsis. Plant Cell 2007;19(1):333-50.
[82] Iwata J-, Ezaki J, Komatsu M, Yokota S, Ueno T, Tanida I, Chiba T, Tanaka K, Kominami E. Excess peroxisomes are degraded by autophagic machinery in mammals. J Biol Chem 2006;281(7):4035-41.
[83] Margittai ÉV, Sitia R. Oxidative protein folding in the secretory pathway and redox signaling across compartments and cells. Traffic 2011;12(1):1-8.
[84] Anelli T, Sitia R. Protein quality control in the early secretory pathway. EMBO J 2008;27(2):315-27.
[85] Grote E, Baba M, Ohsumi Y, Novick PJ. Geranylgeranylated SNAREs are dominant inhibitors of membrane fusion. J Cell Biol 2000;151(2):453-65.
[86] Thoms S, Erdmann R. Dynamin-related proteins and Pex11 proteins in peroxisome division and proliferation. FEBS J 2005;272(20):5169-81.
[87] Kal AJ, Hettema EH, Van Den Berg M, Koerkamp MG, Van Ijlst L, Distel B, Tabak HF. In silicio search for genes encoding peroxisomal proteins in *Saccharomyces cerevisiae*. Cell Biochem Biophys 2000;32:1-8.
[88] McNew JA, Weber T, Engelman DM, Söllner TH, Rothman JE. The length of the flexible SNAREpin juxtamembrane region is a critical determinant of SNARE-dependent fusion. Mol Cell 1999;4(3):415-21.
[89] Novick P, Medkova M, Dong G, Hutagalung A, Reinisch K, Grosshans B. Interactions between rabs, tethers, SNAREs and their regulators in exocytosis. Biochem Soc Trans 2006;34(5):683-6.
[90] Lipschutz JH, Mostov KE. Exocytosis: The many masters of the exocyst. Curr Biol 2002;12(6):R212-214.
[91] Gao X-, Albert S, Tcheperegine SE, Burd CG, Gallwitz D, Bi E. The GAP activity of Msb3p and Msb4p for the rab GTPase Sec4p is required for efficient exocytosis and actin organization. J Cell Biol 2003;162(4):635-46.

[92] Wösten HAB, Moukha SM, Sietsma JH, Wessels JGH. Localization of growth and secretion of proteins in *Aspergillus niger*. J Gen Microbiol 1991;137(8):2017-23.

7

Endocytosis and Exocytosis in Signal Transduction and in Cell Migration

Guido Serini[1,2], Sara Sigismund[3] and Letizia Lanzetti[2,4]
[1]Cell Adhesion Dynamics Laboratory-IRCC, Candiolo
[2]Department of Oncological Sciences University of Torino
[3]1IFOM, Fondazione Istituto FIRC di Oncologia Molecolare, Milan
[4]Membrane Trafficking Laboratory- IRCC, Candiolo
Italy

1. Introduction

Endocytosis is a complex process that is used by eukaryotic cells to internalize fragments of plasma membrane, cell-surface receptors, and various soluble molecules. Many different mechanisms have been developed to achieve internalization of membrane-bound receptors and their ligands and they can be distinguished in clathrin-mediated endocytosis and non-clathrin internalization routes. In the clathrin-mediated endocytosis, receptors bind to the adaptor protein AP2 that, in turn, recruits clathrin to coat the invaginating pits at the plasma membrane. Coated pits are pinched off by the large GTPase dynamin to generate vesicles that traffic from the plasma membrane, undergo uncoating and fuse to the early endosomal compartment. Of note, dynamin is also required in non-clathrin-mediated endocytosis [for detailed recent reviews see (Doherty & McMahon, 2009; Loerke et al, 2009; Mettlen et al, 2009; Traub, 2009)].

From early endosomes vesicles can be re-delivered to the plasma membrane through the exocytic pathway (Grant & Donaldson, 2009). Vesicle budding, uncoating, motility and fusion are controlled by the large family of Rab small GTPases. Rab proteins, in their active GTP-bound form, recruit downstream effectors that, in turn, are responsible for distinct aspects of endosomes function from signal transduction to selection and transport of cargoes. Furthermore, they control vesicular movements on microtubules thus supporting polarized distribution of internalized receptors and signalling molecules [reviewed in (Stenmark, 2009; Zerial & McBride, 2001)]. In this regards, the endo-exocytic processes are profoundly linked with the ability of the cell to elicit receptor-mediated signaling cascades.

Endocytosis has long been considered as an attenuator of signaling as it downregulates receptors at the plasma membrane. However, the ability of internalized receptors to signal from the endosomal compartment and to be recycled to specific regions of the plasma membrane allows signal modulation both in time and in space. Indeed, endocytosis-mediated recycling of receptors is also a major mechanism in the execution of spatially restricted functions, such as cell motility. Moreover, the endo-exocytic cycle of adhesive receptors back and forth from the plasma membrane represents another crucial regulatory aspect played by traffic in the dynamic control of cell-to-cell and cell-to-extracellular matrix contacts.

We, therefore, propose to illustrate the state of the art together with most recent discoveries on the following issues:

1. The signaling endosome: a modality to finely tune persistence of extracellular stimuli inside the cell and to control their re-distribution and compartmentalization. The latter aspect is of extreme relevance for the role of endo-exo membrane trafficking in the execution of cell polarity programs.
2. Involvement of endocytosis and exocytosis in the formation and turnover of cell-to-cell and cell-to-extracellular matrix adhesion. We will review the major findings showing the relevance of membrane trafficking of adhesive receptors, namely cadherins and integrins, and describing the molecular machinery involved that has been identified so far. We will also address recent work indicating that distinct molecular machineries are required for trafficking integrins in active and inactive conformation.
3. Unconventional function of membrane trafficking proteins in mitosis.
   Trafficking molecules also participates to cell cycle progression and to the correct execution of mitosis. We will review the knowledge raised on this issue and discuss how the function of these molecules is related to their established role in membrane trafficking.

2. Endocytosis and signalling

Through endocytosis, active, signalling receptors - such as receptor tyrosine kinases (RTKs) – are removed from the plasma membrane (PM) and destined for degradation and this is crucial to achieve signal extinction and long-term attenuation. Endocytosis is able to remodel the composition of the plasma membrane (PM), thus allowing plasticity in the cellular responses to the microenvironment. Recent evidence, however, has demonstrated that endocytosis has a broader impact on signalling than simply signal extinction (Scita & Di Fiore, 2010; Sorkin & von Zastrow, 2009). Indeed, internalized receptors (and sometimes their ligands) are not only routed to the lysosome for degradation, but, in some cases, can be recycled to specific regions of the PM where polarized signalling is needed for events such as cell migration. Furthermore, signalling might not only occur from the PM, but also could persist along the endocytic pathway as, in the endosomal compartments, signalling receptors are often still bound to their ligands, and continue to be active. More interestingly, signalling receptors in the endosomal compartment could potentially interact with substrates that are not present at the PM. Under this scenario, endocytosis would be a mechanism to sustain signalling and to achieve signal diversification and specificity.

2.1 Signalling elicited by the endocytic compartments

The concept that signalling continues along the endocytic pathway was shown in the case of several signalling receptors, including RTKs and the TGFβR (tumor growth factor β receptor) (Sorkin & von Zastrow, 2009). In all cases, receptors remain bound to their ligand and active once internalized within endosomes, thus sustaining signalling from the intracellular compartments (Burke et al, 2001; Di Guglielmo et al, 1994; Grimes et al, 1996; Haugh et al, 1999; Hayes et al, 2002; Howe et al, 2001; Lai et al, 1989; Pennock & Wang, 2003;

Wada et al, 1992; Wang et al, 2004; Wang et al, 1996). In agreement with this, all the components of the MAPK (mitogen-activated protein kinase) activation cascade can be found in endosomes (Pol et al, 1998; Roy et al, 2002), showing that RTKs signalling persist also after internalization. In this way, sufficient duration and amplitude to signalling is allowed. Furthermore, endosomal-specific proteins have been identified and shown to be required to sustain signalling. One example is represented by P18, which works at the endosomal membrane as an anchor for an ERK-activating scaffold and is required to achieve maximal activation of ERK1/2 (Nada et al, 2009). A similar mechanism occurs in the case of GPCR (G protein-coupled receptor) signalling, where β-arrestin, similarly to P18, acts as a specific scaffold to anchor ERK1/2 to the endosome (DeWire et al, 2007) thus allowing proper signal duration.

A series of genetic evidence support a role for endocytosis in the sustaining of the signalling. Historically, the first proof was provided by the use of a dominant-negative mutant of dynamin that blocks EGF internalization and causes the inhibition of EGF-induced activation of PI3K and ERKs (extracellular signal-regulated kinases) (Vieira et al, 1996). This initial evidence was then reinforced by experiments with siRNAs (small interfering RNAs) targeting proteins involved in internalization, which show that endocytosis is required for ERK activation by several receptor kinases [reviewed in (Sorkin & von Zastrow, 2009)]. Not only endocytosis is crucial to sustain signalling, but it is also required to determine signal specificity and diversification. Indeed, endosomes can support signalling cascades that cannot happen at the PM. The existence of endosome-specific signalling cascades has been shown for different receptor systems, including RTKs, GPCRs and Notch (reviewed in (Scita & Di Fiore, 2010; Sorkin & von Zastrow, 2009)). In the TGFβR pathway, specific signalling proteins are recruited to endosomes through their binding to PI3P (phosphatidylinositol 3-phosphate, which is enriched in endosomal membrane compared to PM) and this allows intracellular-specific signalling. Indeed, the activated TGFβR receptor interacts with the adaptor protein SARA (smad anchor for receptor activation) in early endosomes. SARA is associated with the receptor target SMAD2, and this allows the efficient phosphorylation of SMAD2 by TGFβR in endosomes (Chen et al, 2007; Hayes et al, 2002; Tsukazaki et al, 1998). Once phosphorylated, SMAD2 forms a complex with SMAD4, which translocates to the nucleus to regulate gene transcription.

Importantly, early endosomes are a morphologically and functionally heterogeneous population, characterized by the presence of biochemically distinct membrane subdomains (Lakadamyali et al, 2006; Miaczynska et al, 2004; Sonnichsen et al, 2000; Zoncu et al, 2009).

At the molecular level, small GTPases play a crucial role in determining the different sorting fates of cargoes at these stations, which ultimately impact on the final signalling response [reviewed in (Stenmark, 2009)]. For instance, APPL1-containing endosomes are precursors of early endosomes specifically enriched in Rab5 but lacking EEA1. It has been proposed that the progressive accumulation of PI3P species (through association and activity of phosphatidylinositol 3-kinase, PI3KC3/Vps34) causes the recruitment of EEA1, which competes with APPL1 for Rab5 binding, displacing it from the maturing early endosomes (Zoncu et al, 2009). Importantly, APPL1- but not EEA1-positive endosomes are competent for AKT signalling (Zoncu et al, 2009). This "phosphoinositide switch" is responsible for the maturation of endosomes and it is involved in signalling specification.

A non-canonical example of endosome-specific signalling is provided by the TNFR (tumor necrosis factor receptor) signalling cascade (Schutze et al, 2008) that promotes pro-apoptotic signalling. The components of this pathways are recruited to the ligand-bound TNFR at the plasma membrane (Micheau & Tschopp, 2003). In order for apoptosis to be achieved, the cysteine protease caspase-8 has to be activated by its proteolytic cleavage and this occurs on endosomes (Schneider-Brachert et al, 2004). Although, the mechanisms that prevent caspase-8 recruitment and activation at the PM are not yet known, this represents another example of how endocytosis contributes to signal specificity.

2.2 Regulation of signalling by endosome sorting

Once internalized and sorted to the early endosomes, cargoes can be routed to degradative pathways, terminating signalling, or recycled back to PM, allowing further rounds of activation. Both these mechanisms contribute to regulate signalling in space and time [reviewed in (Marchese et al, 2008; Sorkin & von Zastrow, 2009)].

Transfer of activated receptors to late endosomes/multivesicular bodies (MVB) terminates signalling, either by sequestering receptors in intraluminal vesicles, thus preventing their interaction with signal transducers, or by promoting their lysosomal degradation. Receptor ubiquitination plays a critical role in this process. Indeed, several protein complexes harbouring ubiquitin (Ub)-binding domains recognize ubiquitinated cargoes and escort them along the degradative route to the lysosome (Dikic et al, 2009). These complexes called ESCRT (endosomal sorting complex required for transport) act sequentially at various stations of the degradative route and are involved in MVB inward vesicles budding and cargo sequestration in the intraluminal vesicles of MVBs [for reviews see (Hurley & Hanson, 2010; Raiborg & Stenmark, 2009)].

On the other hand, recycling of internalized receptors to the PM allows the recovery of unoccupied/free receptors to the cell surface and restores receptor sensitivity to extracellular ligands, as is the case for GPCRs. One classical example is represented by β2AR (β2 adrenergic receptor). This class of receptors signals though interaction with PM-resident trimeric G proteins, which transduce signalling from the PM. Upon agonist stimulation, coupling of β2AR trimeric G proteins is inhibited by receptor phosphorylation events [see, for instance, (Benovic et al, 1985; Benovic et al, 1986; Pitcher et al, 1992), reviewed in (Kelly et al, 2008)], which cause functional desensitization of signalling in the absence of endocytosis. However, β-arrestins are recruited to the phosphorylated receptors, triggering their internalization and sorting into a rapid recycling pathway. This step promotes receptor dephosphorylation by an endosome-associated PP2A protein phosphatase, thus ensuring the return of intact receptor for successive rounds of signalling (Pitcher et al, 1995; Vasudevan et al 2011; Yang et al, 1988), a process called "resensitization".

A related example, where the differential trafficking fate determines the duration of the signal, is represented by the EGFR system. When stimulated with TGFα or EGF, EGFR is rapidly internalized. However, while EGF binding to EGFR remains stable at the pH of endosomes, TGFα rapidly dissociates from the receptor. This results in different signalling outputs: EGFR/EGF complex remains stable and active at the endosomal station and is then transported to lysosomes for degradation, allowing signal termination; in contrast, in the case of TGFα, the receptor detaches from ligand at the endosomal station and it is recycled back to the PM, ready to undergo an additional round of activation (Decker, 1990; Ebner & Derynck, 1991; French et al, 1995; Longva et al, 2002). In agreement with this, TGFα is a more potent mitogen than EGF (Waterman et al, 1998). The idea that endosome sorting regulates signalling output as a function of ligand type was shown also in the case of KGFR (keratinocyte growth factor receptor). Indeed, stimulation with two different ligands, KGF or FGF10, targets the receptor to two distinct trafficking routes, degradation vs. recycling, respectively, and this correlates with the higher mitogenic activity exerted by FGF10 on epithelial cells (Belleudi et al, 2007).

The central role of endocytosis in cellular signalling raises the possibility that alteration of this process might contribute to pathological phenotypes in which aberrant signalling is central, such as development and progression of cancer. Several lines of indirect evidence support a role of endocytosis in cancer [reviewed in (Lanzetti & Di Fiore, 2008; Mosesson et al, 2008)]. However, solid proof for a causative role of endocytosis in tumourigenesis is missing. A recent advance in this direction came from a study by Kermorgant's group (Joffre et al, 2011), who investigated the mechanism leading to tumourigenesis of two oncogenic Met mutants (M1268T and D1246N). These mutations cause constitutive Met kinase activity that was originally considered at the basis of their oncogenic potential. By using a combination of *in vitro* and *vivo* approaches, Kermorgant's group showed that endocytosis and intracellular trafficking of these mutants play a crucial role in determining their tumorigenic activity, besides their basal kinase activation. Indeed, these mutants are constitutively internalized and recycled back to the PM at a higher rate compared to WT receptor, and they also show impaired degradation. Importantly, inhibition of internalization with different genetic and pharmaceuticals tools is able to significantly reduce the ability of these mutants of induce transformation *in vitro* and to generate tumours in *ex vivo xenograft* experiments, without altering their activation status. Although the endocytic mechanism used by these mutant receptors is far to be clear (they seem to enter a constitutive pathway that depends on Cbl, Grb2, Clathrin and dynamin and that is independent from receptor kinase activity and ubiquitination), this is the first evidence for a direct involvement of endocytosis and endosome sorting in cancer development.

2.3 Different trafficking routes determine signalling outputs

Different internalization pathways are often associated to distinct intracellular fates. Several signalling receptors, including RTKs, GPCRs, TGFβR, NOTCH and WNT undergo both clathrin-mediated endocytosis (CME) and non-clathrin endocytosis (NCE) and this influences the final signalling output (Le Roy & Wrana, 2005). A mechanism of this kind takes place during internalization and signalling of the EGFR (Sigismund et al, 2005). At low doses of EGF, the EGFR is almost exclusively internalized through CME, which leads to recycling of the receptor and sustains signalling, with only a minor fraction of EGFRs targeted to degradation (Sigismund et al, 2008). At higher doses, about half of the ligand-engaged receptors are then internalized through NCE, a pathway that targets EGFRs to lysosomal degradation causing rapid signal extinction (Sigismund et al, 2008). This dual mechanism perfectly couples with the different EGF concentrations found in body fluids [ranging from 1 to hundreds of ng/ml, reviewed in (Sigismund et al, 2005)]. Indeed, under scarce ligand availability, endocytosis (through CME) preserves the activated receptors from degradation, maximizing the signalling response; at high EGF, the NCE pathway destines the excess of activated EGFR/EGF complex to degradation, protecting cells from overstimulation. This concept has been challenged in other studies, where EGFR was reported to be internalized exclusively through CME at all concentrations of EGF (Kazazic et al, 2006; Rappoport & Simon, 2009). The discrepancy may be due to the different cellular systems used in these studies. It still remains to establish the nature of the NCE pathway used by the EGFR and the molecular mechanism involved [which is still poorly characterized, although it has been shown to be caveolin-independent and to require receptor ubiquitination (Sigismund et al, 2008; Sigismund et al, 2005)].

A similar scenario was previously reported in the case of TGFβR. This receptor is internalized both by CME and NCE and this has profound impact on the final signalling output (Di Guglielmo et al, 2003). Proteins of the TGFβ superfamily signal through the transmembrane Ser-Thr kinase TGFβR type I and type II heteromeric complex (TβRI and TβRII). Ligand-induced assembly of the heteromeric receptor complex activates TβRI, which initiates Smad signalling by phosphorylating the receptor-regulated Smads. The Smad adaptor protein SARA plays a crucial role at this step. Indeed, SARA binds the receptor and contains a FYVE (Fab1p, YOTB, Vac1p and EEA1) domain, which also binds to membranes through specific interactions with phosphatidyl inositol 3' phosphate (PI3P). Receptor internalization through the clathrin pathway is essential for signalling and SARA has been found in the PI(3)P-enriched EEA1-positive endosomes that are downstream of this route (Di Guglielmo et al, 2003). Conversely, receptors that enter cells through NCE are associated with Smad7 and the E3 Ub ligase SMURF; they are ubiquitinated and subjected to degradation (Di Guglielmo et al, 2003).

It is important to note that CME is not always associated to signalling and NCE to degradation, but the opposite is also true, as it was shown in the case of LRP6 [WNT3a-activated low-density receptor-related protein 6, (Yamamoto et al, 2008)]. In the presence of Wnt3a, LPR6 is phosphorylated and internalized into caveolin-positive vesicles, where it can stabilize β-catenin and activates signalling via the CK1g kinase. If LRP6 binds the Wnt3a antagonist Dkk (Dickkopf), it is targeted to the clathrin pathway, which is not competent for signalling but rather directs LRP6 to degradation.

Other examples on how the route of internalization influences the final signalling output have been recently provided in the case of IGF-1R (Martins et al, 2011; Sehat et al, 2008) and PDGFR (De Donatis et al, 2008). In both cases, it has been proposed that they can enter through both clathrin-dependent and -independent pathways depending on the amount of ligand used to stimulate cells, similarly to what has been shown for the EGFR system. This again impacts on the final biological response. For instance, in the case of PDGFR, cells switch from a migrating to a proliferating phenotype in response to an increasing PDGF gradient. It was proposed that the decision to proliferate or migrate relies on the distinct endocytic route followed by the receptor in response to ligand concentration (De Donatis et al, 2008). Although these studies remain at the phenomenological level with no mechanistic insights, they confirm the idea that integration of different internalization pathways is crucial to decode signal information and to specify the signalling response.

2.4 Role of endo-exo membrane trafficking in the execution of cell polarity programs

Endo and exocytosis not only control the persistence and the nature of signals as highlighted above, but also the restricted compartmentalization of the signals. This has profound implications in particular in the establishment of cell polarity, a process that largely relies on the correct localization of protein complexes and signalling platforms at cell-to-cell and cell-to–extracellular matrix contacts. In this regards, a key role in the controlled distribution of signal transducers in restricted areas of the plasma membrane, in response to extracellular cues, is played by small GTPases of the Rab family like Rab5, Rab8 and Rab11.

Rab5 is a master regulator of endocytosis and actin remodelling (Lanzetti et al, 2004; Lanzetti et al, 2000; Palamidessi et al, 2008; Zerial & McBride, 2001). It controls the internalization of a variety of distinct receptors, including the adhesive molecules integrins and cadherins (Palacios et al, 2005; Pellinen et al, 2006), as detailed in paragraph 3, thus participating to the processes of cell-to-cell and cell-to-extracellular matrix adhesion. Importantly, in *Drosophila melanogaster* deletion of Rab5 or disruption of the endocytic protein Syntaxin/Avalanche affects the polarized, restricted apical distribution of the fate-decision receptor Notch and of the polarity determinant Crumbs (Lu & Bilder, 2005). Failure in internalization of Notch and Crumbs causes their accumulation and results in the expansion of the apical membrane domain. Impaired Notch internalization severely impacts on its degradation and signalling and, in turn, this leads to overgrowth of imaginal epithelial tissues (Lu & Bilder, 2005) indicating that endocytosis may also control epithelial tissue proliferation.

Rab8 participates in polarized transport of molecules to the basolateral membrane (Huber et al, 1993) and also in cilia (Nachury et al, 2007). Genetic deletion of Rab8 in mice has been found to affect the distribution of apical proteins to the surface of intestinal epithelial cells resulting in accumulation of vacuoles containing apical hydrolases and microvilli with the final outcome of animal death by starvation (Sato et al, 2007). Thus, Rab8 has been proposed to play a crucial role in the biogenesis of the apical membrane, a process that is profoundly influenced also by another Rab protein involved in recycling routes: Rab11 [reviewed in (Hoekstra et al, 2004)]. Indeed trafficking *via* the recycling endosomes is required for the establishment or rearrangement of cell polarity in various settings including cellularization, cell–to-cell boundary rearrangement, asymmetric cell division, and cell migration (Assaker et al, 2010; Bryant et al, 2010; Emery et al, 2005; Xu et al, 2011). Furthermore, it provides a very efficient mechanism to reinforce polarity by feedback loops (Assaker et al, 2010).

In addition to these GTPases, the endocytic protein Numb has also been implicated in the establishment of apical-basolateral polarity. Numb participates to cadherin endocytosis by interacting with the E-cadherin/p120 complex and promotes E-cadherin endocytosis. Impairment of Numb induces mislocalization of E-cadherin from the lateral to the apical membrane. This function of Numb appears to rely on its phosphorylation by Atypical protein kinase C (aPKC), a member of the PAR complex, as it prevents association of phosphorylated Numb with p120 and α-adaptin thereby attenuating E-cadherin endocytosis (Sato et al, 2011).

Beside the involvement of endo-exocytosis in apical-basolateral polarity, these trafficking routes are also required in the establishment of planar cell polarity (PCP) [for a detailed reviews on membrane trafficking in cell polarity see (Nelson, 2009)]. Intracellular membrane trafficking has emerged as a crucial regulator of PCP in the *Drosophila* wing where inhibition of dynamin or Rab11 disrupts PCP-dependent hexagonal repacking (Classen et al, 2005). More recently, Rab5 has been found to bind to Go and to participates in PCP and in Wingless signal transduction, pathways initiated by G-protein coupled receptors of the

Frizzled (Fz) family. Additionally, Rab4 and Rab11 function in Fz- and Go-mediated signaling to favor PCP over canonical Wingless signaling (Purvanov et al, 2010). Furthermore, the Rab5-effector Rabenosyn-5 is required for the polarized distribution of PCP proteins at the apical cell boundaries aiding the establishment of planar polarity (Mottola et al, 2010).

The requirement for regulation of clathrin-mediated endocytosis in planar cell polarity also emerges from the study showing that the planar polarized RhoGEF2 controls the function of Dia and Myosin II which, in turn, are responsible for the initiation of E-cadherin endocytosis by regulating their lateral clustering (Levayer et al, 2011).

Another relevant instance of the involvement of endo/exocytosis in the execution of polarized function is directed cell migration. Also in this case important lessons come from the *Drosophila* model. In the fruit fly, endocytosis of motogenic receptors and their recycling to the plasma membrane serve to maintain their polarized distribution at the leading edge of migrating cells, thus promoting directional motility (Jekely et al, 2005; McDonald et al, 2006; McDonald et al, 2003; Montell, 2003; Wang et al, 2006). This is achieved *via* a tight control of endocytosis and recycling in restricted areas of the cell membrane through the regulation of a subset of molecules such as the endocytic E3 ligase Cbl, or the Rab5 GEF Sprint (Jekely et al, 2005).

Collectively, these observations provide genetic evidence that one physiological role of endocytosis is to ensure localized intracellular responses to extracellular cues, i.e. the spatial restriction of signalling. Similar circuitries are also exploited in mammalian cells to achieve and maintain cell polarity and also to execute polarized functions such as directed cell migration (Balasubramanian et al, 2007; Caswell & Norman, 2008; Jones et al, 2006; Palamidessi et al, 2008; Riley et al, 2003; Schlunck et al, 2004). Of note, directed cell migration in mammalian cells has been found to require Rab proteins like Rab25 and Rab5 (Caswell et al, 2007; Palamidessi et al, 2008). Rab25 promotes the extension of long pseudopodia in 3D matrices, by regulating the recycling of a pool of □5□1 (Caswell et al., 2007; detailed in paragraph 3), Instead, Rab5-dependent endocytosis allows for the activation of Rac, induced by motogenic stimuli, on early endosomes. Subsequent recycling of Rac to the plasma membrane ensures localized formation of actin-based migratory protrusions (Palamidessi et al, 2008).

3. Regulation of cell adhesion dynamics by trafficking adhesive receptors

The acquisition of key molecular strategies that support social cell functions, such as intercellular communication and adhesion either to other cells or to the surrounding environment, represented a tenet in the evolution from simple unicellular to complex multicellular organisms on the Earth (Rokas, 2008). Indeed, the appearance of genes encoding for adhesion receptors is likely to have represented a major driving force of the so called Cambrian explosion during which, around 500 million years ago, the appearance on our planet of multicellular organisms, aka metazoans, and an astonishingly wide exploration of most of their possible morphological organizations took place (Abedin & King, 2010). In mammalians, the ability of dynamically regulating cell adhesion in space and time is crucial for several physiological and pathological phenomena, such as embryonic development (Hynes, 2007), tissue and organ morphogenesis and repair (Insall & Machesky,

2009), leukocyte extravasation (Hogg et al, 2011), platelet aggregation (Tao et al, 2010), and cancer cell metastatic dissemination throughout the body (Roussos et al, 2011). Cadherins (Takeichi, 2011) and integrins represent the main classes of transmembrane receptors respectively mediating cell-to-cell and cell-to-extracellular matrix (ECM) adhesion in mammals. A dynamic control of cell adhesion can be accomplished by regulation of either conformation or endo-exocytic traffic of adhesion receptors. Cadherin and integrin conformational activation can be triggered by the binding of either extracellular divalent cations, e.g. Ca^{2+} for cadherins (Takeichi, 2011) or Mg^{2+} for integrins (Tiwari et al, 2011), or cytosolic proteins, such as talin and kindlin in the case of integrins (Moser et al, 2009). The mechanisms that directly supersede to the control of cadherin (Gumbiner, 2005; Niessen et al, 2011; Takeichi, 2011) and integrin conformation (Moser et al, 2009; Shattil et al, 2010) have been extensively described elsewhere. Here, we will instead review the emerging evidence of how cell adhesion and migration critically depends on cadherin and integrin traffic.

3.1 Role of E-cadherin traffic in adherens junction maintenance and remodeling

Normal epithelial tissues are hold together by adherens junctions (AJs), *i.e.* cell-to-cell adhesion sites that originate after the dimerization *in trans* of epithelial (E)-cadherin molecules (Gumbiner, 2005; Niessen et al, 2011; Takeichi, 2011). E-cadherin-dependent assembly of AJs is required to assemble and maintain the apico-basal polarity of functional epithelia (Rodriguez-Boulan et al, 2005).

In *Drosophila* and in mammals, the maintenance of both AJs and epithelial polarity depends on a complex formed by the small GTPase Cdc42 and its partner PAR6 that binds aPKC (Goldstein & Macara, 2007; Iden & Collard, 2008; McCaffrey & Macara, 2009).

Interestingly, Cdc42, PAR6, and aPKC are required for the activation of a signaling pathway responsible for the dynamin-driven pinch-off of vesicles during E-cadherin endocytosis from *Drosophila* AJs (Baum & Georgiou, 2011; Georgiou et al, 2008; Leibfried et al, 2008) and a genome wide siRNA screen in *C. elegans* also identified Cdc42, PAR6, and aPKC as key regulators of endocytosis (Balklava et al, 2007). In addition, pharmacological inhibition of dynamin coupled to two-photon FRAP microscopy demonstrated that in mammalian cells E-cadherin engaged at mature stationary AJs turns over by endocytosis and not by free diffusion through the PM (de Beco et al, 2009). *Drosophila* Cdc42 interacting protein 4 (Cip4), aka transducer of Cdc42-dependent actin assembly 1 (TOCA-1) in mammals, displays both an FCH-Bin–Amphiphysin–Rvs (F-BAR) and a Src homology 3 (SH3) domains that respectively bind curved membranes and dynamin (Fricke et al, 2009). Of note, Cip4 knockdown causes AJ and E-cadherin endocytosis defects identical to those caused by the lack of components of the Cdc42/PAR6/aPKC apical complex (Baum & Georgiou, 2011; Leibfried et al, 2008).

Once internalized, E-cadherin is first trafficked to Rab5 containing early endosomes and from there to a Rab11-positive recycling compartment (Emery & Knoblich, 2006; Harris & Tepass, 2010; Wirtz-Peitz & Zallen, 2009). Sec10 and Sec15 proteins then directly bind and interconnect the β-catenin-bound endosomal E-cadherin to the exocyst complex located at the PM, hence favoring the recycling of the adhesion receptor (Langevin et al, 2005).

There is now a mounting consensus that the maintenance of stable AJs requires the continuous and local traffic of E-cadherin back and forth from the PM (Baum & Georgiou,

2011; Emery & Knoblich, 2006; Harris & Tepass, 2010; Wirtz-Peitz & Zallen, 2009). Therefore, endless cycles of polarized endocytosis and recycling of E-cadherin are responsible for the existence in space and time of AJs that warrant an efficient intercellular adhesion in stable epithelia.

This would suggest that in living cells, because of the intrinsic physical and biochemical properties of its molecular components, what appears as a stable adhesion site is nothing but an almost continuous and swift spatio-temporal succession of short-lived adhesive events. In this framework, endocytosis could be required either to remove and then replenish *via* recycling the adhesive material or to provide a substantial fraction of the force required to maintain adhesion.

Moreover, the incessant turnover of E-cadherins would allow cells to rapidly adapt the structure of their AJs in response to extracellular signals during tissue reshaping. Indeed, during embryonic development, cancer cell metastatization, and tissue fibrosis epithelial cells activate the epithelial-mesenchymal transition (EMT) program during which they lose their AJs and become motile (Kalluri & Weinberg, 2009; Thiery & Sleeman, 2006). For example, in epithelial cells, hepatocyte growth factor (HGF), acting through the MET tyrosine kinase receptor, activates H-Ras that, by stimulating the Rab5 guanosine exchange factor Ras and Rab interactor 2 (RIN2), induces E-cadherin endocytosis (Kimura et al, 2006). In addition, HGF signals *via* Src and generates a tyrosine phosphosite on E-cadherin where the E3-ubiquitin ligase Hakai docks to trigger the ubiquitination and lysosomal degradation of E-cadherin (Fujita et al, 2002; Palacios et al, 2005).

3.2 Combined regulation of integrin function by conformation and traffic

Integrin heterodimers can switch from low (inactive) to high affinity (active) conformation for their ECM ligands (Hynes, 2002). Conformational activation of integrins can be due to the interaction of their cytoplasmic tails with different proteins acting as positive (e.g. talin and kindlin) (Moser et al, 2009; Shattil et al, 2010) or negative (e.g. mammary-derived growth inhibitor, MDGI) modulators (Nevo et al, 2010). Due to their ability to mechanosense the surrounding ECM environment and mediate the interactions that support cell adhesion and migration (Parsons et al, 2011), active integrins are key regulators of several important adhesion dependent functions, such as assembly and morphogenetic movements of tissues and organs or migration of isolated/clustered cells through the body (e.g. immune or cancer cells). For example, the remodeling of immature vascular networks that occurs during embryonic, but not tumor angiogenesis, depends on the ability of endothelial cells (ECs) to instantaneously mechanotransduce variations in fluid shear stress (Hahn & Schwartz, 2009).

Integrin traffic is increasingly recognized as a key determinant in the dynamic control of cell adhesion to the ECM (Caswell et al, 2009; Pellinen et al, 2006). Integrins can be internalized in a clathrin-dependent as well as in a clathrin-independent way. For example, α5β1 integrin, the major fibronectin (FN) receptor, can be endocytosed into clathrin-coated vesicles (CCVs) (Pellinen et al, 2008) or by a caveolin-mediated pathway (Shi & Sottile, 2008). It was initially hypothesized that in migrating cells integrins can be preferentially endocytosed in ECM-adhesion sites located at the trailing edge and then recycled back *en masse* toward the leading edge (Bretscher, 1989). More recently, such a theoretical long range model has been challenged by an experimental short range model that showed how in cells migrating in 3D matrices a spatially restricted subpopulation of α5β1 integrin is instead internalized from the PM of ECM-adhesions located at the cell front and quickly recycled back to the same or proximal adhesive structures (Caswell & Norman, 2008; Caswell et al, 2007; Caswell et al, 2009). The Rab11 subfamily member Rab25, which resides in a vesicular compartment located in close proximity of the tips of invading pseudopods, physically interacts with the β1 subunit of the internalized integrins and promotes tumor cell invasion, likely by favoring the localized recycling of α5β1 integrin (Caswell & Norman, 2008; Caswell et al, 2007; Caswell et al, 2009). Another key regulator of integrin traffic in motile cells is the Rab11 effector Rab-coupling protein (RCP), which binds with β3 integrin and, when αvβ3 integrin is inhibited, switches to the cytodomain of β1 integrin, connecting α5β1 integrin with Rab11 and thus favoring its recycling to the PM (Caswell et al, 2008; Caswell et al, 2009). Of note, RCP also associates with EGFR and, upon αvβ3 inhibition, the recycling to the PM of endocytosed EGFR is enhanced in coordination with that of α5β1, finally resulting in an increased EGFR auto-phosphorylation and downstream activation of AKT (Caswell et al, 2008; Caswell et al, 2009).

In the last couple of years, the new concept that endocytosis of active and inactive integrins could be mediated by different sorting machineries started emerging. Neuropilin 1 (Nrp1) is a transmembrane protein, initially identified in neurons, that is also expressed in ECs, where it works as a co-receptor for both pro- and anti-angiogenic factors, such as vascular endothelial growth factor (VEGF)-A165 and semaphorin 3A (SEMA3A) respectively (Bussolino et al, 2006; Neufeld & Kessler, 2008; Serini & Bussolino, 2004). The very C-terminal SEA motif of Nrp1 cytodomain binds the endocytic adaptor GAIP interacting protein C terminus 1 (GIPC1)/synectin (Cai & Reed, 1999) that can also bind to the motor Myosin VI (Myo6) (Reed et al, 2005). Nrp1, *via* its cytodomain, controls EC adhesion to FN in a way that does not depend on its function as co-receptor for either VEGF-A or SEMA3A, but rather on its ability to promote the GIPC1/synectin- and Myo6-dependent endocytosis of the active, but not inactive conformation of α5β1 integrin from ECM adhesions (Valdembri et al, 2009). Remarkably, Nrp1 silencing does not affect the ratio between active and inactive α5β1 integrin, indicating that not only the conformational switch of integrins, but also the regulation of active integrin traffic and distribution constitutes an equally crucial parameter in the control of EC adhesion to the ECM (Valdembri et al, 2009). It has hence been proposed a model in which, upon FN binding, active α5β1 integrin associates with Nrp1 at the PM. GIPC1/synectin and Myo6 then favor the rapid internalization of the active α5β1/Nrp1 complex into Rab5-positive early endosomes, from which (active) α5β1 is then recycled back to the PM, likely in newly forming ECM-adhesion sites.

The described endo-exocytic cycle of active integrins back and forth from ECM adhesions is remarkably similar to the traffic dependent E-cadherin dynamics observed in AJs of epithelial cells (*see above*). It is therefore tempting to speculate that also an ECM adhesion site could result from a rapid sequence of localized and exceptionally brief adhesive events, during which traffic could be crucial either to endocytose and then immediately recycle active integrins or to generate the force that has to be applied on ECM-bound active integrins to allow cell adhesion. Likely because GIPC1/synectin also binds the C-terminal SDA motif of the α5 integrin subunit cytodomain (Tani & Mercurio, 2001), while in ECs Nrp1 and Myo6 are specifically dedicated to the endocytosis of active α5β1 integrin, GIPC1 controls inactive α5β1 internalization as well (Valdembri et al, 2009). The different molecular composition of the machineries that control active *vs.* inactive integrin traffic could imply that higher amounts of endocytic proteins are required to effectively internalize ECM-bound integrins. Accordingly, the force-generating retrograde motor Myo6 (Spudich & Sivaramakrishnan, 2010) participates to endocytosis, transport of endosomal vesicles along F-actin (Hasson, 2003), and active integrin internalization (Valdembri et al, 2009) as well.

Clathrin coats exist either as classical curved clathrin-coated pits or as flat clathrin-coated plaques that depend on the presence of the actin cytoskeleton and occur only at ECM-adherent surfaces, indicating that integrin-mediated adhesion of cells to the ECM likely control the organization of the different clathrin-based endocytic structures (Kirchhausen, 2009; Saffarian et al, 2009). The potential role of cell-to-ECM adhesion in regulating clathrin-mediated endocytosis is further supported by the recent experimental observation that the closer clathrin-coated pits are to integrin-containing adhesion sites the slower are their internalization dynamics (Batchelder & Yarar, 2010). It is indeed possible that the binding of integrins to the ECM could give rise to forces that counteract the pulling forces required to deform and curve the PM to finally allow clathrin-based internalization. Such a hypothesis could also account for the requirement of different molecular complexes for active *vs.* inactive integrin internalization.

To date, only few proteins have been selectively involved in inactive, but not active, integrin traffic and the degree of specificity for the bent/inactive integrin conformation is still matter of debate. A prominent example is represented by the endocytic adaptor protein disabled 2 (DAB2), that is able to directly bind the cytodomain of integrin β subunits (Calderwood et al, 2003), and was recently found to selectively promote the internalization of inactive β1 integrins (Teckchandani et al, 2009). However, during ECM-adhesion disassembly experiments, Chao and Kunz, by incubating living cells with the anti-active β1 integrin monoclonal antibody 12G10, found that active β1 integrins could be endocytosed in a DAB2-dependent manner as well (Chao & Kunz, 2009). However, since incubation of living cells with function activating or blocking antibodies represents a significant bias in the study of integrin activation physiology, further work is needed to better characterize the role of DAB2.

4. Unconventional function of membrane trafficking proteins in mitosis

Recent findings have shown that clathrin-mediated endocytosis is active throughout mitosis, while the recycling pathway slows down from prophase until the completion of anaphase (Boucrot & Kirchhausen, 2007). These data have been generated by monitoring the changes in plasma membrane area during mitosis in living cells with a membrane-impermeant dye that becomes fluorescent upon binding to the outer leaflet of the plasma membrane. Since the dye cannot flip to the inner leaflet, only endocytic vesicles generated by internalization of the plasma membrane can be visualized. At metaphase, these plasma membrane-derived vesicles are not delivered back at the surface resulting in a net decrease of the cell area. In turn, this translates in cell detachment and round up from prophase to anaphase. The recycling pathway recovers at telophase when the forming daughter cells start to spread again (Boucrot & Kirchhausen, 2007).

Interestingly, mitotic phosphorylation of Rab4, a GTPase required for recycling from early endosomes to the plasma membrane (van der Sluijs et al, 1992), prevents its localization at endosomal membranes (Ayad et al, 1997). During mitosis, phosphorylated Rab4 is in the cytosol complexed with the peptidyl-prolyl isomerase Pin1 and it is no longer able to recruit downstream effectors on endosomes (Gerez et al, 2000). Thus an appealing possibility is that Rab4 phosphorylation might participates in the inhibition of the recycling pathway measured by Boucrot and Kirchhausen during the early steps of mitosis.

Of note, fusion of early endosomes in mitosis is blocked via cdc2-dependent phosphorylation events (Tuomikoski et al, 1989). This might represent an additional mechanism to inhibit vesicles recycling at the plasma membrane by altering the homeostasis of the endosomal compartment and affecting the generation of exocytic vesicles. Inhibition of homotypic fusion of early endosomes at mitosis is also caused by decreased residence time of the early endosome-tethering molecule EEA1 on endosomal membranes (Bergeland et al, 2008). It would be interesting to define how the acceleration of the EEA1 cycle between cytosol and membranes is achieved in mitotic cells.

Endocytic/trafficking proteins are also emerging as important factors required for the proper execution of cell division. Beside the involvement of trafficking molecules in membrane delivery to the cleavage furrow at cytokinesis [for recent reviews see (McKay & Burgess, 2011; Montagnac et al, 2008)], some of these proteins also display specific functions in mitosis. Here we will review knowledge rising on this issue.

One of the best-characterized endocytic molecules showing a distinct role in mitosis is the clathrin heavy chain. The clathrin complex is organized in a triskelion made of three heavy chains each with an associated light chain (ter Haar et al, 1998). At metaphase, clathrin also localizes to kinetochore fibers (spindle microtubules connecting kinetochores to spindle poles) of the spindle apparatus (Royle et al, 2005). Here it stabilizes spindle microtubules aiding congression of chromosomes on the metaphase plate. Depletion of clathrin heavy chain by RNA interference causes failure in the correct attachment of chromosomes to kinetochore fibers resulting in misaligned chromosomes and in persistent activation of the mitotic checkpoint thus prolonging mitosis (Royle et al, 2005). More recently, some advances in understanding clathrin function at the spindle have been made. Clathrin heavy chain has been found to bind to TACC3, phosphorylated on serine 558 by Aurora A, and to recruit it to the spindle. In turn, TACC3 is responsible for localization of ch-TOG, a protein that promotes microtubule assembly and spindle stability, to spindles (Lin et al, 2010). In agreement, functional ablation of clathrin heavy chain causes loss of ch-TOG from spindles and destabilizes kinetochore fibers affecting chromosome congression. Based on electron microscopy data, it has been proposed that TACC3/ch-TOG/clathrin heavy chain complex works as an inter-microtubules bridge that stabilizes kinetochore fibers by physical crosslinking reducing the rate of microtubule catastrophe (Booth et al, 2011).

Another important endocytic player is epsin, an adaptor molecule that binds and deforms membranes driving curvature of clathrin-coated pits (Ford et al, 2002). At mitosis, epsin participates in spindle morphogenesis indirectly through its ability to regulate mitotic membrane organization (Liu & Zheng, 2009). In cells depleted of epsin, by RNAi-mediated silencing, the membrane network that uniformly surrounds the chromosomes is distorted with uneven membrane distribution frequently showing layers of membrane whorls. This, in turn, alters spindle morphology resulting in splayed spindle poles and multipolar spindles (Liu & Zheng, 2009).

Huntingtin-interacting protein 1-related (HIP1r) functions in clathrin-mediated endocytosis and links endocytosis to the actin cytoskeleton (Engqvist-Goldstein et al, 2001). HIP1r also localizes to the spindle and its depletion by RNA interference causes chromosome misalignment and activation of the spindle checkpoint (Park, 2011).

In addition, is worth to mention that Rab6A', a GTPase that regulates trafficking between the Golgi and post-Golgi membrane compartments, is also required for spindle stability (Mallard et al, 2002). At mitosis, depletion of Rab6A' arrests cells at metaphase (Miserey-Lenkei et al, 2006). Aligned chromosomes, in Rab6A'-depleted cells, show increased amount of p150^{Glued}, a subunit of the dynein/dynactin complex, and of Mad2 at kinetochores. p150^{Glued} takes part in the release of the checkpoint protein Mad2 from kinetochores thus switching off the mitotic checkpoint, an operation required for the transition of cells from metaphase to anaphase. The inability of Rab6A'-silenced cells to progress mitosis might be the consequence of defective p150^{Glued}-mediated transport of Mad2 out of kinetochores resulting in the failure to turn off the checkpoint. Thus Rab6A', by regulating the dynamics of the dynein/dynactin complex at the kinetochores, cooperates to the inactivation of the Mad2-spindle checkpoint.

Some trafficking proteins have also been found to act at the centrosome which is part of the mitotic machinery that ensure proper chromosome segregation. One of these proteins is dynamin. In addition to its membrane localization, dynamin is at the centrosome throughout the cell cycle and localizes to the spindle midzone and to the cleavage furrow during cytokinesis (Thompson et al, 2004; Thompson et al, 2002). Depletion of dynamin by RNA interference causes centrosome separation indicating a role for dynamin in the maintenance of centrosome cohesion (Thompson et al, 2004).

The Autosomal Recessive Hypercholesterolemia (ARH) protein provides another example. ARH is a cargo-specific adaptor that functions in clathrin-mediated endocytosis of receptors of the LDLR family (Shin et al, 2001). It displays a complex subcellular localization being on endocytic vesicles and at the centrosome in interphase. During mitosis, it also localizes to kinetochores, spindle poles and midbody. The suggested function for ARH is in centrosome assembly, as ARH-/- embryonic fibroblasts show smaller centrosomes. Since ARH binds to the dynein motor protein it could cooperate in the transport of components to the centrosome. Of note, functional ablation of ARH also strongly delays cytokinesis (Lehtonen et al, 2008).

In addition, the Rab-GAP protein RN-tre is phosphorylated at mitosis and dephosphorylated by the dual-specificity phosphatase Cdc14A (Lanzetti et, 2007). Cdc14A controls key mitotic events and it is also implicated in centrosome function in human cells (Mailand et al, 2002). Mitotic phosphorylation on RN-tre modulates its GAP activity establishing an additional link between endocytosis and the machinery working at mitosis (Lanzetti et al, 2007).

Finally, Rab5 is required for nuclear membrane breakdown at mitosis, as depletion of this GTPase in *C. elegans* delays nuclear envelope disassembly and the release of nuclear envelope and lamina components (Audhya et al, 2007). The activity of Rab5 in nuclear envelope disassembly appears to result from its involvement in structuring the ER, of which the nuclear membrane represents a functional district (Audhya et al, 2007). Rab5 participates to mitotic ER clustering and to disassembly of the nuclear envelope also in mammalian cells

(Audhya et al, 2007; Serio et al, 2011). Although the molecular mechanisms are unclear, it has been proposed that Rab5 might act *in trans*, while localized on endosomes, by interacting with effectors on the ER membrane to induce their homotypic fusion. Furthermore, recent findings have shown that, at mitosis, Rab5 is required for proper chromosome alignment both in human cells and in the *Drosophila* system (Serio et al, 2011; Capalbo et al, 2011).

One relevant question is whether the modality of function for these molecules at mitosis is distinct from their role in membrane trafficking during interphase. A couple of observations argue in favor of this possibility.

First, some of these proteins display a subcellular localization in mitosis distinct from trafficking membranes. For instance, the globular N-terminal domain of the clathrin heavy chain is responsible for clathrin localization to kinetochore fibers and a number of assays, including labeling of intracellular membranes, electron microscopy analysis and mass spectrometry, revealed that it does not coat membranes at the spindle but it rather bind to microtubules or to microtubules-associated proteins (Royle et al, 2005). Localization of dynamin to centrosome, which is a non-membranous organelle, is dynamic and occurs through its middle domain in a microtubules-independent manner (Thompson et al, 2004).

Second, these molecules appear to interact with binding partners distinct from those involved in vesicular trafficking pathways and such interactions seem to be relevant during cell division. Indeed, clathrin has been reported to bind and stabilize spindle microtubules (Royle et al, 2005) while dynamin interacts with the centrosomal protein γ-tubulin (Thompson et al, 2004). In addition, the β2-adaptin subunit of the clathrin adaptor AP2 associates, at least in vitro, with a component of the mitotic spindle checkpoint, the kinase BubR1. Although the physiological meaning of this interaction is unknown, it might provide a link between endocytic proteins and the mitotic checkpoint machinery (Cayrol et al, 2002). Of note, two accessory components of clathrin coated pits, epsin and Eps15 are phosphorylated at mitosis and such modification reduces their binding to the α-adaptin subunit of AP2 (Chen et al, 1999). Among the different hypothesis that can be envisioned, one appealing possibility is that mitotic phosphorylation of epsin and Eps15 alters their binding capabilities promoting formation of protein complexes working at mitosis and involving partners distinct from AP2. Importantly, epsin has been shown to facilitate spindle organization independently from its endocytic function by using cell-free spindle assembly assays. In these assays, *Xenopus* egg extracts, lacking the membrane cortex, have been depleted of epsin and reconstituted with purified epsin or with epsin lacking the membrane-bending domain. Only full length epsin was able to rescue the spindle defects demonstrating that the membrane curvature activity of epsin is required for the establishment of spindle morphology independently from endocytosis (Liu & Zheng, 2009). This study nicely extends the concept that endocytic proteins have a role in mitosis distinct from the one exerted during interphase.

Given that endocytosis is active throughout the cell cycle and that, at mitosis, some endocytic proteins are also involved in pathways different from internalization, these molecules might play two distinct functions simultaneously thus coordinating membrane traffic with the execution of mitotic events.

Genetic instability is a driving force in tumourigenesis and it is prompted by alteration in centrosome function and in spindle assembly (Lengauer et al, 1998; Lingle et al, 2002; Orr-Weaver & Weinberg, 1998). Since endocytic proteins participate in the regulation of mitotic events, this could represent a novel, previously unrecognized, link between endocytosis and cancer.

5. Conclusions

Endo and exocytosis are well-known mechanisms that regulate signal transduction and the execution of different cellular programs. The number of players, their crosstalk and the networks that they generate is continuously growing adding novel layers of complexity and definition to the current picture.

Intriguingly, increasing evidence shows that signalling itself can control and modulate endocytic pathways (Collinet et al, 2010). Activated receptors elicit a variety of signals that directly or indirectly control endocytosis by several means including phospho-modification of downstream effectors involved in endocytosis, control of protein synthesis and also modulation of actin cytoskeleton dynamics, a process that aids clathrin-mediated endocytosis. This is an emerging view in the trafficking field that will certainly disclose new areas of investigation.

6. Acknowledgments

Work in Letizia Lanzetti's lab is supported by grants from: Associazione Italiana per la Ricerca sul Cancro (START UP program), the Association for International Cancer Research, the Fondazione Piemontese per la Ricerca sul Cancro - ONLUS – Intramural Grant 2010 and Regione Piemonte - Ricerca Sanitaria Finalizzata 2008, and 2009. Work in the Lab of Guido Serini is supported by grants from: Telethon Italy; Compagnia di San Paolo - Neuroscience Program Multicentre Projects; Associazione Italiana per la Ricerca sul Cancro; Fondazione Piemontese per la Ricerca sul Cancro - ONLUS – Intramural Grant 2010; Ministero della Salute - Programma Ricerca Oncologica 2006; Regione Piemonte - Ricerca Sanitaria Finalizzata 2006, 2008, and 2009, Ricerca industriale e sviluppo precompetitivo 2006: grant SPLASERBA; Associazione Augusto per la Vita.

7. References

Abedin M, King N (2010) Diverse evolutionary paths to cell adhesion. *Trends Cell Biol* 20: 734-742

Assaker G, Ramel D, Wculek SK, González-Gaitán M, Emery G (2010) Spatial restriction of receptor tyrosine kinase activity through a polarized endocytic cycle controls border cell migration. *Proc Natl Acad Sci U S A* 107: 22558-22563

Audhya A, Desai A, Oegema K (2007) A role for Rab5 in structuring the endoplasmic reticulum. *J Cell Biol* 178: 43-56

Ayad N, Hull M, Mellman I (1997) Mitotic phosphorylation of rab4 prevents binding to a specific receptor on endosome membranes. *EMBO J* 16: 4497-4507

Balasubramanian N, Scott DW, Castle JD, Casanova JE, Schwartz MA (2007) Arf6 and microtubules in adhesion-dependent trafficking of lipid rafts. *Nat Cell Biol* 9: 1381-1391

Balklava Z, Pant S, Fares H, Grant BD (2007) Genome-wide analysis identifies a general requirement for polarity proteins in endocytic traffic. *Nat Cell Biol* 9: 1066-1073

Batchelder EM, Yarar D (2010) Differential requirements for clathrin-dependent endocytosis at sites of cell-substrate adhesion. *Mol Biol Cell* 21: 3070-3079

Baum B, Georgiou M (2011) Dynamics of adherens junctions in epithelial establishment, maintenance, and remodeling. *J Cell Biol* 192: 907-917

Belleudi F, Leone L, Nobili V, Raffa S, Francescangeli F, Maggio M, Morrone S, Marchese C, Torrisi MR (2007) Keratinocyte growth factor receptor ligands target the receptor to different intracellular pathways. *Traffic* 8: 1854-1872

Benovic JL, Pike LJ, Cerione RA, Staniszewski C, Yoshimasa T, Codina J, Caron MG, Lefkowitz RJ (1985) Phosphorylation of the mammalian beta-adrenergic receptor by cyclic AMP-dependent protein kinase. Regulation of the rate of receptor phosphorylation and dephosphorylation by agonist occupancy and effects on coupling of the receptor to the stimulatory guanine nucleotide regulatory protein. *J Biol Chem* 260: 7094-7101

Benovic JL, Strasser RH, Caron MG, Lefkowitz RJ (1986) Beta-adrenergic receptor kinase: identification of a novel protein kinase that phosphorylates the agonist-occupied form of the receptor. *Proc Natl Acad Sci U S A* 83: 2797-2801

Bergeland T, Haugen L, Landsverk OJ, Stenmark H, Bakke O (2008) Cell-cycle-dependent binding kinetics for the early endosomal tethering factor EEA1. *EMBO Rep* 9: 171-178

Booth DG, Hood FE, Prior IA, Royle SJ (2011) A TACC3/ch-TOG/clathrin complex stabilises kinetochore fibres by inter-microtubule bridging. *Embo J* 30(5):906-919

Boucrot E, Kirchhausen T (2007) Endosomal recycling controls plasma membrane area during mitosis. *Proc Natl Acad Sci U S A* 104: 7939-7944

Bretscher MS (1989) Endocytosis and recycling of the fibronectin receptor in CHO cells. *EMBO J* 8: 1341-1348

Bryant DM, Datta A, Rodríguez-Fraticelli AE, Peränen J, Martín-Belmonte F, Mostov KE (2010) A molecular network for de novo generation of the apical surface and lumen. *Nat Cell Biol* 12: 1035-1045

Burke P, Schooler K, Wiley HS (2001) Regulation of epidermal growth factor receptor signaling by endocytosis and intracellular trafficking. *Mol Biol Cell* 12: 1897-1910

Bussolino F, Valdembri D, Caccavari F, Serini G (2006) Semaphoring Vascular Morphogenesis. *Endothelium* 13: 81-91

Cai H, Reed RR (1999) Cloning and characterization of neuropilin-1-interacting protein: a PSD-95/Dlg/ZO-1 domain-containing protein that interacts with the cytoplasmic domain of neuropilin-1. *J Neurosci* 19: 6519-6527

Calderwood DA, Fujioka Y, de Pereda JM, García-Alvarez B, Nakamoto T, Margolis B, McGlade CJ, Liddington RC, Ginsberg MH (2003) Integrin β cytoplasmic domain interactions with phosphotyrosine-binding domains: A structural prototype for diversity in integrin signaling. *Proceedings of the National Academy of Sciences of the United States of America* 100: 2272-2277

Capalbo L, D'Avino PP, Archambault V, Glover DM. (2011) Rab5 GTPase controls chromosome alignment through Lamin disassembly and relocation of the NuMA-like protein Mud to the poles during mitosis. *Proc Natl Acad Sci U S A*. 108(42):17343-8.

Caswell P, Norman J (2008) Endocytic transport of integrins during cell migration and invasion. *Trends Cell Biol* 18: 257-263

Caswell PT, Chan M, Lindsay AJ, McCaffrey MW, Boettiger D, Norman JC (2008) Rab-coupling protein coordinates recycling of alpha5beta1 integrin and EGFR1 to promote cell migration in 3D microenvironments. *J Cell Biol* 183: 143-155

Caswell PT, Spence HJ, Parsons M, White DP, Clark K, Cheng KW, Mills GB, Humphries MJ, Messent AJ, Anderson KI, McCaffrey MW, Ozanne BW, Norman JC (2007) Rab25 associates with alpha5beta1 integrin to promote invasive migration in 3D microenvironments. *Dev Cell* 13: 496-510

Caswell PT, Vadrevu S, Norman JC (2009) Integrins: masters and slaves of endocytic transport. *Nat Rev Mol Cell Biol* 10: 843-853

Cayrol C, Cougoule C, Wright M (2002) The beta2-adaptin clathrin adaptor interacts with the mitotic checkpoint kinase BubR1. *Biochem Biophys Res Commun* 298: 720-730

Chao WT, Kunz J (2009) Focal adhesion disassembly requires clathrin-dependent endocytosis of integrins. *FEBS Lett* 583: 1337-1343

Chen H, Slepnev VI, Di Fiore PP, De Camilli P (1999) The interaction of epsin and Eps15 with the clathrin adaptor AP-2 is inhibited by mitotic phosphorylation and enhanced by stimulation-dependent dephosphorylation in nerve terminals. *J Biol Chem* 274: 3257-3260

Chen YG, Wang Z, Ma J, Zhang L, Lu Z (2007) Endofin, a FYVE domain protein, interacts with Smad4 and facilitates TGF-beta signaling. *J Biol Chem*

Classen AK, Anderson KI, Marois E, Eaton S (2005) Hexagonal packing of Drosophila wing epithelial cells by the planar cell polarity pathway. *Dev Cell* 9: 805-817

Collinet C, Stöter M, Bradshaw CR, Samusik N, Rink JC, Kenski D, Habermann B, Buchholz F, Henschel R, Mueller MS, Nagel WE, Fava E, Kalaidzidis Y, Zerial M (2010) Systems survey of endocytosis by multiparametric image analysis. *Nature* 464: 243-249

de Beco S, Gueudry C, Amblard Fo, Coscoy S (2009) Endocytosis is required for E-cadherin redistribution at mature adherens junctions. *Proceedings of the National Academy of Sciences* 106: 7010-7015

De Donatis A, Comito G, Buricchi F, Vinci MC, Parenti A, Caselli A, Camici G, Manao G, Ramponi G, Cirri P (2008) Proliferation versus migration in platelet-derived growth factor signaling: the key role of endocytosis. *J Biol Chem* 283: 19948-19956

Decker SJ (1990) Epidermal growth factor and transforming growth factor-alpha induce differential processing of the epidermal growth factor receptor. *Biochem Biophys Res Commun* 166: 615-621

DeWire SM, Ahn S, Lefkowitz RJ, Shenoy SK (2007) Beta-arrestins and cell signaling. *Annu Rev Physiol* 69: 483-510

Di Guglielmo GM, Baass PC, Ou WJ, Posner BI, Bergeron JJ (1994) Compartmentalization of SHC, GRB2 and mSOS, and hyperphosphorylation of Raf-1 by EGF but not insulin in liver parenchyma. *EMBO J* 13: 4269-4277

Di Guglielmo GM, Le Roy C, Goodfellow AF, Wrana JL (2003) Distinct endocytic pathways regulate TGF-beta receptor signalling and turnover. *Nat Cell Biol* 5: 410-421

Dikic I, Wakatsuki S, Walters KJ (2009) Ubiquitin-binding domains - from structures to functions. *Nat Rev Mol Cell Biol* 10: 659-671

Doherty GJ, McMahon HT (2009) Mechanisms of endocytosis. *Annu Rev Biochem* 78: 857-902

Ebner R, Derynck R (1991) Epidermal growth factor and transforming growth factor-alpha: differential intracellular routing and processing of ligand-receptor complexes. *Cell Regul* 2: 599-612

Emery G, Hutterer A, Berdnik D, Mayer B, Wirtz-Peitz F, Gaitan MG, Knoblich JA (2005) Asymmetric Rab 11 endosomes regulate delta recycling and specify cell fate in the Drosophila nervous system. *Cell* 122: 763-773

Emery G, Knoblich JA (2006) Endosome dynamics during development. *Curr Opin Cell Biol* 18: 407-415

Engqvist-Goldstein AE, Warren RA, Kessels MM, Keen JH, Heuser J, Drubin DG (2001) The actin-binding protein Hip1R associates with clathrin during early stages of endocytosis and promotes clathrin assembly in vitro. *J Cell Biol* 154: 1209-1223

Ford MG, Mills IG, Peter BJ, Vallis Y, Praefcke GJ, Evans PR, McMahon HT (2002) Curvature of clathrin-coated pits driven by epsin. *Nature* 419: 361-366

French AR, Tadaki DK, Niyogi SK, Lauffenburger DA (1995) Intracellular trafficking of epidermal growth factor family ligands is directly influenced by the pH sensitivity of the receptor/ligand interaction. *J Biol Chem* 270: 4334-4340

Fricke R, Gohl C, Dharmalingam E, Grevelhörster A, Zahedi B, Harden N, Kessels M, Qualmann B, Bogdan S (2009) Drosophila Cip4/Toca-1 integrates membrane trafficking and actin dynamics through WASP and SCAR/WAVE. *Curr Biol* 19: 1429-1437

Fujita Y, Krause G, Scheffner M, Zechner D, Leddy HEM, Behrens J, Sommer T, Birchmeier W (2002) Hakai, a c-Cbl-like protein, ubiquitinates and induces endocytosis of the E-cadherin complex. *Nat Cell Biol* 4: 222-231

Georgiou M, Marinari E, Burden J, Baum B (2008) Cdc42, Par6, and aPKC regulate Arp2/3-mediated endocytosis to control local adherens junction stability. *Curr Biol* 18: 1631-1638

Gerez L, Mohrmann K, van Raak M, Jongeneelen M, Zhou XZ, Lu KP, van Der Sluijs P (2000) Accumulation of rab4GTP in the cytoplasm and association with the peptidyl-prolyl isomerase pin1 during mitosis. *Mol Biol Cell* 11: 2201-2211

Goldstein B, Macara IG (2007) The PAR proteins: fundamental players in animal cell polarization. *Dev Cell* 13: 609-622

Grant BD, Donaldson JG (2009) Pathways and mechanisms of endocytic recycling. *Nat Rev Mol Cell Biol* 10: 597-608

Grimes ML, Zhou J, Beattie EC, Yuen EC, Hall DE, Valletta JS, Topp KS, LaVail JH, Bunnett NW, Mobley WC (1996) Endocytosis of activated TrkA: evidence that nerve growth factor induces formation of signaling endosomes. *J Neurosci* 16: 7950-7964

Gumbiner BM (2005) Regulation of cadherin-mediated adhesion in morphogenesis. *Nat Rev Mol Cell Biol* 6: 622-634

Hahn C, Schwartz MA (2009) Mechanotransduction in vascular physiology and atherogenesis. *Nat Rev Mol Cell Biol* 10: 53-62

Harris TJC, Tepass U (2010) Adherens junctions: from molecules to morphogenesis. *Nat Rev Mol Cell Biol* 11: 502-514

Hasson T (2003) Myosin VI: two distinct roles in endocytosis. *J Cell Sci* 116: 3453-3461

Haugh JM, Huang AC, Wiley HS, Wells A, Lauffenburger DA (1999) Internalized epidermal growth factor receptors participate in the activation of p21(ras) in fibroblasts. *J Biol Chem* 274: 34350-34360

Hayes S, Chawla A, Corvera S (2002) TGF beta receptor internalization into EEA1-enriched early endosomes: role in signaling to Smad2. *J Cell Biol* 158: 1239-1249

Hoekstra D, Tyteca D, van IJzendoorn SC (2004) The subapical compartment: a traffic center in membrane polarity development. *J Cell Sci* 117: 2183-2192

Hogg N, Patzak I, Willenbrock F (2011) The insider's guide to leukocyte integrin signalling and function. *Nat Rev Immunol* 11: 416-426

Howe CL, Valletta JS, Rusnak AS, Mobley WC (2001) NGF signaling from clathrin-coated vesicles: evidence that signaling endosomes serve as a platform for the Ras-MAPK pathway. *Neuron* 32: 801-814

Huber LA, Pimplikar S, Parton RG, Virta H, Zerial M, Simons K (1993) Rab8, a small GTPase involved in vesicular traffic between the TGN and the basolateral plasma membrane. *J Cell Biol* 123: 35-45

Hurley JH, Hanson PI (2010) Membrane budding and scission by the ESCRT machinery: it's all in the neck. *Nat Rev Mol Cell Biol* 11: 556-566

Hynes RO (2002) Integrins: bidirectional, allosteric signaling machines. *Cell* 110: 673-687.

Hynes RO (2007) Cell-matrix adhesion in vascular development. *J Thromb Haemost* 5 Suppl 1: 32-40

Iden S, Collard JG (2008) Crosstalk between small GTPases and polarity proteins in cell polarization. *Nat Rev Mol Cell Biol* 9: 846-859

Insall RH, Machesky LM (2009) Actin dynamics at the leading edge: from simple machinery to complex networks. *Dev Cell* 17: 310-322

Jekely G, Sung HH, Luque CM, Rorth P (2005) Regulators of endocytosis maintain localized receptor tyrosine kinase signaling in guided migration. *Dev Cell* 9: 197-207

Joffre C, Barrow R, Menard L, Calleja V, Hart IR, Kermorgant S (2011) A direct role for Met endocytosis in tumorigenesis. *Nat Cell Biol* 13(7): 827-837

Jones MC, Caswell PT, Norman JC (2006) Endocytic recycling pathways: emerging regulators of cell migration. *Curr Opin Cell Biol* 18: 549-557

Kalluri R, Weinberg RA (2009) The basics of epithelial-mesenchymal transition. *The Journal of Clinical Investigation* 119: 1420-1428

Kazazic M, Roepstorff K, Johannessen LE, Pedersen NM, van Deurs B, Stang E, Madshus IH (2006) EGF-induced activation of the EGF receptor does not trigger mobilization of caveolae. *Traffic* 7: 1518-1527

Kelly E, Bailey CP, Henderson G (2008) Agonist-selective mechanisms of GPCR desensitization. *Br J Pharmacol* 153 Suppl 1: S379-388

Kimura T, Sakisaka T, Baba T, Yamada T, Takai Y (2006) Involvement of the Ras-Ras-activated Rab5 Guanine Nucleotide Exchange Factor RIN2-Rab5 Pathway in the Hepatocyte Growth Factor-induced Endocytosis of E-cadherin. *Journal of Biological Chemistry* 281: 10598-10609

Kirchhausen T (2009) Imaging endocytic clathrin structures in living cells. *Trends Cell Biol* 19: 596-605

Lai WH, Cameron PH, Doherty JJ, 2nd, Posner BI, Bergeron JJ (1989) Ligand-mediated autophosphorylation activity of the epidermal growth factor receptor during internalization. *J Cell Biol* 109: 2751-2760

Lakadamyali M, Rust MJ, Zhuang X (2006) Ligands for clathrin-mediated endocytosis are differentially sorted into distinct populations of early endosomes. *Cell* 124: 997-1009

Langevin J, Morgan MJ, Rossé C, Racine V, Sibarita J-B, Aresta S, Murthy M, Schwarz T, Camonis J, Bellaïche Y (2005) Drosophila Exocyst Components Sec5, Sec6, and Sec15 Regulate DE-Cadherin Trafficking from Recycling Endosomes to the Plasma Membrane. *Developmental cell* 9: 365-376

Lanzetti L, Di Fiore PP (2008) Endocytosis and cancer: an 'insider' network with dangerous liaisons. *Traffic* 9: 2011-2021

Lanzetti L, Margaria V, Melander F, Virgili L, Lee MH, Bartek J, Jensen S (2007) Regulation of the Rab5 GTPase-activating protein RN-tre by the dual specificity phosphatase Cdc14A in human cells. *J Biol Chem* 282: 15258-15270

Lanzetti L, Palamidessi A, Areces L, Scita G, Di Fiore PP (2004) Rab5 is a signalling GTPase involved in actin remodelling by receptor tyrosine kinases. *Nature* 429: 309-314

Lanzetti L, Rybin V, Malabarba MG, Christoforidis S, Scita G, Zerial M, Di Fiore PP (2000) The Eps8 protein coordinates EGF receptor signalling through Rac and trafficking through Rab5. *Nature* 408: 374-377.

Le Roy C, Wrana JL (2005) Clathrin- and non-clathrin-mediated endocytic regulation of cell signalling. *Nat Rev Mol Cell Biol* 6: 112-126

Lehtonen S, Shah M, Nielsen R, Iino N, Ryan JJ, Zhou H, Farquhar MG (2008) The endocytic adaptor protein ARH associates with motor and centrosomal proteins and is involved in centrosome assembly and cytokinesis. *Mol Biol Cell* 19: 2949-2961

Leibfried A, Fricke R, Morgan MJ, Bogdan S, Bellaiche Y (2008) Drosophila Cip4 and WASp define a branch of the Cdc42-Par6-aPKC pathway regulating E-cadherin endocytosis. *Curr Biol* 18: 1639-1648

Lengauer C, Kinzler KW, Vogelstein B (1998) Genetic instabilities in human cancers. *Nature* 396: 643-649

Levayer R, Pelissier-Monier A, Lecuit T (2011) Spatial regulation of Dia and Myosin-II by RhoGEF2 controls initiation of E-cadherin endocytosis during epithelial morphogenesis. *Nat Cell Biol* 13: 529-540

Lin CH, Hu CK, Shih HM (2010) Clathrin heavy chain mediates TACC3 targeting to mitotic spindles to ensure spindle stability. *J Cell Biol* 189 (7):1097-1105

Lingle WL, Barrett SL, Negron VC, D'Assoro AB, Boeneman K, Liu W, Whitehead CM, Reynolds C, Salisbury JL (2002) Centrosome amplification drives chromosomal instability in breast tumor development. *Proc Natl Acad Sci U S A* 99: 1978-1983

Liu Z, Zheng Y (2009) A requirement for epsin in mitotic membrane and spindle organization. *J Cell Biol* 186: 473-480

Loerke D, Mettlen M, Yarar D, Jaqaman K, Jaqaman H, Danuser G, Schmid SL (2009) Cargo and dynamin regulate clathrin-coated pit maturation. *PLoS Biol* 7: e57

Longva KE, Blystad FD, Stang E, Larsen AM, Johannessen LE, Madshus IH (2002) Ubiquitination and proteasomal activity is required for transport of the EGF receptor to inner membranes of multivesicular bodies. *J Cell Biol* 156: 843-854

Lu H, Bilder D (2005) Endocytic control of epithelial polarity and proliferation in Drosophila. *Nat Cell Biol* 7: 1232-1239

Mailand N, Lukas C, Kaiser BK, Jackson PK, Bartek J, Lukas J (2002) Deregulated human Cdc14A phosphatase disrupts centrosome separation and chromosome segregation. *Nat Cell Biol* 4: 317-322

Mallard F, Tang BL, Galli T, Tenza D, Saint-Pol A, Yue X, Antony C, Hong W, Goud B, Johannes L (2002) Early/recycling endosomes-to-TGN transport involves two SNARE complexes and a Rab6 isoform. *J Cell Biol* 156: 653-664

Marchese A, Paing MM, Temple BR, Trejo J (2008) G protein-coupled receptor sorting to endosomes and lysosomes. *Annu Rev Pharmacol Toxicol* 48: 601-629

Martins AS, Ordonez JL, Amaral AT, Prins F, Floris G, Debiec-Rychter M, Hogendoorn PC, de Alava E (2011) IGF1R Signaling in Ewing Sarcoma Is Shaped by Clathrin-/Caveolin-Dependent Endocytosis. *PLoS One* 6: e19846

McCaffrey LM, Macara IG (2009) Widely conserved signaling pathways in the establishment of cell polarity. *Cold Spring Harb Perspect Biol* 1: a001370

McDonald JA, Pinheiro EM, Kadlec L, Schupbach T, Montell DJ (2006) Multiple EGFR ligands participate in guiding migrating border cells. *Dev Biol* 296: 94-103

McDonald JA, Pinheiro EM, Montell DJ (2003) PVF1, a PDGF/VEGF homolog, is sufficient to guide border cells and interacts genetically with Taiman. *Development* 130: 3469-3478

McKay HF, Burgess DR (2011) 'Life is a highway': membrane trafficking during cytokinesis. *Traffic* 12: 247-251

Mettlen M, Pucadyil T, Ramachandran R, Schmid SL (2009) Dissecting dynamin's role in clathrin-mediated endocytosis. *Biochem Soc Trans* 37: 1022-1026

Miaczynska M, Pelkmans L, Zerial M (2004) Not just a sink: endosomes in control of signal transduction. *Curr Opin Cell Biol* 16: 400-406

Micheau O, Tschopp J (2003) Induction of TNF receptor I-mediated apoptosis via two sequential signaling complexes. *Cell* 114: 181-190

Miserey-Lenkei S, Couedel-Courteille A, Del Nery E, Bardin S, Piel M, Racine V, Sibarita JB, Perez F, Bornens M, Goud B (2006) A role for the Rab6A' GTPase in the inactivation of the Mad2-spindle checkpoint. *Embo J* 25: 278-289

Montagnac G, Echard A, Chavrier P (2008) Endocytic traffic in animal cell cytokinesis. *Curr Opin Cell Biol* 20: 454-461

Montell DJ (2003) Border-cell migration: the race is on. *Nat Rev Mol Cell Biol* 4: 13-24

Moser M, Legate KR, Zent R, Fassler R (2009) The tail of integrins, talin, and kindlins. *Science* 324: 895-899

Mosesson Y, Mills GB, Yarden Y (2008) Derailed endocytosis: an emerging feature of cancer. *Nat Rev Cancer* 8: 835-850

Mottola G, Classen AK, González-Gaitán M, Eaton S, Zerial M (2010) A novel function for the Rab5 effector Rabenosyn-5 in planar cell polarity. *Development* 137: 2353-2364

Nachury MV, Loktev AV, Zhang Q, Westlake CJ, Peränen J, Merdes A, Slusarski DC, Scheller RH, Bazan JF, Sheffield VC, Jackson PK (2007) A core complex of BBS proteins cooperates with the GTPase Rab8 to promote ciliary membrane biogenesis. *Cell* 129: 1201-1213

Nada S, Hondo A, Kasai A, Koike M, Saito K, Uchiyama Y, Okada M (2009) The novel lipid raft adaptor p18 controls endosome dynamics by anchoring the MEK-ERK pathway to late endosomes. *Embo J* 28: 477-489

Nelson WJ (2009) Remodeling epithelial cell organization: transitions between front-rear and apical-basal polarity. *Cold Spring Harb Perspect Biol* 1: a000513

Neufeld G, Kessler O (2008) The semaphorins: versatile regulators of tumour progression and tumour angiogenesis. *Nature Reviews Cancer* 8: 632-645

Nevo J, Mai A, Tuomi S, Pellinen T, Pentikainen OT, Heikkila P, Lundin J, Joensuu H, Bono P, Ivaska J (2010) Mammary-derived growth inhibitor (MDGI) interacts with integrin alpha-subunits and suppresses integrin activity and invasion. *Oncogene* 29: 6452-6463

Niessen CM, Leckband D, Yap AS (2011) Tissue organization by cadherin adhesion molecules: dynamic molecular and cellular mechanisms of morphogenetic regulation. *Physiol Rev* 91: 691-731

Orr-Weaver TL, Weinberg RA (1998) A checkpoint on the road to cancer. *Nature* 392: 223-224

Palacios F, Tushir JS, Fujita Y, D'Souza-Schorey C (2005) Lysosomal targeting of E-cadherin: a unique mechanism for the down-regulation of cell-cell adhesion during epithelial to mesenchymal transitions. *Mol Cell Biol* 25: 389-402

Palamidessi A, Frittoli E, Garre M, Faretta M, Mione M, Testa I, Diaspro A, Lanzetti L, Scita G, Di Fiore PP (2008) Endocytic trafficking of Rac is required for the spatial restriction of signaling in cell migration. *Cell* 134: 135-147

Park SJ (2011) Huntingtin-interacting protein 1-related is required for accurate congression and segregation of chromosomes. *BMB Rep* 43: 795-800

Parsons JT, Horwitz AR, Schwartz MA (2011) Cell adhesion: integrating cytoskeletal dynamics and cellular tension. *Nat Rev Mol Cell Biol* 11: 633-643

Pellinen T, Arjonen A, Vuoriluoto K, Kallio K, Fransen JA, Ivaska J (2006) Small GTPase Rab21 regulates cell adhesion and controls endosomal traffic of beta1-integrins. *J Cell Biol* 173: 767-780

Pellinen T, Tuomi S, Arjonen A, Wolf M, Edgren H, Meyer H, Grosse R, Kitzing T, Rantala JK, Kallioniemi O, Fassler R, Kallio M, Ivaska J (2008) Integrin trafficking regulated by Rab21 is necessary for cytokinesis. *Dev Cell* 15: 371-385

Pennock S, Wang Z (2003) Stimulation of cell proliferation by endosomal epidermal growth factor receptor as revealed through two distinct phases of signaling. *Mol Cell Biol* 23: 5803-5815

Pitcher J, Lohse MJ, Codina J, Caron MG, Lefkowitz RJ (1992) Desensitization of the isolated beta 2-adrenergic receptor by beta-adrenergic receptor kinase, cAMP-dependent protein kinase, and protein kinase C occurs via distinct molecular mechanisms. *Biochemistry* 31: 3193-3197

Pitcher JA, Payne ES, Csortos C, DePaoli-Roach AA, Lefkowitz RJ (1995) The G-protein-coupled receptor phosphatase: a protein phosphatase type 2A with a distinct subcellular distribution and substrate specificity. *Proc Natl Acad Sci U S A* 92: 8343-8347

Pol A, Calvo M, Enrich C (1998) Isolated endosomes from quiescent rat liver contain the signal transduction machinery. Differential distribution of activated Raf-1 and Mek in the endocytic compartment. *FEBS Lett* 441: 34-38

Purvanov V, Koval A, Katanaev VL (2010) A direct and functional interaction between Go and Rab5 during G protein-coupled receptor signaling. *Sci Signal* 3: ra65

Raiborg C, Stenmark H (2009) The ESCRT machinery in endosomal sorting of ubiquitylated membrane proteins. *Nature* 458: 445-452

Rappoport JZ, Simon SM (2009) Endocytic trafficking of activated EGFR is AP-2 dependent and occurs through preformed clathrin spots. *J Cell Sci* 122: 1301-1305

Reed BC, Cefalu C, Bellaire BH, Cardelli JA, Louis T, Salamon J, Bloecher MA, Bunn RC (2005) GLUT1CBP(TIP2/GIPC1) Interactions with GLUT1 and Myosin VI:

Evidence Supporting an Adapter Function for GLUT1CBP 10.1091/mbc.E04-11-0978. *Mol Biol Cell* 16: 4183-4201

Riley KN, Maldonado AE, Tellier P, D'Souza-Schorey C, Herman IM (2003) Betacap73-ARF6 interactions modulate cell shape and motility after injury in vitro. *Mol Biol Cell* 14: 4155-4161

Rodriguez-Boulan E, Kreitzer G, Musch A (2005) Organization of vesicular trafficking in epithelia. *Nat Rev Mol Cell Biol* 6: 233-247

Rokas A (2008) The origins of multicellularity and the early history of the genetic toolkit for animal development. *Annu Rev Genet* 42: 235-251

Roussos ET, Condeelis JS, Patsialou A (2011) Chemotaxis in cancer. *Nat Rev Cancer* 11: 573-587

Roy S, Wyse B, Hancock JF (2002) H-Ras signaling and K-Ras signaling are differentially dependent on endocytosis. *Mol Cell Biol* 22: 5128-5140

Royle SJ, Bright NA, Lagnado L (2005) Clathrin is required for the function of the mitotic spindle. *Nature* 434: 1152-1157

Saffarian S, Cocucci E, Kirchhausen T (2009) Distinct dynamics of endocytic clathrin-coated pits and coated plaques. *PLoS Biol* 7: e1000191

Sato K, Watanabe T, Wang S, Kakeno M, Matsuzawa K, Matsui T, Yokoi K, Murase K, Sugiyama I, Ozawa M, Kaibuchi K (2011) Numb controls E-cadherin endocytosis through p120 catenin with aPKC. *Mol Biol Cell*

Sato T, Mushiake S, Kato Y, Sato K, Sato M, Takeda N, Ozono K, Miki K, Kubo Y, Tsuji A, Harada R, Harada A (2007) The Rab8 GTPase regulates apical protein localization in intestinal cells. *Nature* 448: 366-369

Schlunck G, Damke H, Kiosses WB, Rusk N, Symons MH, Waterman-Storer CM, Schmid SL, Schwartz MA (2004) Modulation of Rac localization and function by dynamin. *Mol Biol Cell* 15: 256-267

Schneider-Brachert W, Tchikov V, Neumeyer J, Jakob M, Winoto-Morbach S, Held-Feindt J, Heinrich M, Merkel O, Ehrenschwender M, Adam D, Mentlein R, Kabelitz D, Schutze S (2004) Compartmentalization of TNF receptor 1 signaling: internalized TNF receptosomes as death signaling vesicles. *Immunity* 21: 415-428

Schutze S, Tchikov V, Schneider-Brachert W (2008) Regulation of TNFR1 and CD95 signalling by receptor compartmentalization. *Nat Rev Mol Cell Biol* 9: 655-662

Scita G, Di Fiore PP (2010) The endocytic matrix. *Nature* 463: 464-473

Sehat B, Andersson S, Girnita L, Larsson O (2008) Identification of c-Cbl as a new ligase for insulin-like growth factor-I receptor with distinct roles from Mdm2 in receptor ubiquitination and endocytosis. *Cancer Res* 68: 5669-5677

Serini G, Bussolino F (2004) Common cues in vascular and axon guidance. *Physiology (Bethesda)* 19: 348-354

Serio G, Margaria V, Jensen S, Oldani A, Bartek J, Bussolino F, Lanzetti L. (2011) Small GTPase Rab5 participates in chromosome congression and regulates localization of the centromere-associated protein CENP-F to kinetochores. *Proc Natl Acad Sci U S A*. 108(42):17337-42.

Shattil SJ, Kim C, Ginsberg MH (2010) The final steps of integrin activation: the end game. *Nat Rev Mol Cell Biol* 11: 288-300

Shi F, Sottile J (2008) Caveolin-1-dependent {beta}1 integrin endocytosis is a critical regulator of fibronectin turnover. *Journal of Cell Science* 121: 2360-2371

Shin D, Garcia-Cardena G, Hayashi S, Gerety S, Asahara T, Stavrakis G, Isner J, Folkman J, Gimbrone MA, Jr., Anderson DJ (2001) Expression of ephrinB2 identifies a stable genetic difference between arterial and venous vascular smooth muscle as well as endothelial cells, and marks subsets of microvessels at sites of adult neovascularization. *Dev Biol* 230: 139-150

Sigismund S, Argenzio E, Tosoni D, Cavallaro E, Polo S, Di Fiore PP (2008) Clathrin-mediated internalization is essential for sustained EGFR signaling but dispensable for degradation. *Dev Cell* 15: 209-219

Sigismund S, Woelk T, Puri C, Maspero E, Tacchetti C, Transidico P, Di Fiore PP, Polo S (2005) Clathrin-independent endocytosis of ubiquitinated cargos. *Proc Natl Acad Sci U S A* 102: 2760-2765

Sonnichsen B, De Renzis S, Nielsen E, Rietdorf J, Zerial M (2000) Distinct membrane domains on endosomes in the recycling pathway visualized by multicolor imaging of Rab4, Rab5, and Rab11. *J Cell Biol* 149: 901-914

Sorkin A, von Zastrow M (2009) Endocytosis and signaling: intertwining molecular networks. *Nat Rev Mol Cell Biol* In press

Spudich JA, Sivaramakrishnan S (2010) Myosin VI: an innovative motor that challenged the swinging lever arm hypothesis. *Nat Rev Mol Cell Biol* 11: 128-137

Stenmark H (2009) Rab GTPases as coordinators of vesicle traffic. *Nat Rev Mol Cell Biol* 10: 513-525

Takeichi M (2011) Self-organization of animal tissues: cadherin-mediated processes. *Dev Cell* 21: 24-26

Tani TT, Mercurio AM (2001) PDZ interaction sites in integrin a subunits. TIP-2/GIPC binds to a type I recognition sequence in a6A/a5 and a novel sequence in a6B. *J Biol Chem* 276: 36535–36542

Tao L, Zhang Y, Xi X, Kieffer N (2010) Recent advances in the understanding of the molecular mechanisms regulating platelet integrin αIIbβ3 activation. *Protein Cell* 1: 627-637

Teckchandani A, Toida N, Goodchild J, Henderson C, Watts J, Wollscheid B, Cooper JA (2009) Quantitative proteomics identifies a Dab2/integrin module regulating cell migration. *The Journal of Cell Biology* 186: 99-111

ter Haar E, Musacchio A, Harrison SC, Kirchhausen T (1998) Atomic structure of clathrin: a beta propeller terminal domain joins an alpha zigzag linker. *Cell* 95: 563-573

Thiery JP, Sleeman JP (2006) Complex networks orchestrate epithelial-mesenchymal transitions. *Nat Rev Mol Cell Biol* 7: 131-142

Thompson HM, Cao H, Chen J, Euteneuer U, McNiven MA (2004) Dynamin 2 binds gamma-tubulin and participates in centrosome cohesion. *Nat Cell Biol* 6: 335-342

Thompson HM, Skop AR, Euteneuer U, Meyer BJ, McNiven MA (2002) The large GTPase dynamin associates with the spindle midzone and is required for cytokinesis. *Curr Biol* 12: 2111-2117

Tiwari S, Askari JA, Humphries MJ, Bulleid NJ (2011) Divalent cations regulate the folding and activation status of integrins during their intracellular trafficking. *J Cell Sci* 124: 1672-1680

Traub LM (2009) Tickets to ride: selecting cargo for clathrin-regulated internalization. *Nat Rev Mol Cell Biol* 10: 583-596

Tsukazaki T, Chiang TA, Davison AF, Attisano L, Wrana JL (1998) SARA, a FYVE domain protein that recruits Smad2 to the TGFbeta receptor. *Cell* 95: 779-791

Tuomikoski T, Felix MA, Dorée M, Gruenberg J (1989) Inhibition of endocytic vesicle fusion in vitro by the cell-cycle control protein kinase cdc2. *Nature* 342: 942-945

Valdembri D, Caswell PT, Anderson KI, Schwarz JP, Konig I, Astanina E, Caccavari F, Norman JC, Humphries MJ, Bussolino F, Serini G (2009) Neuropilin-1/GIPC1 signaling regulates alpha5beta1 integrin traffic and function in endothelial cells. *PLoS Biol* 7: e25

van der Sluijs P, Hull M, Huber LA, Mâle P, Goud B, Mellman I (1992) Reversible phosphorylation--dephosphorylation determines the localization of rab4 during the cell cycle. *EMBO J* 11: 4379-4389

Vasudevan NT, Mohan ML, Gupta MK, Hussain AK, Naga Prasad SV (2011) Inhibition of protein phosphatase 2A activity by PI3Kgamma regulates beta-adrenergic receptor function. *Mol Cell* 41: 636-648

Vieira AV, Lamaze C, Schmid SL (1996) Control of EGF receptor signaling by clathrin-mediated endocytosis. *Science* 274: 2086-2089

Wada I, Lai WH, Posner BI, Bergeron JJ (1992) Association of the tyrosine phosphorylated epidermal growth factor receptor with a 55-kD tyrosine phosphorylated protein at the cell surface and in endosomes. *J Cell Biol* 116: 321-330

Wang X, Bo J, Bridges T, Dugan KD, Pan TC, Chodosh LA, Montell DJ (2006) Analysis of cell migration using whole-genome expression profiling of migratory cells in the Drosophila ovary. *Dev Cell* 10: 483-495

Wang Y, Pennock SD, Chen X, Kazlauskas A, Wang Z (2004) Platelet-derived growth factor receptor-mediated signal transduction from endosomes. *J Biol Chem* 279: 8038-8046

Wang Z, Tung PS, Moran MF (1996) Association of p120 ras GAP with endocytic components and colocalization with epidermal growth factor (EGF) receptor in response to EGF stimulation. *Cell Growth Differ* 7: 123-133

Waterman H, Sabanai I, Geiger B, Yarden Y (1998) Alternative intracellular routing of ErbB receptors may determine signaling potency. *J Biol Chem* 273: 13819-13827

Wirtz-Peitz F, Zallen JA (2009) Junctional trafficking and epithelial morphogenesis. *Current Opinion in Genetics & Development* 19: 350-356

Xu J, Lan L, Bogard N, Mattione C, Cohen RS (2011) Rab11 is required for epithelial cell viability, terminal differentiation, and suppression of tumor-like growth in the Drosophila egg chamber. *PLoS One* 6: e20180

Yamamoto H, Sakane H, Michiue T, Kikuchi A (2008) Wnt3a and Dkk1 regulate distinct internalization pathways of LRP6 to tune the activation of beta-catenin signaling. *Dev Cell* 15: 37-48

Yang SD, Fong YL, Benovic JL, Sibley DR, Caron MG, Lefkowitz RJ (1988) Dephosphorylation of the beta 2-adrenergic receptor and rhodopsin by latent phosphatase 2. *J Biol Chem* 263: 8856-8858

Zerial M, McBride H (2001) Rab proteins as membrane organizers. *Nat Rev Mol Cell Biol* 2: 107-117

Zoncu R, Perera RM, Balkin DM, Pirruccello M, Toomre D, De Camilli P (2009) A phosphoinositide switch controls the maturation and signaling properties of APPL endosomes. *Cell* 136: 1110-1121

8

Making the Final Cut – The Role of Endosomes During Mitotic Cell Division

Rytis Prekeris
Department of Cell and Developmental Biology, School of Medicine, Anschutz Medical Campus, University of Colorado, Denver, Aurora, CO USA

1. Introduction

The last step of cell division is the physical separation of two daughter cells via a process known as cytokinesis (Barr and Gruneberg, 2007; Prekeris and Gould, 2008). After replication of the genetic material, the mother cell divides by the formation of a cleavage furrow that constricts the cytoplasm, thus leaving two daughter cells connected by a thin intracellular bridge (ICB). The resolution of this bridge, abscission, results in a physical separation of the two daughter cells and usually occurs on either, or both, sides of the midbody within the ICB. This abscission event occurs on either or both sides of the midbody within the intracellular bridge (ICB). While earlier studies thought this abscission event was simply a continuation of the constriction placed on the cleavage furrow by the actomyosin contractile ring, it was later discovered that abscission is a highly complex and organized event consisting of much more than a simple actin and non-muscle myosin constricting ring. Endocytic membrane transport, ESCRT protein complex function, and cytoskeletal reorganization were all shown to contribute to cytokinesis and abscission (Barr and Gruneberg, 2007; Prekeris and Gould, 2008). The goal of this review is provide an overview the newest advances in our understanding about the dynamics and roles of endosomal transport during cytoskeletal re-arrangements and ESCRT complex assembly throughout cytokinesis.

2. Post-Golgi transport is required for the completion of cytokinesis

Classically, cytokinesis in animal cells was thought to be mediated by the formation and contraction of the actomyosin contractile ring, which assembles at the equator of the dividing cells and is regulated by a variety of molecules, including the recruitment and activation of a RhoA GTPase-dependent signaling cascade (Glotzer, 2005). In contrast, plant cytokinesis was demonstrated to depend on the transport of post-Golgi organelles, which assemble and fuse to form an organelle, known as a phragmoplast (Jurgens, 2005). This organelle ultimately fuses with the plasma membrane, bisecting the plant cell into two daughter cells. These mechanisms in animal and plant cell division were thought to have differentially evolved due to the fact that plant cells have a rigid cell wall, while animal cells only need to separate a very dynamic and "bendable" plasma membrane. Interestingly, during recent years, multiple studies have challenged this dogma and suggested that animal cell cytokinesis may not be entirely different from plant cell cytokinesis. The first clues, that membrane transport may play a role in animal cytokinesis, came from the observations that in large embryonic cells, such as amphibian eggs, the addition of endo-membranes is required for expanding the plasma membrane during the formation and ingression of the cleavage furrow (Bluemink and de Laat, 1973). Later studies have shown that similar membrane addition is required during cellularization of *Drosophila melanogaster* embryos (Albertson et al., 2008; Hickson et al., 2003). In addition, multiple genetics and proteomics screens in *Caenorhabditis elegans* and *Drosophila melanogaster* have identified many known endocytic membrane trafficking proteins, such as dynamin, SNAREs and Rab GTPases as factors required for the successful completion of cytokinesis (Low et al., 2003; Pelissier et al., 2003; Riggs et al., 2003). Based on this work, it has become widely accepted that endosomes are specifically targeted to the forming cleavage furrow, and appear to mediate the late step(s) of cell division. Finally, recent studies have suggested that these endosomes are not a homogeneous pool of endocytic membranes, but in fact consist of several post-Golgi organelles with different transport dynamics and distinct functions (Dambournet et al., 2011; Nezis et al., 2010; Schiel et al., 2011).

2.1 TGN-derived secretory vesicles

Some of the first organelles that were shown to be targeted to the forming cleavage furrow were trans-Golgi Network (TGN)-derived secretory organelles. Elegant work from Steve Doxsey and colleagues has demonstrated that secretory organelles can be targeted to the midbody of the cleavage furrow by binding to the centriolin-exocyst protein complex (Gromley et al., 2005). Centriolin was originally described as a protein that associates with the mother centriole. However, during late cytokinesis, centriolin also associates with a novel structure in the midbody, known as the midbody ring. This midbody-associated centriolin was shown to act as a scaffolding factor by binding to and recruiting the Exocyst protein complex to the midbody (Gromley et al., 2005). The exocyst complex was originally described as a tethering factor for secretory vesicles in budding yeast (TerBush et al., 1996), and later was shown to play a similar role in mammalian cells (Hsu et al., 1999). The exocyst is a multi-protein complex that is comprised of Sec3, Sec5, Sec6, Sec8, Sec10, Sec15, Exo70 and Exo84 subunits (Hurley, 2010). The exocyst complex was already shown to localize to the neck of the bud in budding yeast and to the midbody of the dividing animal cells (TerBush et al., 1996). As the result, these data provided a tantalizing possibility that the centriolin-exocyst complex may serve as a tethering factor for the targeting of secretory vesicles to the forming cleavage furrow. Indeed, the knock-down of various exocyst subunits leads to defects in late cytokinesis (Fielding et al., 2005; Gromley et al., 2005). In addition to the exocyst complex, centriolin was also shown to bind and recruit a SNARE-associated protein, SNAPIN (Gromley et al., 2005), thus providing another link between centriolin and membrane transport/fusion. Consistent with these data, SNAPIN also appears to be required for the successful completion of cytokinesis (Gromley et al., 2005). Taken together, it was proposed that the compound fusion of these secretory vesicles with each other and with the furrow plasma membrane may lead to the final scission of the intracellular bridge (ICB) connecting daughter cells. Since this model was introduced, high-resolution microscopy and tomography studies have questioned whether compound secretory vesicle fusion can mediate abscission (Elia et al., 2011; Schiel et al., 2011). It is clear that secretory vesicles are delivered to the forming cleavage furrow during early-to-mid telophase, the function of these secretory vesicles remains unknown.

2.2 Rab11-endosomes

In addition to secretory vesicles, recycling endosomes also have emerged as important players in mediating abscission. Several reports have demonstrated that pronounced changes occur in endocytic recycling during mitosis, and that these changes are required for the successful completion of cytokinesis (Boucrot and Kirchhausen, 2007). Additionally, VAMP8, a known endocytic SNARE, also was shown to be present in the cleavage furrow and is required for mitotic cell division (Schiel et al., 2011). Originally it was proposed that recycling endosomes, just like secretory vesicles, initiate abscission by fusing with each other and the plasma membrane, thus building a separating membrane in a manner similar to the formation of a phragmoplast in plant cells. However, recent data indicates that fusion of recycling endosomes instead mediates formation of the "secondary ingression" (Figure 2, Endosome fusion model), although it remains unclear how these recycling endosomes mechanistically induce this secondary ingression (Schiel et al., 2011). Rab11 is a small monomeric GTPase that plays a major role in the trafficking of recycling endosomes during interphase (Prekeris, 2003). As the result, it was proposed that Rab11 may play a role in targeting recycling endosomes to the cleavage furrow during mitosis. Indeed, work in several model organisms, such as *Caenorhabditis elegans* and *Drosophila mellanogaster* have shown that Rab11 is required for cytokinesis (Hickson et al., 2003; Pelissier et al., 2003; Skop et al., 2001). Similarly, recent work from several laboratories has shown that Rab11 also mediates late cytokinesis in mammalian cells (Hickson et al., 2003; Horgan et al., 2004; Wilson et al., 2004). Rab GTPases work by binding and recruiting distinct effector proteins to membranes. Because of the ability of the Rabs to bind multiple effector proteins, considerable effort within the last decade has been dedicated to the identification and characterization of Rab effector proteins and their roles in various cellular pathways. Work from several laboratories has identified Rab11-FIPs (**Rab11** **F**amily **I**nteracting **P**roteins) as Rab11 effector proteins that serve as "targeting complexes" in specific recycling endosome transport pathways (Hales et al., 2001; Prekeris et al., 2001). Rab11-FIPs (henceforth referred to as FIPs) consists of five members, which bind to Rab11 with a very high affinity (~100-200nM) (Junutula et al., 2004). All FIPs contain a highly conserved C-terminally located Rab11-binding domain (RBD) (Prekeris, 2003). Recently solved crystal structures of FIP3/Rab11 and FIP2/Rab11 complexes have shown that FIPs form a parallel coiled-coil homo-dimer and bind to two Rab11-GTP molecules via Rab11-switch regions (Eathiraj et al., 2006; Jagoe et al., 2006). Interestingly, FIP binding induces a conformational change in Rab11, perhaps explaining the very high affinity of Rab11 and FIP binding (Eathiraj et al., 2006). Based on their structure and sequence similarity, the FIPs are divided into two classes (Figure 1). Class I FIPs (FIP1, FIP2 and FIP5) contain a N-terminally located phospholipid binding C2 domain and were shown to regulate endocytic recycling during interphase (Prekeris, 2003). Class II FIPs (FIP3 and FIP4) lack a C2 domain, but contain a N-terminally located proline-rich domain (PRD) and two calcium binding EF hands (Prekeris, 2003), although the functions of these domains remain unclear. FIP3, a Class II FIP, has emerged as a key regulator of recycling endosome targeting to the cleavage furrow during cytokinesis. In mammalian cells, knock-down of FIP3 results in failed cytokinesis, leading to the formation of bi-nucleate or multi-nucleate cells (Wilson et al., 2004). Similarly, nuclear fallout protein (nuf), a *Drosophila* orthologue of FIP3, is required for the cellularization of *Drosophila* embryos (Riggs et al., 2007). FIP3 also makes an excellent marker of furrow-associated recycling endosomes, because during cytokinesis FIP3 is present exclusively at this site in the cell (Schiel et al., 2011; Simon et al., 2008).

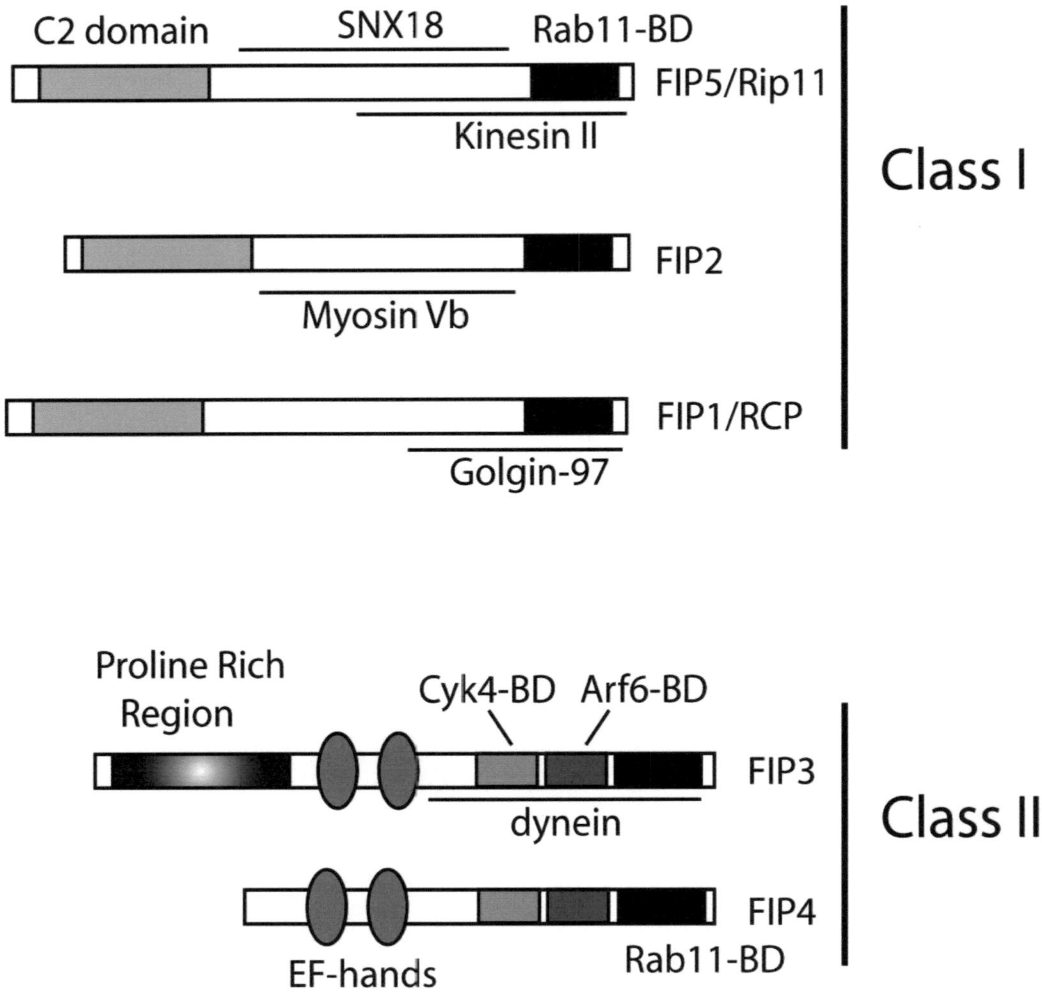

Fig. 1. The schematic representation of mammalian Rab11-FIP family members. Rab11-BD stands for Rab11-binding domain. Lines indicate binding domains mediating interactions with various FIP-binding proteins.

As a result, recycling endosomes enriched at the cleavage furrow, are often referred to as FIP3-endosomes (Schiel et al., 2011). Recent work from many laboratories on FIP3-endosome dynamics during cell division has identified the machinery that allows the targeting and accumulation of FIP3-endosomes close to the midbody of the dividing cells. It was shown that FIP3-endosomes are delivered to the cleavage furrow along central spindle microtubules in a kinesin-dependent manner. An elegant study by Dr. Chavrier and colleagues showed that the directionality of recycling endosome transport along central spindle microtubules depends on the differential association of endosomes with either the Kinesin I or dynein molecular motors (Montagnac et al., 2009). Interestingly, some evidence suggests that FIP3 may directly bind to dynein via an association with the dynein light intermediate chain 2 (Horgan et al., 2010). Once delivered to the cleavage furrow, FIP3-endosomes accumulate at close proximity to the midbody, a step that is required for the completion of cytokinesis (Wilson et al., 2004). This accumulation depends on two distinct tethering mechanisms. It was indicated that FIP3 can directly bind to Cyk4/MgcRacGAP, and that this interaction is required for the efficient targeting of FIP3-endosomes (Simon et al., 2008). Cyk4 is a subunit of the Centralspindlin complex that is localized at the midbody during cytokinesis (Glotzer, 2005). In addition, FIP3 also was shown to bind Arf6 GTPase, an endocytic protein that is known to be required for cytokinesis (Hickson et al., 2003; Schonteich et al., 2007). Interestingly, Arf6 also binds the Sec10 subunit of the exocyst complex (Prigent et al., 2003). Rab11 was also shown to bind the exocyst complex via its Sec15 subunit (Wu et al., 2005). Thus Rab11/FIP3-containing endosomes can interact with the Exocyst tethering complex in at least two distinct binding interfaces. Why do FIP3-endosomes need a multiple protein-protein interactions to be targeted to the midbody? One possibility is that all of these targeting proteins work as the 'belt and braces' to ensure the fidelity of FIP3-endosome transport and targeting.

2.3 Rab35-endosomes

In addition to Rab11, Rab35 also recently emerged as a potential regulator of endosomes during mitotic cell division. The possible involvement of Rab35 in regulating cytokinesis was first uncovered during a non-biased Rab siRNA library screen for cell division defects in *Drosophila* tissue culture cells, and later was confirmed to also be required for cytokinesis in mammalian cells as well (Kouranti et al., 2006). Just like Rab11, Rab35 appears to act during the late stages of cytokinesis, since cells expressing the dominant-negative Rab35 mutant can still form and ingress a cleavage furrow, but fail to undergo abscission (Kouranti et al., 2006). The function of Rab35 still remains to be fully understood, mostly due to the fact that Rab35 effector proteins are only beginning to be identified. Rab35 has been implicated in regulating endocytic transport during interphase as it is localized to the plasma membranes, clathrin coated pits and endosomes (Chesneau et al., 2012). Consequently, Rab35 was shown to bind EHD1 and regulate fast endocytic recycling of MHC class I and II, components of the immunological synapse, and some synaptic proteins (Allaire et al., 2010; Walseng et al., 2008). Rab35 was also shown to bind fascin, a known actin cross-linking protein (Zhang et al., 2009). Consistent with this finding, Rab35 was suggested to regulate actin polymerization/bundling during neurite outgrowth, phagocytosis and cell motility (Chevallier et al., 2009; Egami et al., 2011; Kanno et al., 2010). The exact function of Rab35 during cytokinesis remains to be fully understood, however, it was proposed that it also may regulate the actin cytockeleton within the cleavage furrow (see below) (Dambournet et al., 2011; Kouranti et al., 2006).

3. The role of endosomes in regulating PI3P, PI(4,5)P2 and the actin cytoskeleton during cell division

Phosphoinositides (PIs) are well-established regulators of cytokinesis. It has been demonstrated by work in many laboratories that phosphoinositide 4-5 bisphosphate (PI(4,5)P2) is enriched at the ingressing cleavage furrow (Field et al., 2005). Inhibition of PI(4,5)P2 production by overexpressing a kinase-dead PI4P5-kinase leads to an increase in multinucleation, an indication of failed cytokinesis (Emoto and Umeda, 2000; Field et al., 2005). Two enzymes, phosphatase and tensin homologue on chromosome 10 (PTEN) and PI3-kinase, which regulate PI(4,5)P2 and PI(3,4,5)P3 levels, are required for cytokinesis

(Janetopoulos and Devreotes, 2006; Nezis et al., 2010). Interestingly, endosomes have recently emerged as important modulators of PI(4,5)P2 during cell division. For example, Vps34 (PI3-kinase C3) is delivered to the cleavage furrow by associating with Rab11-containing endosomes (Nezis et al., 2010). Several PI3K-III accessory proteins, such as Becklin 1, Vps15, UVRAG and BIF1 were all shown to be associated with endosomes and required for cytokinesis (Nezis et al., 2010). PI4P5-kinase was also shown to bind and be activated by Arf6, a protein which is targeted to the midbody by binding to the exocyst complex and FIP3/Rab11 (Hickson et al., 2003; Schonteich et al., 2007). Finally, Rab35 was shown to bind and recruit to the furrow a protein called phosphoinositide 5-phosphatase OCRL, that is responsible for a genetic disease oculocerebral syndrome of Lowe (Dambournet et al., 2011). This evidence clearly demonstrates that the cellular levels of PIs at the furrow are tightly regulated, and relay on several independent pathways that must coordinate with each other during mitotic progression. The actin cytoskeleton is one of the key players during cytokinesis. Classically, cytokinesis is defined as occurring through the formation and contraction of the actomyosin contractile ring at the equator of the cell. Initial actomyosin ring assembly and activation is regulated by RhoA GTPase and appears to be independent of endocytic transport, since the inhibition of either Rab11 or Rab35-dependent membrane delivery does not prevent the initial cleavage furrow ingression (Dambournet et al., 2011; Wilson et al., 2004). In late cytokinesis, the role of the actin cytoskeleton is less clearly defined. Actin appears to be required for the initial stabilization of the intracellular bridge, presumably by binding to a septin network via anillin adaptor proteins (Piekny and Maddox, 2010). Actin filaments also associate with the plasma membrane at the furrow by binding to the ezrin/radixin/moesin family of proteins, which bind directly with the plasma membrane (Kunda et al., 2008; Kunda et al., 2011). Finally, it is generally accepted that the actin cytoskeleton needs to be disassembled for the final abscission step to take place, however the machinery that regulates this final disassembly remains unclear. Interestingly, Rab35 appears to mediate the delivery of OCRL to the furrow only during late cytokinesis (Dambournet et al., 2011). Thus, it was postulated that OCRL may be responsible for the final disassembly of the actin cytoskeleon by depleting PI(4,5)P2 levels at the furrow. Consistent with this, it was demonstrated that a delay in abscission resulting from the depletion of OCRL or Rab35 can be rescued by incubating dividing cells with low concentrations of the actin depolymerizing agent, latrunculin-A (Dambournet et al., 2011). In addition to Rab35, Rab11 and FIP3 also appear to play a role in regulating the actin cytoskeleton during cytokinesis. Inhibition of *Drosophila* homologue of FIP3, nuclear fallout protein (Nuf), resulted in a very dramatic reorganization of the actin cytoskeleton within cleavage furrows (Riggs et al., 2007; Riggs et al., 2003). While the mechanisms that mediate Rab11 and FIP3 effects on actin remains to be determined, new data suggest that FIP3-containing endosomes deliver p50RhoGAP to the cleavage furrow (unpublished data). Since p50RhoGAP inactivates RhoA GTPase (Barrett et al., 1997; Sirokmany et al., 2006; Zhou et al., 2010), it is tempting to speculate that the targeting of FIP3 endosomes to the furrow may result in the inactivation of RhoA, and the subsequent depolymerization of the actin cytoskeleton.

4. ESCRTs and cytokinesis

Recently work from several laboratories has implicated the endosomal sorting complex required for transport (ESCRT) proteins in mediating the abscission step of cytokinesis. The

ESCRT proteins were originally identified as proteins involved in protein sorting to the lysosomes and in the formation of intraluminal vesicles during the maturation of multivesicular bodies. The ESCRT complex consists of four different protein complexes: ESCRT-0, ESCRT-I, ESCRT-II and ESCRT-III. While ESCRT-0 and ESCRT-I are involved in cargo recognition and sorting, ESCRT-II and ESCRT-III seem to play a role in initial vesicle formation (Wollert and Hurley, 2010). Finally, ESCRT-III appears to mediate the final scission of intraluminal vesicles (Wollert and Hurley, 2010). Since intraluminal vesicle scission is topologically similar to the scission of the intracellular bridge during cytokinesis, it was proposed that ESCRT proteins may mediate the abscission step of cytokinesis (Carlton et al., 2008; Carlton and Martin-Serrano, 2007).

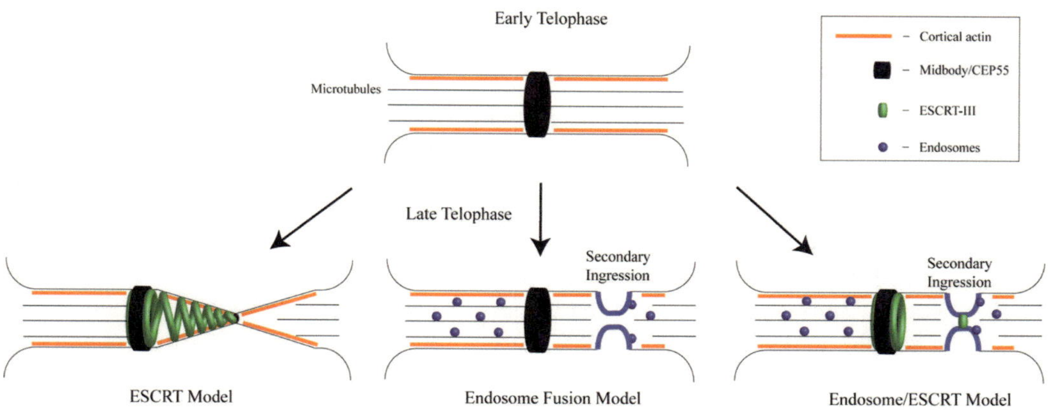

Fig. 2. Models for ESCRT and endosome roles during abscission.

Indeed, knock-down of various ESCRT-II and ESCRT-III members result in the inhibition of abscission, while having no effect on initial cleavage furrow ingression (Caballe and Martin-Serrano, 2011). Interestingly, ESCRT-dependent abscission does not require ESCRT-0 and ESCRT-I complexes (Hurley, 2010; Samson et al., 2008). Instead, ESCRT-III is recruited to the midbody of the cleavage furrow by binding to ALIX or Tsg101 proteins, which accumulate at the midbody by interacting with well-established midbody protein CEP55 (Caballe and Martin-Serrano, 2011). The actual scission step is mediated by the ESCRT-III complex subunit CHMP4. CHMP4 has the ability to form ~5 nm filaments and tubulate liposomes *in vitro* (Ghazi-Tabatabai et al., 2008; Lata et al., 2008). Since abscission usually occurs outside the midbody, it was postulated that CHMP4 is recruited to the midbody, where it polymerizes to form continuous spiral filaments that induces a gradual decrease in the diameter of the ICB and eventually leads to the scission of the membranes (Elia et al., 2011; Guizetti et al., 2011) (Figure 2, ESCRT model). While the discovery of ESCRT involvement in abscission does provide the novel conceptual framework for the mechanism of abscission, many questions remain to be answered. ESCRT complexes usually mediate scission of the membrane tubes that are about 24-50 nm in diameter. Consistent with this, it was shown that the ESCRT-III complex associate with the high-curvature (>100 nm) membranes (Fyfe et al., 2011). Thus, it remains unclear how the ESCRT-III polymers can form and contract intracellular bridges that are 2-3 mm in size (Schiel et al., 2011). Furthermore, the stability of the ICB is maintained by a complex cortical network consisting of actin and septin filaments, which are cross-linked to each other and to the plasma membrane by anillin. How the

ESCRT-III "spirals" are formed in the presence of these cytoskeletal elements remains a mystery. Some of these questions may be explained by recent findings that endosome delivery and fusion with the plasma membrane of the intracellular bridge may induce the disassembly of the actin cytoskeleton and lead to the secondary ingression that decreases the diameter of the intracellular bridge to ~100-200 nm (Dambournet et al., 2011; Schiel et al., 2011). Perhaps this endosome-dependent secondary ingression initiates the "de novo" recruitment of activated ESCRT-III to the abscission site by removing the actin cytoskeleton and narrowing the ICB to a smaller size (Figure 2, Endosome/ESCRT model) (Schiel and Prekeris, 2011). Indeed, while accumulation of the ESCRT-III at the abscission site can be readily detected, most recent studies have failed to observe an ESCRT "spiral" emanating from the midbody and continuing to the abscission site (Elia et al., 2011; Schiel et al., 2011).

5. Conclusions and future objectives

Work from multiple laboratories in the last few years has significantly advanced our understanding about the core machinery of the abscission step of cytokinesis. All these data have demonstrated that cell abscission is an immensely complicated event that involves coordinated changes in membrane transport, microtubules, the actin cytoskeleton, septin filaments and the ESCRT complexes. How all these components are regulated, and what the mechanisms of the cross-talk between them may be, remain completely unknown and will be the focus of future studies. One of the biggest problems in studying the spatiotemporal dynamics of various cellular components during cell division has been the inability to visualize the individual organelles or cytoskeletal elements within the intracellular bridge, due to the resolution limits of the light microscopy. The emergence of novel super-resolution imaging techniques, such as photo-activated localization microscopy (PALM), stimulated emission depletion microscopy (STED) and correlation high-resolution tomography will allow us to begin addressing some of these questions and testing/combining multiple competing abscission models.

6. Acknowledgments

I am grateful to John Schiel (UC SOM) and Carly Willenborg (UC SOM) for a critical reading of the manuscript. I apologize to all colleagues for not being able to cite all work related to cytokinesis, due to the focused nature of this review. Work in the Prekeris laboratory has been funded by the National Institute of Health (DK064380), Susan G. Komens Breast Cancer Research Foundation and the Cancer League of Colorado Foundation.

7. References

Albertson, R., Cao, J., Hsieh, T.S., and Sullivan, W. (2008). Vesicles and actin are targeted to the cleavage furrow via furrow microtubules and the central spindle. J Cell Biol 181, 777-790.

Allaire, P.D., Marat, A.L., Dall'Armi, C., Di Paolo, G., McPherson, P.S., and Ritter, B. (2010). The Connecdenn DENN domain: a GEF for Rab35 mediating cargo-specific exit from early endosomes. Mol Cell 37, 370-382.

Barr, F.A., and Gruneberg, U. (2007). Cytokinesis: placing and making the final cut. Cell 131, 847-860.

Barrett, T., Xiao, B., Dodson, E.J., Dodson, G., Ludbrook, S.B., Nurmahomed, K., Gamblin, S.J., Musacchio, A., Smerdon, S.J., and Eccleston, J.F. (1997). The structure of the GTPase-activating domain from p50rhoGAP. Nature 385, 458-461.

Bluemink, J.G., and de Laat, S.W. (1973). New membrane formation during cytokinesis in normal and cytochalasin B-treated eggs of Xenopus laevis. I. Electron microscope observations. J Cell Biol 59, 89-108.

Boucrot, E., and Kirchhausen, T. (2007). Endosomal recycling controls plasma membrane area during mitosis. Proc Natl Acad Sci U S A 104, 7939-7944.

Caballe, A., and Martin-Serrano, J. (2011). ESCRT machinery and cytokinesis: the road to daughter cell separation. Traffic 12, 1318-1326.

Carlton, J.G., Agromayor, M., and Martin-Serrano, J. (2008). Differential requirements for Alix and ESCRT-III in cytokinesis and HIV-1 release. Proc Natl Acad Sci U S A 105, 10541-10546.

Carlton, J.G., and Martin-Serrano, J. (2007). Parallels between cytokinesis and retroviral budding: a role for the ESCRT machinery. Science 316, 1908-1912.

Chesneau, L., Dambournet, D., Machicoane, M., Kouranti, I., Fukuda, M., Goud, B., and Echard, A. (2012). An ARF6/Rab35 GTPase Cascade for Endocytic Recycling and Successful Cytokinesis. Curr Biol.

Chevallier, J., Koop, C., Srivastava, A., Petrie, R.J., Lamarche-Vane, N., and Presley, J.F. (2009). Rab35 regulates neurite outgrowth and cell shape. FEBS Lett 583, 1096-1101.

Dambournet, D., Machicoane, M., Chesneau, L., Sachse, M., Rocancourt, M., El Marjou, A., Formstecher, E., Salomon, R., Goud, B., and Echard, A. (2011). Rab35 GTPase and OCRL phosphatase remodel lipids and F-actin for successful cytokinesis. Nat Cell Biol 13, 981-988.

Eathiraj, S., Mishra, A., Prekeris, R., and Lambright, D.G. (2006). Structural Basis for Rab11-mediated Recruitment of FIP3 to Recycling Endosomes. J Mol Biol 364, 121-135.

Egami, Y., Fukuda, M., and Araki, N. (2011). Rab35 regulates phagosome formation through recruitment of ACAP2 in macrophages during FcgammaR-mediated phagocytosis. J Cell Sci 124, 3557-3567.

Elia, N., Sougrat, R., Spurlin, T.A., Hurley, J.H., and Lippincott-Schwartz, J. (2011). Dynamics of endosomal sorting complex required for transport (ESCRT) machinery during cytokinesis and its role in abscission. Proc Natl Acad Sci U S A 108, 4846-4851.

Emoto, K., and Umeda, M. (2000). An essential role for a membrane lipid in cytokinesis. Regulation of contractile ring disassembly by redistribution of phosphatidylethanolamine. J Cell Biol 149, 1215-1224.

Field, S.J., Madson, N., Kerr, M.L., Galbraith, K.A., Kennedy, C.E., Tahiliani, M., Wilkins, A., and Cantley, L.C. (2005). PtdIns(4,5)P2 functions at the cleavage furrow during cytokinesis. Curr Biol 15, 1407-1412.

Fielding, A.B., Schonteich, E., Matheson, J., Wilson, G., Yu, X., Hickson, G.R., Srivastava, S., Baldwin, S.A., Prekeris, R., and Gould, G.W. (2005). Rab11-FIP3 and FIP4 interact with Arf6 and the exocyst to control membrane traffic in cytokinesis. Embo J 24, 3389-3399.

Fyfe, I., Schuh, A.L., Edwardson, J.M., and Audhya, A. (2011). Association of the endosomal sorting complex ESCRT-II with the Vps20 subunit of ESCRT-III generates a curvature-sensitive complex capable of nucleating ESCRT-III filaments. J Biol Chem 286, 34262-34270.

Ghazi-Tabatabai, S., Saksena, S., Short, J.M., Pobbati, A.V., Veprintsev, D.B., Crowther, R.A., Emr, S.D., Egelman, E.H., and Williams, R.L. (2008). Structure and disassembly of filaments formed by the ESCRT-III subunit Vps24. Structure 16, 1345-1356.

Glotzer, M. (2005). The molecular requirements for cytokinesis. Science 307, 1735-1739.

Gromley, A., Yeaman, C., Rosa, J., Redick, S., Chen, C.T., Mirabelle, S., Guha, M., Sillibourne, J., and Doxsey, S.J. (2005). Centriolin anchoring of exocyst and SNARE complexes at the midbody is required for secretory-vesicle-mediated abscission. Cell 123, 75-87.

Guizetti, J., Schermelleh, L., Mantler, J., Maar, S., Poser, I., Leonhardt, H., Muller-Reichert, T., and Gerlich, D.W. (2011). Cortical constriction during abscission involves helices of ESCRT-III-dependent filaments. Science 331, 1616-1620.

Hales, C.M., Griner, R., Hobdy-Henderson, K.C., Dorn, M.C., Hardy, D., Kumar, R., Navarre, J., Chan, E.K., Lapierre, L.A., and Goldenring, J.R. (2001). Identification and characterization of a family of Rab11-interacting proteins. J Biol Chem 276, 39067-39075.

Hickson, G.R.X., Matheson, J., Riggs, B., Maier, V.H., Fielding, A.B., Prekeris, R., Sullivan, W., Barr, F.A., and Gould, G.W. (2003). Arfophilins are dual Arf/Rab11 binding proteins that regulate recycling endosome distribution and are related to Drosophila nuclear fallout. Molecular Biology of the Cell 14, 2908-2920.

Horgan, C.P., Hanscom, S.R., Jolly, R.S., Futter, C.E., and McCaffrey, M.W. (2010). Rab11-FIP3 links the Rab11 GTPase and cytoplasmic dynein to mediate transport to the endosomal-recycling compartment. J Cell Sci 123, 181-191.

Horgan, C.P., Walsh, M., Zurawski, T.H., and McCaffrey, M.W. (2004). Rab11-FIP3 localises to a Rab11-positive pericentrosomal compartment during interphase and to the cleavage furrow during cytokinesis. Biochem Biophys Res Commun 319, 83-94.

Hsu, S.C., Hazuka, C.D., Foletti, D.L., and Scheller, R.H. (1999). Targeting vesicles to specific sites on the plasma membrane: the role of the sec6/8 complex. Trends Cell Biol 9, 150-153.

Hurley, J.H. (2010). The ESCRT complexes. Crit Rev Biochem Mol Biol 45, 463-487.

Jagoe, W.N., Lindsay, A.J., Read, R.J., McCoy, A.J., McCaffrey, M.W., and Khan, A.R. (2006). Crystal structure of rab11 in complex with rab11 family interacting protein 2. Structure 14, 1273-1283.

Janetopoulos, C., and Devreotes, P. (2006). Phosphoinositide signaling plays a key role in cytokinesis. J Cell Biol 174, 485-490.

Junutula, J.R., Schonteich, E., Wilson, G.M., Peden, A.A., Scheller, R.H., and Prekeris, R. (2004). Molecular characterization of Rab11 interactions with members of the family of Rab11-interacting proteins. J Biol Chem 279, 33430-33437.

Jurgens, G. (2005). Plant cytokinesis: fission by fusion. Trends Cell Biol 15, 277-283.

Kanno, E., Ishibashi, K., Kobayashi, H., Matsui, T., Ohbayashi, N., and Fukuda, M. (2010). Comprehensive screening for novel rab-binding proteins by GST pull-down assay using 60 different mammalian Rabs. Traffic 11, 491-507.

Kouranti, I., Sachse, M., Arouche, N., Goud, B., and Echard, A. (2006). Rab35 regulates an endocytic recycling pathway essential for the terminal steps of cytokinesis. Curr Biol 16, 1719-1725.

Kunda, P., Pelling, A.E., Liu, T., and Baum, B. (2008). Moesin controls cortical rigidity, cell rounding, and spindle morphogenesis during mitosis. Curr Biol 18, 91-101.

Kunda, P., Rodrigues, N.T., Moeendarbary, E., Liu, T., Ivetic, A., Charras, G., and Baum, B. (2011). PP1-Mediated Moesin Dephosphorylation Couples Polar Relaxation to Mitotic Exit. Curr Biol.

Lata, S., Schoehn, G., Jain, A., Pires, R., Piehler, J., Gottlinger, H.G., and Weissenhorn, W. (2008). Helical structures of ESCRT-III are disassembled by VPS4. Science 321, 1354-1357.

Low, S.H., Li, X., Miura, M., Kudo, N., Quinones, B., and Weimbs, T. (2003). Syntaxin 2 and endobrevin are required for the terminal step of cytokinesis in mammalian cells. Dev Cell 4, 753-759.

Montagnac, G., Sibarita, J.B., Loubery, S., Daviet, L., Romao, M., Raposo, G., and Chavrier, P. (2009). ARF6 Interacts with JIP4 to control a motor switch mechanism regulating endosome traffic in cytokinesis. Curr Biol 19, 184-195.

Nezis, I.P., Sagona, A.P., Schink, K.O., and Stenmark, H. (2010). Divide and ProsPer: the emerging role of PtdIns3P in cytokinesis. Trends Cell Biol 20, 642-649.

Pelissier, A., Chauvin, J.P., and Lecuit, T. (2003). Trafficking through Rab11 endosomes is required for cellularization during Drosophila embryogenesis. Curr Biol 13, 1848-1857.

Piekny, A.J., and Maddox, A.S. (2010). The myriad roles of Anillin during cytokinesis. Semin Cell Dev Biol 21, 881-891.

Prekeris, R. (2003). Rabs, Rips, FIPs, and Endocytic Membrane Traffic. ScientificWorldJournal 3, 870-880.

Prekeris, R., and Gould, G.W. (2008). Breaking up is hard to do - membrane traffic in cytokinesis. J Cell Sci 121, 1569-1576.

Prekeris, R., Davies, J.M., and Scheller, R.H. (2001). Identification of a novel Rab11/25 binding domain present in Eferin and Rip proteins. J Biol Chem 276, 38966-38970.

Prigent, M., Dubois, T., Raposo, G., Derrien, V., Tenza, D., Rosse, C., Camonis, J., and Chavrier, P. (2003). ARF6 controls post-endocytic recycling through its downstream exocyst complex effector. J Cell Biol 163, 1111-1121.

Riggs, B., Fasulo, B., Royou, A., Mische, S., Cao, J., Hays, T.S., and Sullivan, W. (2007). The concentration of Nuf, a Rab11 effector, at the microtubule-organizing center is cell cycle regulated, dynein-dependent, and coincides with furrow formation. Mol Biol Cell 18, 3313-3322.

Riggs, B., Rothwell, W., Mische, S., Hickson, G.R., Matheson, J., Hays, T.S., Gould, G.W., and Sullivan, W. (2003). Actin cytoskeleton remodeling during early Drosophila furrow formation requires recycling endosomal components Nuclear-fallout and Rab11. J Cell Biol 163, 143-154.

Samson, R.Y., Obita, T., Freund, S.M., Williams, R.L., and Bell, S.D. (2008). A role for the ESCRT system in cell division in archaea. Science 322, 1710-1713.

Schiel, J.A., and Prekeris, R. (2011). ESCRT or endosomes?: Tales of the separation of two daughter cells. Commun Integr Biol 4, 606-608.

Schiel, J.A., Park, K., Morphew, M.K., Reid, E., Hoenger, A., and Prekeris, R. (2011). Endocytic membrane fusion and buckling-induced microtubule severing mediate cell abscission. J Cell Sci 124, 1411-1424.

Schonteich, E., Pilli, M., Simon, G.C., Matern, H.T., Junutula, J.R., Sentz, D., Holmes, R.K., and Prekeris, R. (2007). Molecular characterization of Rab11-FIP3 binding to ARF GTPases. Eur J Cell Biol 86, 417-431.

Simon, G.C., Schonteich, E., Wu, C.C., Piekny, A., Ekiert, D., Yu, X., Gould, G.W., Glotzer, M., and Prekeris, R. (2008). Sequential Cyk-4 binding to ECT2 and FIP3 regulates cleavage furrow ingression and abscission during cytokinesis. Embo J.

Sirokmany, G., Szidonya, L., Kaldi, K., Gaborik, Z., Ligeti, E., and Geiszt, M. (2006). Sec14 homology domain targets p50RhoGAP to endosomes and provides a link between Rab and Rho GTPases. J Biol Chem 281, 6096-6105.

Skop, A.R., Bergmann, D., Mohler, W.A., and White, J.G. (2001). Completion of cytokinesis in C. elegans requires a brefeldin A-sensitive membrane accumulation at the cleavage furrow apex. Curr Biol 11, 735-746.

TerBush, D.R., Maurice, T., Roth, D., and Novick, P. (1996). The Exocyst is a multiprotein complex required for exocytosis in Saccharomyces cerevisiae. Embo J 15, 6483-6494.

Walseng, E., Bakke, O., and Roche, P.A. (2008). Major histocompatibility complex class II-peptide complexes internalize using a clathrin- and dynamin-independent endocytosis pathway. J Biol Chem 283, 14717-14727.

Wilson, G.M., Fielding, A.B., Simon, G.C., Yu, X., Andrews, P.D., Peden, A.A., Gould, G.W., and Prekeris, R. (2004). The FIP3-Rab11 Protein Complex Regulates Recycling Endosome Targeting to the Cleavage Furrow during Late Cytokinesis. Mol Biol Cell.

Wollert, T., and Hurley, J.H. (2010). Molecular mechanism of multivesicular body biogenesis by ESCRT complexes. Nature 464, 864-869.

Wu, S., Mehta, S.Q., Pichaud, F., Bellen, H.J., and Quiocho, F.A. (2005). Sec15 interacts with Rab11 via a novel domain and affects Rab11 localization in vivo. Nat Struct Mol Biol 12, 879-885.

Zhang, J., Fonovic, M., Suyama, K., Bogyo, M., and Scott, M.P. (2009). Rab35 controls actin bundling by recruiting fascin as an effector protein. Science 325, 1250-1254.

Zhou, Y.T., Chew, L.L., Lin, S.C., and Low, B.C. (2010). The BNIP-2 and Cdc42GAP homology (BCH) domain of p50RhoGAP/Cdc42GAP sequesters RhoA from inactivation by the adjacent GTPase-activating protein domain. Mol Biol Cell 21, 3232-3246.

9

Enzymology and Regulation of ArfGAPs and ArfGEFs

Peng Zhai, Xiaoying Jian, Ruibai Luo and Paul A. Randazzo
National Cancer Institute, National Institutes of Health
USA

1. Introduction

Arf family GTP-binding proteins, a subfamily of the Ras superfamily, are critical regulators of membrane traffic and actin remodeling (Kahn, 2009; Kahn et al., 2006; Gillingham and Munro, 2007; Donaldson and Jackson, 2011). The Arf family contains six Arf proteins in most mammals (five in humans) that are divided into three classes based on primary sequence and phylogenetic considerations (Kahn et al., 2006). The function of the Arf proteins requires switching between GDP and GTP bound forms. The accessory proteins that mediate the transitions between Arf•GDP and Arf•GTP function as enzymes and can be studied using the formalisms of enzymology.

Like other GTP binding proteins, switching between Arf•GDP and Arf•GTP is achieved by a controlled cycle of GTP binding and hydrolysis. The two steps are catalyzed by distinct enzymes. The conversion of Arf•GDP to Arf•GTP is accomplished through nucleotide exchange, with the apo form of Arf as an intermediate. Nucleotide, however, binds tightly to Arf, resulting in very slow intrinsic exchange rates, and the apo form of Arf is unstable. Guanine nucleotide exchange factors (GEFs) are enzymes that accelerate the reaction, by decreasing affinity for nucleotide and stabilizing the apo form of Arf (Casanova, 2007; Gillingham and Munro, 2007; Renault et al., 2003). Arf proteins are unusual among Ras superfamily proteins in having no detectable GTPase activity. Conversion of Arf•GTP to Arf•GDP is catalyzed by GTPase-activating proteins (GAPs;Gillingham and Munro, 2007; Donaldson and Jackson, 2011; Ha et al., 2008b; Kahn et al., 2008; Randazzo and Hirsch, 2004; Spang et al., 2010). The GEFs and GAPs are both large families of proteins with diverse structural features. The control of binding and hydrolysis of GTP by Arf is thought to be achieved by regulation of the ArfGAPs and ArfGEFs. The study of the ArfGAPs and ArfGEFs as allosterically controlled enzymes is providing valuable information about their regulation and insights into the roles in cell physisology.

2. ArfGAP family of proteins

Thirty-one genes encode proteins with Arf GAP domains in humans (Kahn et al., 2008). The proteins are divided into 10 groups (Figure. 1) based on domain structure and phylogenetic analysis (Kahn et al., 2008). Six groups have the ArfGAP catalytic domain at the extreme N-terminus of the protein. In the other four groups, which comprise 20 proteins, the ArfGAP domain is sandwiched between a PH and Ankyrin repeat domains. Other than having the common ArfGAP catalytic domain, the groups are structurally diverse. Some of the structural differences are thought to contribute to differential regulation of the catalytic activity of the GAPs through lipid or protein binding, which has been tested for several ArfGAPs. The molecular mechanism for catalysis by ArfGAPs remains unclear. The first reported structure of ArfGAP/Arf complex is for the ArfGAP domain of ArfGAP1 in complex with Arf1 bound to GDP, the product of the reaction (Goldberg). This structure argued against the general "Arginine-finger" mechanism for the GAPs that has been described for GAPs for other Ras superfamily members (Scheffzek et al., 1998). Studies of the enzymology of ArfGAP1 and ASAP1 first revealed that the available crystal structure may have to be reinterpreted.

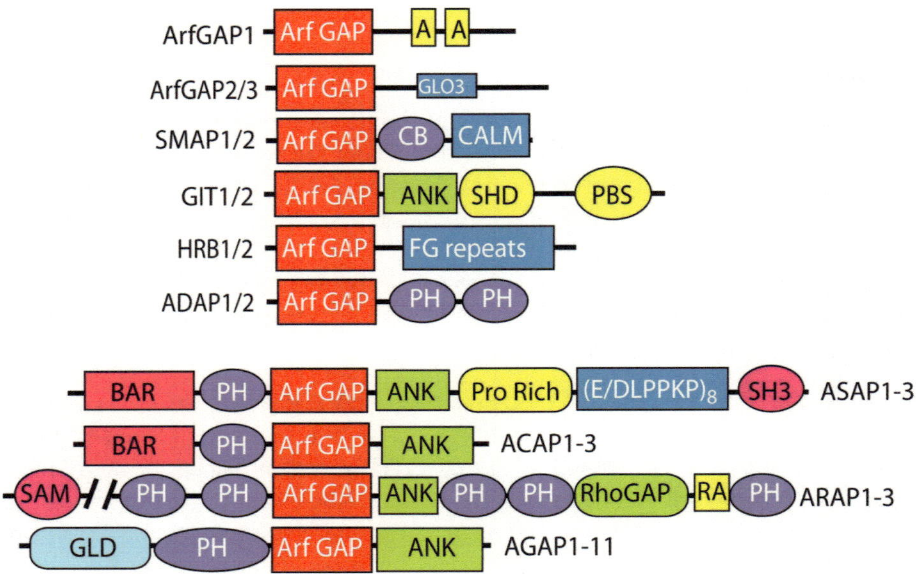

Fig. 1. Schematic of human ArfGAPs. Abbreviations-A, Ankyrin repeat; BAR, Bin/Amphiphysin/Rvs; PBS, paxillin-binding site; PH, pleckstrin homology; SAM, sterile alpha motif; SH3, Src-homology 3; SHD, Spa2 homology; CB, clathrin box; CALM, CALM-binding domain; GLD, GTP-binding protein-like domain; RA, Ras association domain; and GLO3, GLO3 motif.

The ArfGAPs contained all necessary elements for efficiently inducing GTP hydrolysis (Luo et al., 2007; Jian et al., 2009). In addition, arginine 497 of ASAP1 and arginine 50 of ArfGAP1 likely serve in the catalytic capacity described for the arginine finger in Ras GAPs. Indeed, recent structural studies of another ArfGAP/Arf complex reveals an arginine that is catalytic (Ismail et al., 2010). The enzymology indicates that our current understanding of the catalytic mechanism is still incomplete. The proteins used for the crystal (Ismail et al., 2010) have $1/10^5$ the optimal activity of the ASAP3. Thus, in addition to providing insights into the cellular functions of ArfGAPs, which will be described in more detail below, the enzymology has provided important information about the molecular basis of catalysis. Here, we discuss general considerations about the enzymology of these proteins and then discuss two specific examples.

2.1 General considerations in the kinetic analysis of ArfGAPs

ArfGAPs catalyze what can be considered a single substrate reaction (the second substrate is water), schematized as

$$\mathrm{Arf \bullet GTP} \xrightarrow{\mathrm{GAP}} \mathrm{Arf \bullet GDP} + \mathrm{P_i} \tag{1}$$

For more complex schemes, we simplify the notation. If we let E =GAP, S=Arf•GTP and P=Arf•GDP, then the scheme is

$$\mathrm{E + S} \underset{k_{-1}}{\overset{k_1}{\rightleftharpoons}} \mathrm{ES} \underset{k_{-2}}{\overset{k_2}{\rightleftharpoons}} \mathrm{EP} \underset{k_{-3}}{\overset{k_3}{\rightleftharpoons}} \mathrm{E + P} \tag{2}$$

If we consider that ES and EP rapidly isomerize, then

$$\mathrm{E + S} \underset{k_{-1}}{\overset{k_1}{\rightleftharpoons}} \mathrm{ES} \underset{k_{-2}}{\overset{k_2}{\rightleftharpoons}} \mathrm{E + P} \tag{3}$$

The kinetics can be complex for several reasons. Excluding other factors such as dimerization of the ArfGAP and allosteric modifiers, the first possible additional complexity of the simple E+S→ ES scheme is that the substrate is restricted to a surface and the reaction occurs on the same membrane surface. The restriction to the surface is important for two reasons related to analyzing GAP activity. First, if the enzyme is also restricted to the surface, the collision rate of enzyme and substrate will be determined by the surface concentration, i.e. mass/area, rather than mass/volume. Second, the quality of the surface is important. In the few cases examined, the quality of the surface is a more important consideration than the total surface area (Jian et al., 2009). With Arf GAPs that reversibly associate with surfaces, surface dilution does not seem to affect reaction rate so long as surface area is about 5 fold greater than the surface occupied by the maximum amount of Arf present. The quality of surface, however, is critical with different parameters determined when using mixed micelles of Triton X-100, LUVs containing all saturated acyl groups in the phospholipids or LUVs containing unsaturated lipids. Catalysis and regulation of the ArfGAPs can be analyzed without invoking surface dilution kinetics but comparisons between proteins are only valid when the same quantity and composition of lipid or detergent are used as the hydrophobic surface to support the reaction. Keeping the surface constant, the initial velocity equation is

$$v_i = \frac{V_{max} \cdot \mathrm{Arf \bullet GTP}}{K_m + \mathrm{Arf \bullet GTP}}, \tag{4}$$

the Michaelis-Menten equation for initial reaction velocity. Using the symbol S for the substrate Arf•GTP gives the familiar notation used in the equation.

$$v_i = \frac{V_{max} \cdot S}{K_m + S} \tag{5}$$

The effect of an allosteric modifier is schematized

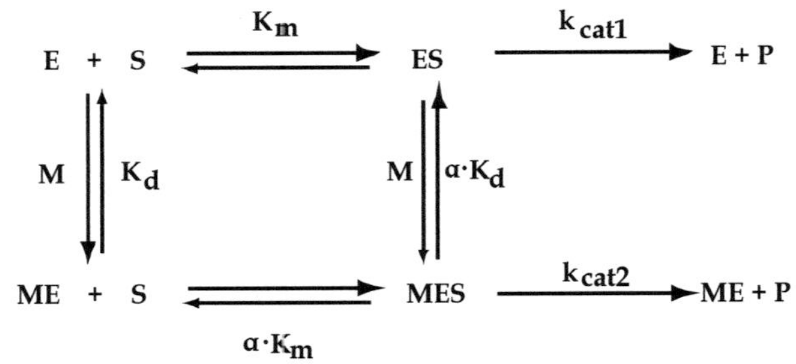

(6)

The initial velocity equation is:

$$v_i = \frac{\dfrac{E_t \cdot k_{cat1} \cdot S}{K_m} + \dfrac{E_t \cdot k_{cat2} \cdot M \cdot S}{\alpha \cdot K_m \cdot K_d}}{1 + \dfrac{S}{K_m} + \dfrac{M}{K_d} + \dfrac{S \cdot M}{\alpha \cdot K_m \cdot K_d}} \qquad (7)$$

Where E is the GAP (enzyme), S is Arf•GTP and M is the allosteric modifier. K_m is the concentration of substrate at which the enzyme proceeds with half maximal velocity, k_{cat} is the turnover number, α is the effect of the modifier on substrate binding, E_t is the total GAP in the reaction, and $E_t \cdot k_{cat} = V_{max}$.

As briefly discussed for ASAP1 in this chapter, a second consideration for ArfGAPs is the potential role of dimerization in regulating the reaction. We are not aware of any description, to date, of an ArfGAP that requires the consideration of dimerization to explain the kinetics, but only a few ArfGAPs have been analyzed.

2.2 ASAP1: Examination of putative lipid binding domain leads to model of activation by two signals

ASAP1 is of interest to cell biologists because it has been implicated as one critical factor for cancer cell invasion and metastasis (Ha et al., 2008b; Sabe et al., 2006) with the most compelling evidence coming from studies of uveal melanoma (Ehlers et al., 2005). Consistent with a potential role in cancer, ASAP1 affects cellular adhesions and cell migration (Randazzo et al., 2000; Onodera et al., 2005; Bharti et al., 2007; Ha et al., 2008b; Ha et al., 2008a). Recently, additional interest in ASAP1 comes from the study of ciliogenesis (Ward et al., 2011;Mazelova et al., 2009). ASAP1 is necessary for the delivery of proteins to primary cilia. The molecular bases for the contribution of ASAP1 to the pathologic behavior of cells or the physiological cellular function are not known. The mechanisms by which catalytic activity is regulated are being defined with the hope of furthering the understanding of the role of ASAP1 in cell physiology.

The regulation of ArfGAPs by phosphoinositides was an early finding that led to the purification of ASAP family Arf GAPs (Randazzo and Kahn, 1994). ASAPs contain, from the N-terminus, a BAR, PH, ArfGAP, Ank repeat, Proline rich, E/DLPPKP repeat and SH3 domains, while ASAP1 also has a 30 amino acid extension on the N-terminus of the BAR domain (Jian et al., 2009). Although many Arf GAPs have PH domains, which can bind to phosphoinositides, the hypothesis that phosphoinositide binding to the PH domain regulates catalysis by the GAP domain has been most extensively examined for ASAP1.

The PH domain is one of two potential lipid binding domains in ASAP1, the other being the BAR domain. The PH domain of ASAP1 has some consensus with PIP_2 binding PH domains and was found to bind to phosphatidylinositol 4,5-bisphosphate ($PI4,5P_2$). Initial analysis revealed that the PH domain was important for catalytic activity. Recombinant protein with the PH domain had 3 - 4 orders of magnitude greater activity than recombinant protein lacking the PH domain (Kam et al., 2000; Che et al., 2005; Luo et al., 2008). Extrapolating from other PH domain proteins, the function of the PH domain was thought to be recruitment of the protein to PIP_2-containing membranes, which also contained the substrate, Arf•GTP; however, in the case of ASAP1, recruitment could be uncoupled from activation (Che et al., 2005). PIP_2 binding was found to cause conformational changes in the PH domain in ASAP1 and in more than 100-fold stimulation of catalytic activity (Che et al., 2005). Changing residues in the PH domain that reduced PIP_2 binding resulted in a change in both K_m and k_{cat} for the reaction consistent with the concept that PIP_2 binding induces conformational changes in the protein leading to increased activity. A simple recruitment mechanism would lead to an isolated change in K_m for the enzymatic reaction. In addition, although ASAP1 is recruited to membrane ruffles, the recruitment is independent of the PH domain. PIP_2 binding to the PH domain of ASAP1 may be necessary for enzymatic activity, but it may not be sufficient to regulate the protein. There must be a signal to at least recruit ASAP1 to the site of action. Studies examining the BAR domain reveal regulation may be more complex.

BAR domains are bundles of 3 α helices that homodimerize to form banana shaped structures (Zhu et al., 2007). The function of some BAR domains is related to binding membranes where they are thought to either induce or sense curvature (Habermann, 2004; McMahon and Gallop, 2005; McMahon and Gallop, 2005) (Figure 2). This hypothesis was tested for ASAP1. The BAR domain of ASAP1 dimerizes with a dissociation constant of less than 10 nM (Nie et al., 2006). The isolated BAR was not found to be stable, but the isolated BAR-PH tandem was stable and could induce membrane curvature (Nie et al., 2006). Any

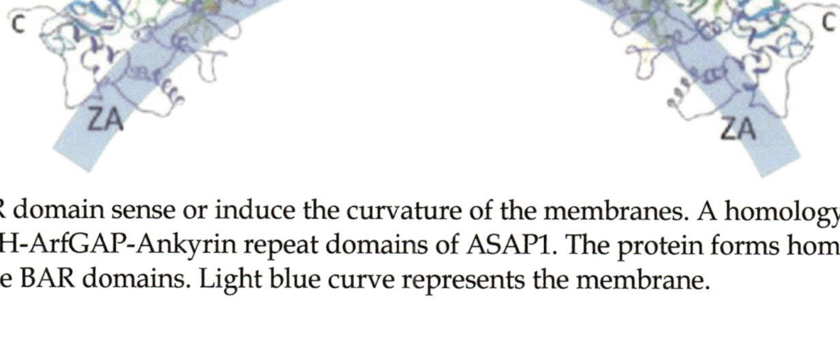

Fig. 2. BAR domain sense or induce the curvature of the membranes. A homology model of the BAR-PH-ArfGAP-Ankyrin repeat domains of ASAP1. The protein forms homodimer through the BAR domains. Light blue curve represents the membrane.

additional domains, e.g. a construct comprised of the BAR, PH, ArfGAP and Ank repeat domains did not induce membrane curvature nor sense membrane curvature. Comparisons of recombinant proteins derived from ASAP1 containing or lacking the BAR domain revealed that the extension from the BAR domain inhibited GAP activity, presumably acting in trans within the homodimer of ASAP1 (Jian et al., 2009) (Figure. 3). Neither lipid composition nor curvature of vesicles affected the autoinhibition. This leads to the still untested hypothesis that proteins that bind to the BAR domain of ASAP1 may stimulate activity by relieving the autoinhibition. The current model for regulation of ASAP1 is that simultaneous binding of protein to the BAR domain and to the PH domain leads to activation.

The contribution of the domains C-terminal of the ankyrin repeats in the regulation of GAP activity has not been extensively examined. Plausible models include the SH domain interacting with PXXP motifs in the N-terminus of the molecule or molecules that bind to the PXXP motifs c-terminal of the ank repeat domains interacting with one of the N-terminal domain. There is support for the idea that in mouse ASAP1, src binding to the PXXP motif can phosphorylate residues near the PH domain resulting is reduced GAP activity (Kruljac-Letunic et al., 2003).

The ACAP subfamily is similar to the ASAPs having a structure comprised of BAR,PH, ArfGAP and Ank repeat domains. Similar to ASAPs, the ACAPs are regulated by phosphoinositides, which was expected for the PH domain. Different than ASAPs (Figure. 3), the ACAPs do not contain the N-terminal extension of the BAR domain that has an autoinhibitory function in ASAPs. The ACAPs, therefore, are likely to be regulated by distinct mechanisms from those used by the ASAPs. Inhibition by proteins that associate with the BAR-PH domain is one possibility.

	Inhibitory
ASAP1	MRSSASRLSSFSSRDSLWNRMPDQI SVSEFIAETTEDYNSPTTSSFTTRLHNCRN
ASAP2	MPD QI SVS EFVAETHEDYKAPTASSFTTRTAQCRN
ASAP3	MPE QF SVAEFLAVTAEDLSSPAGAAAFAAKMPRYR
ACAP1	MKMTVDFEECLKDSPRFR
ACAP2	MTVKLDFEECLKDSPRFR

Fig. 3. Alignment of the N-termini of ASAPs and ACAPs. The autoinhibitory fragment identified in ASAP1 is indicated as "Inhibitory." Identities among ASAPs are indicated by gray shading. Loci numbers: ASAP1, NP_060952; ASAP2, NP_003878; ASAP3, NP_060177; ACAP1, NP_055531; ACAP2, NP_036419.

Other examples of ArfGAPs regulated by phosphoinositides include ARAPs, ACAPs and AGAPs. ARAPs contain, from the N-terminus, SAM, two PH, ArfGAP, two PH, RhoGAP RA, and PH domains (total of 5 PH domains). PH domains 1 and 3 (see schematic in Figure 1) have consensus for PIP3 binding PH domains and PIP3-binding to PH domain 1 stimulates GAP activity (Campa et al., 2009). Like the PH domain for ASAP1, the PH domain does not mediate recruitment to the membrane surface containing Arf•GTP. AGAPs are also stimulated by phosphoinositides but specificity among the phosphoinositides is not apparent.

2.3 AGAP1: Allosteric regulation through a GTP-binding protein like domain

The AGAP proteins are comprised of a GTP-binding protein-like domain (GLD), split PH, ArfGAP and ankyrin repeat domains (Nie et al., 2002). Two AGAPs bind clathrin adaptor proteins and have effects on endocytic membrane traffic (Nie et al., 2003; Nie et al., 2005). One of these, AGAP2, has also been implicated in the progression of glioblastoma (Ye and Snyder, 2004). Defining the regulation of GAP activity is of significance both to understanding membrane traffic and cancer.

Initial examination of regulation of the AGAPs focused on phosphoinositides and later, with the discovery that clathrin adaptor proteins bind to the PH domain, on the adaptors (Nie et al., 2005; Nie et al., 2002). GLD was at first discounted as a regulator of the ArfGAP catalytic activity because deletion of the domain did not affect activity. Recombinant AGAPs, with or without the GLD have less than 1 % of the activity of ASAP1 and no apparent substrate specificity, indicating that there was likely a means of increasing AGAP catalytic activity, as explained below. Two-hybrid screening revealed that the GLD is a protein binding site. When a complex is formed, the GAP activity is increased for Arf1 and decreased for Arf6. The mechanistic basis still needs to be determined (Luo et al., submitted).

The additional value of the work on AGAP1 is that it illustrated two concepts important to studying enzymes that regulate proteins. First, turnover number, or at least catalytic power, is relevant. Low activity could indicate a poorly folded protein, in which case the data obtained may not be physiologically relevant. On the other hand, if the protein is properly folded, as was the case for AGAP1, low activity may indicate that positive regulatory mechanisms remain to be discovered. Second, studying an inactive enzyme could be misleading. Although enough activity may be present to make measurements, the enzyme may not optimally recognize the physiological substrate. Other proteins with similarity to the physiological substrate may be fortuitously used. These properties of the GAPs need to be considered when expressing the proteins in cells, since activators may be titrated away, and the bulk of the GAP may be relatively inactive.

2.4 ArfGAP1, 2 and 3: Control by two interacting proteins

ArfGAP1 was the first identified ArfGAP and the first GAP found to regulate membrane traffic, although its precise role remains unknown (Hsu, 2011; Hsu et al., 2009;Kahn, 2009; East and Kahn, 2011; Kahn, 2011;Beck et al., 2011; Weimer et al., 2008; Beck et al., 2009b; Shiba et al., 2011; Spang et al., 2010). The protein is approximately 50 kDa with the Arf GAP domain at the N-terminus and a unique C-terminus that contains two ALPS motifs that are described below. ArfGAP2 and 3 have a similar overall structure but in place of the ALPS motifs contain a Glo3 homology domain (Figure 1). ArfGAP1, ArfGAP2 and ArfGAP3 localize to the Golgi where they regulate Golgi-to-ER membrane traffic. Early work reported that PIP_2 could activate ArfGAP1, despite lack of a PH domain, but that result was later found to result from the use of nonmyristoylated Arf as a substrate, which is recruited to membrane surfaces by PIP_2 (Randazzo, 1997). Later, diacyglycerol was found to activate ArfGAP1. The effect was attributed to effects on lipid packing in the vesicles used in the experiments (Antonny et al., 1997). Subsequently, increasing vesicle curvature, which also results in loosened packing of the lipid head groups, was found to increase activity (Bigay et al., 2003). The effect depends on two ArfGAP Lipid Packing Sensor (ALPS) motifs in

ArfGAP1. The model based on this result was embraced as it could explain the timing of GTP hydrolysis on Arf during coated vesicle formation (for review see Nie and Randazzo, 2006). However, the idea of curvature sensitivity of ArfGAP1 has been difficult to reconcile with current models of coatomer and ArfGAP1 function, and regulation by interaction with proteins may be a plausible alternative model for the regulation of this protein. Furthermore, the curvature sensing model has not been tested in vivo and there is little kinetic support of the model. It is not known, for instance, how the change in curvature affects enzymatic parameters K_m and k_{cat}.

A second model for the regulation of ArfGAP1 has been proposed, which also seems to apply to ArfGAP2 and ArfGAP3. ArfGAP1 binds to the vesicle coat protein coatomer and to cargo proteins (Hsu et al., 2009; Lee et al., 2005). Coatomer is a protein that drives formation of transport intermediates that carry material between the Golgi apparatus and the endoplasmic reticulum (ER) and cargo proteins are the material carried in the transport intermediates. ArfGAP2 and ArfGAP3 were subsequently discovered and found to also bind coatomer and cargo (Frigerio et al., 2007; Kliouchnikov et al., 2009; Weimer et al., 2008). As early as 1999, coatomer was found to stimulate GAP activity of ArfGAP1 (Goldberg, 1999; Goldberg, 2000). The following model was formulated to help analyze the kinetics (note that we abbreviate coatomer as C in the schematic, instead of M which is used for allosteric modifier in other sections of this chapter).

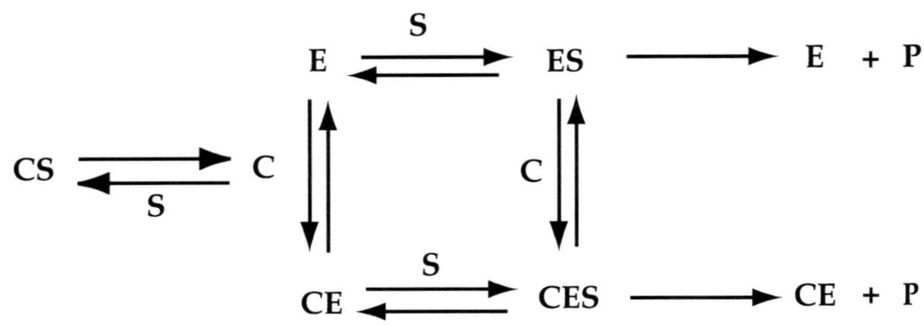

(8)

Coatomer is unusual as an allosteric modifier because it could bind the substrate Arf•GTP independently of ArfGAP1. Therefore, the possibility of substrate sequestration by the allosteric modifier had to be taken into account when analyzing kinetics. If coatomer activated ArfGAP1 by direct interaction but was also able to sequester Arf•GTP, titration of coatomer would result in a biphasic curve as was observed, described by the following initial rate equation:

$$v_i = \frac{K_{cs} \cdot (V_{max1} \cdot K_{m2} \cdot K_c + V_{max2} \cdot K_{m1} \cdot C) \cdot S}{K_{m1} \cdot K_{m2} \cdot (K_{cs} + C) \cdot (K_c + C) + K_{cs} \cdot (K_{m2} \cdot K_c + K_{m1} \cdot C) \cdot S} \quad (9)$$

Where K_{m1} is the Michaelis constant for the Arf with GAP in the absence of coatomer, K_{m2} is the Michaelis constant for Arf with GAP in complex with coatomer, V_{max1} is is the limiting rate of the reaction in the absence of coatomer, V_{max2} is the limiting rate for the GAP in complex with coatomer, K_c is the affinity for coatomer in the absence of Arf, K_{cs} is the affinity for coatomer in the presence of Arf, C is coatomer, E is GAP and S is Arf•GTP. Consistent with the prediction of the equation, Luo and colleagues (Luo and Randazzo, 2008; Luo et al., 2009) found a biphasic coatomer dependence under conditions of limiting substrate. At low substrate concentration, the sequestration effect is dominant at high coatmer concentration, while at the high substrate concentration, the activation effect is dominant.

ArfGAP2 and ArfGAP3 were found to be similar in that GAP activity depended on binding to coatomer. Titration revealed coatomer affected the K_m for both ArfGAP1 and ArfGAP2. ArfGAP2 and ArfGAP3 bind coatomer more tightly than does ArfGAP1. Consequently, the coatomer•ArfGAP2/3 complex could be formed at concentrations of coatomer low enough for substrate sequestration to be ignored. For this reason, ArfGAP2 was used for subsequent studies examining the effect of cargo on GAP activity. In the experiments a peptide from cargo was used as a model of cargo because of the challenges of expressing recombinant transmembrane proteins in bacteria. Cargo was found to act as an allosteric modifier, increasing the k_{cat} of the reaction. The effect depended on the presence of coatomer. ArfGAP1 was also stimulated by cargo in the presence of coatomer. Therefore, rather than curvature, ArfGAPs that function with coatomer are stimulated by the coat-cargo complex. These results have implications important to our understanding of membrane traffic. Previously Arf was thought to function as a bridge between coat proteins and membranes. Arf•GTP, in this model, is required to hold coat on the membrane through the process of trapping cargo and forming a vesicle. GTP hydrolysis would trigger the dissociation of coat necessary after a vesicle is formed. The curvature sensing model fit this paradigm, since the GAP would be most active on the highly curved vesicle and would have little activity on the flat surface on which the vesicle is formed. However, coat-cargo complex is formed prior to making a vesicle, so that activation of the GAP would also occur prior to vesicle formation. The competing models of the role of Arf and ArfGAPs for the formation of coated vesicles are discussed in more detail in a series of papers published from 2009 to 2011 (Shiba et al., 2011; Hsu, 2011; Hsu et al., 2009; Beck et al., 2009b; Beck et al., 2009a). Importantly, the enzymology has been found valuable to gain insights into biological processes.

3. The ArfGEF family of proteins

There are at least 16 proteins with Arf GEF, also called sec7, domains in humans (Casanova, 2007; Donaldson and Jackson, 2011). They are divided into 5 groups: BIG1/2 and GBF; Brag;

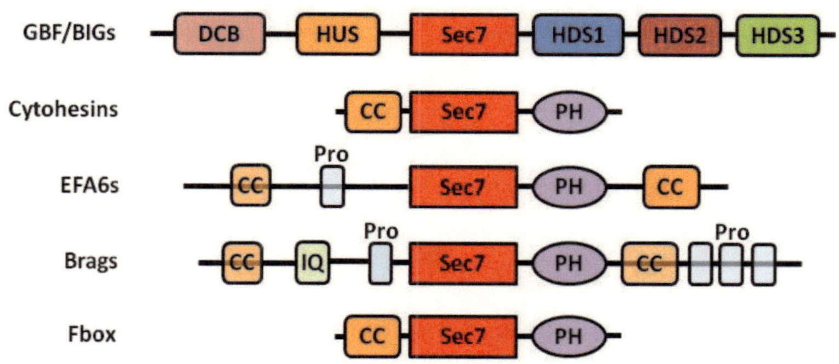

Fig. 4. Schematic of ArfGEFs. Abbreviations-CC, coiled-coil ; DCB, dimerization and cyclophilin-binding domain; HDS, Homology Downstream of Sec7 domain; HUS, Homology Upstream of Sec7 domain; IQ, IQ motif; PH, pleckstrin homology; Pro, Proline-rich. The semi-transparent means not universally present in all subfamily members.

Cytohesins/ARNO, EFA6 and Fbox. Like the ArfGAPs, the GEFs are a family of structurally diverse and complex proteins (Figure 4).

3.1 General considerations in the kinetic analysis of ArfGEFs

The exchange factors catalyze the exchange reaction by a bi bi ping pong mechanism, referring to two substrates (Arf•GDP and GTP), two products (GDP and Arf•GTP) and a reaction that proceeds with binding of the first substrate (Arf•GDP), followed by release of the first product (GDP) and formation of a distinct enzyme intermediate (EA as shown in scheme 10, or F in scheme 11 as it is often presented) prior to binding of the second substrate (GTP) and release of the second product (Arf•GTP). The essential elements of the reaction are schematized as

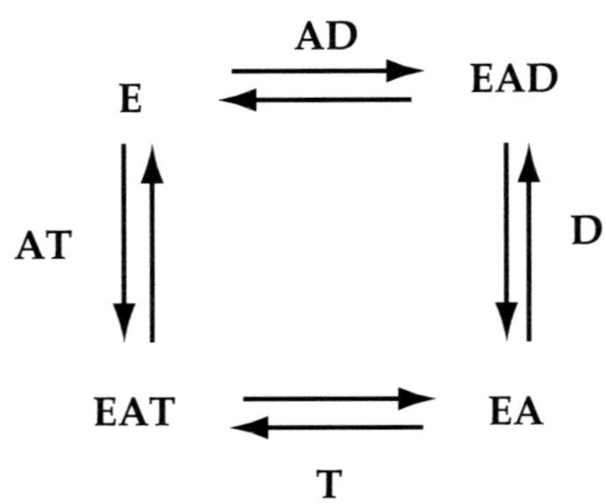

(10)

E = ArfGEF; AD=Arf1•GDP; AT = Arf1•GTP; T = GTP; D = GDP

The general scheme often shown for bi bi ping pong is

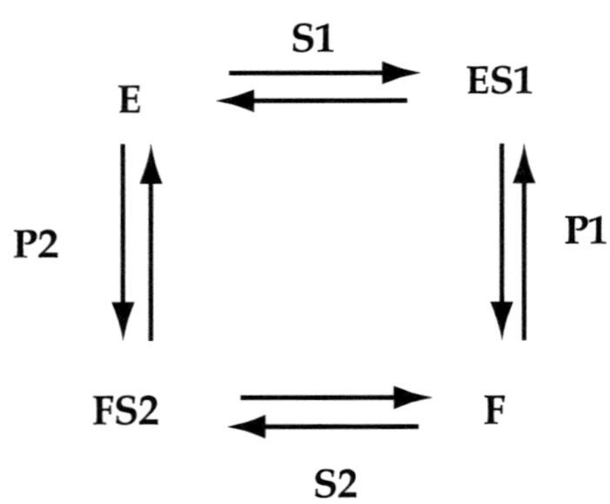

(11)

In the forward direction, both substrates are soluble but the ES1 and EP2 complexes are membrane restricted as is the second product (Arf•GTP). When studying initial rates, the reaction can be treated as a soluble system, without invoking surface dilution kinetics. Nevertheless, the quality of the surface may affect stability of the enzyme complexes starting with ES1 through FS2 and the relaxation of the transition state of the enzyme to the ground state. A similar situation was found for ASAP1, an ArfGAP (Jian et al.,2009; Luo et al., 2007).

The initial velocity of the forward reaction based on the scheme above, excluding the presence of products, is:

$$v_i = \frac{V_1 \cdot S1 \cdot S2}{K_{S2} \cdot S1 + K_{S1} \cdot S2 + S1 \cdot S2} \quad (12)$$

where S1 is Arf•GDP and S2 is GTP. Holding S1 or S2 constant at saturating concentrations gives the following two equations:

$$v_i = \frac{V_{max,app1} \cdot S2}{K_{m,app1} + S2} \quad (13)$$

$$v_i = \frac{V_{max,app2} \cdot S1}{K_{m,app2} + S1} \quad (14)$$

Note that these equations do not account for allosterism. The kinetic consequences of dimerization have not been examined. However, other allosteric modifiers have been considered and have been treated as described for GAPs, represented in schematic 6 and equation 7.

As for the GAPs, the allosteric modifier could potentially modify the K_m or the k_{cat} for one or both substrates. The full equation for allosteric modification, examining a single substrate (e.g. Arf•GDP), holding the second substrate constant and saturating, is:

$$v_i = \frac{\frac{\alpha \cdot K_d}{\alpha \cdot K_d + M} \cdot V_{max1} \cdot Arf1 \bullet GDP + \frac{M}{\alpha \cdot K_d + M} \cdot V_{max2} \cdot Arf1 \bullet GDP}{\frac{\alpha \cdot K_m \cdot (K_d + M)}{\alpha \cdot K_d + M} + Arf1 \bullet GDP} \quad (15)$$

where M is the modifier.

The potential effect on k_{cat} is given by the equation:

$$k_{cat,obs} = \frac{\alpha \cdot K_d \cdot k_{cat1} + M \cdot k_{cat2}}{\alpha \cdot K_d + M} \quad (16)$$

And the potential effect on the K_m is given by:

$$K_{m,app} = \frac{K_m \cdot \alpha \cdot (K_d + M)}{\alpha \cdot K_d + M} \quad (17)$$

Note, in this latter case, the K_d for the modifier should be affected by the substrate described in following equation,

$$K_{d,app} = \frac{K_d \cdot \alpha \cdot (K_m + S1)}{\alpha \cdot K_m + S1} \quad (18)$$

which provides an additional test of the model for allostery. We described this below for Brag2.

The enzymology of two ArfGEFs has been examined using some of these principles.

3.2 ARNO/cytohesin/Grp1: Example of ArfGEF regulated by relief of autoinhibition

ARNO proteins are comprised of coiled coil, sec7 (catalytic), PH and polybasic (PB) domains. The ARNO group of proteins has roles in diverse cellular processes: regulation of cell adhesion and migration (Goldfinger et al., 2003;Santy and Casanova, 2001;Nagel et al., 1998; Geiger et al., 2000;Hernandez-Deviez et al., 2004); insulin signaling (Fuss et al., 2006;Hafner et al., 2006); and; vesicle transport (Hurtado-Lorenzo et al., 2006; Merkulova et al., 2010; Merkulova et al., 2011;Caumon et al., 2000). The regulation described for ARNO is complex. The effect of protein and lipid binding to the PH domain is discussed here. Protein binding to the coiled-coil domain (Esteban et al., 2006;Goldfinger et al., 2003) and PKC mediated phosphorylation (DiNitto et al., 2007) also contribute to regulating ARNO.

The molecular basis for the effect of protein and lipid binding to the PH domain has been examined in some detail (DiNitto et al., 2007; Cohen et al., 2007). ARNO is autoinhibited by the linker region between the sec7 and PH domains and a C-terminal amphipathic helix, which physically block the Arf binding site. Binding of Arl4•GTP, Arf6•GTP and phosphoinositides to the PH domain has two functions. One is to recruit ARNO to the membrane surface on which it is active and the second to induce a conformational change in the PH domain that relieves autoinhibition. Phosphorylation of ARNO by protein kinase C (PKC) also alleviates autoinhibition (DiNitto et al., 2007; Frank et al., 1998). The characterization was done primarily with a truncated form of Arf, lack an N-terminal extension that is unique to the Arf family of GTP binding proteins. A possible function of N-terminus - Arno interaction in regulation will be interesting to examine.

3.3 Brag2: PIP_2 acts as allosteric modifier binding to the PH domain

The Brag subgroup of GEF proteins has three members characterized by the presence of IQ, sec7, PH and coiled-coil domains (Casanova, 2007). Brag1 and 3 are found primarily in brain. Brag2, although enriched in brain, is ubiquitously expressed. Brag2 affects endocytosis of cell adhesion molecules, including cadherins and integrins, and has been implicated in antiangiogenic signaling in endothelial cells and invasion of breast cancer cells.

Recent work examining Brag2 supports the idea that PIP_2 allosterically modifies activity by binding to the PH domain (Jian and Randazzo, manuscript in preparation). The work was an extension of work examining signaling by semaphorin. Sema3E is an antiangiogenic factor that binds to Plexin D1 resulting in recruitment of PIP kinase and increased Arf6 exchange factor activity mediated by Brag2 (Sakurai et al., 2010; Sakurai et al., 2011). Brag2 was found to bind to PIP_2, which stimulated exchange factor activity in vitro. Subsequent work identified residues within the PH domain that bound to PIP_2. PH domains are thought to be recruitment domains, but the two substrates for Brag2, Arf•GDP and GTP, are soluble, so recruitment to a membrane by itself would not result in increased activity. PIP_2 was found to increase both the K_m and k_{cat} for the exchange reaction, and, consistent with behavior as an allosteric modifier with an effect on K_m, the substrate Arf•GDP affected the

K_d for the ligand PIP_2. Based on these results, PIP_2 must induce some rearrangement of the catalytic pocket. One possibility is that PIP_2 stabilizes the transition state, which is restricted to the membrane. Given that PIP_2 has also been found to stabilize the apo form of Arf (Terui et al., 1994), it is possible that PIP_2 binds to Arf, in addition to the PH domain of Brag2, within the transition complex.

4. Conclusions

The knowledge of kinetic parameters is limited to a few GAPs and GEFs. The information available has provided a number of insights into the biological function of the proteins and potential regulation. For instance, the effect of cotaomer and cargo on ArfGAP1 led to the idea that it may act prior to transport vesicle formation rather than after vesicle formation as has been generally accepted. Activation of ARNO by Arf6 and Arl4 has led to the idea of sequential signaling functions of Arf proteins.

Other aspects of the known enzymology of GAPs and GEFs, such as the discrepant turnover numbers among the GAPs, are intriguing. The slow turnover number could result from a lack of understanding of optimal conditions for a particular ArfGAP, including potential allosteric modifiers that may stimulate activity. Also possible, the different turnover numbers may be related to the biological process being controlled. In addition to examination of additional GAPs and GEFs and further characterization of individual proteins to find optimal conditions for enzymatic activity, identification of GEF/GAP pairs will be important for understanding the function of the Arf proteins in biological processes.

5. Acknowledgment

The work was supported by the intramural program of the National Cancer Institute, National Institutes of Health, USA.

6. References

Antonny,B., Huber,I., Paris,S., Chabre,M., and Cassel,D. (1997). Activation of ADP-ribosylation factor 1 GTPase-activating protein by phosphatidylcholine-derived diacylglycerols. J. Biol. Chem. *272*, 30848-30851.

Beck,R., Adolf,F., Weimer,C., Bruegger,B., and Wieland,F.T. (2009a). ArfGAP1 Activity and COPI Vesicle Biogenesis. Traffic *10*, 307-315.

Beck,R., Brugger,B., and Wieland,F. (2011). GAPs in the context of COPI. Cellular Logistics *1, 52-54.*

Beck,R., Ravet,M., Wieland,F.T., and Cassel,D. (2009b). The COPI system: Molecular mechanisms and function. FEBS Lett *583*, 2701-2709.

Bharti,S., Inoue,H., Bharti,K., Hirsch,D.S., Nie,Z., Yoon,H.Y., Artym,V., Yamada,K.A., Mueller,S.C., Barr,V.A., and Randazzo,P.A. (2007). Src-dependent phosphorylation of ASAP1 regulates podosomes. Mol. Cell. Biol. *27*, 8271-8283.

Bigay,J., Gounon,P., Robineau,S., and Antonny,B. (2003). Lipid packing sensed by ArfGAP1 couples COPI coat disassembly to membrane bilayer curvature. Nature *426*, 563-566.

Campa,F., Yoon,H.Y., Ha,V.L., Szentpetery,Z., Balla,T., and Randazzo,P.A. (2009). A PH Domain in the Arf GTPase-activating Protein (GAP) ARAP1 Binds Phosphatidylinositiol 3,4,5-Trisphosphate and Regulates Arf GAP Activity Independently of Recruitment to the Plasma Membranes. J. Biol. Chem. *284*, 28069-28083.

Casanova,J.E. (2007). Regulation of arf activation: the sec7 family of guanine nucleotide exchange factors. Traffic *8*, 1476-1485.

Caumon,A.S., Vitale,N., Gensse,M., Galas,M.C., Casanova,J.E., and Bader,M.F. (2000). Identification of a plasma membrane-associated guanine nucleotide exchange factor for ARF6 in chromaffin cells - Possible role in the regulated exocytotic pathway. J. Biol. Chem. *275*, 15637-15644.

Che,M.M., Boja,E.S., Yoon,H.-Y., Gruschus,J., Jaffe,H., Stauffer,S., Schuck,P., Fales,H.M., and Randazzo,P.A. (2005). Regulation of ASAP1 by phospholipids is dependent on the interface between the PH and Arf GAP domains. Cellular Signalling *17*, 1276-1288.

Cohen,L.A., Honda,A., Varnai,P., Brown,F.D., Balla,T., and Donaldson,J.G. (2007). Active Arf6 recruits ARNO/cytohesin GEFs to the PM by binding their PH domain. Mol. Biol. Cell *18*, 2244-2253.

DiNitto,J.P., Delprato,A., Lee,M.T.G., Cronin,T.C., Huang,S., Guilherme,A., Czech,M.P., and Lambright,D.G. (2007). Structural basis and mechanism of autoregulation in 3-phosphoinositide-dependent Grp1 family Arf GTPase exchange factors. Mol. Cell *28*, 569-583.

Donaldson,J.G. and Jackson,C.L. (2011). ARF family G proteins and their regulators: roles in membrane transport, development and disease. Nature Rev Mol Cell Biol *12*, 362-375.

East,M.P. and Kahn,R.A. (2011). Models for the functions of Arf GAPs. Sem Cell & Dev. Biol. *22*, 3-9.

Ehlers,J.P., Worley,L., Onken,M.D., and Harbour,J.W. (2005). DDEF1 is located in an amplified region of chromosome 8q and is overexpressed in uveal melanoma. Clin Canc Res *11*, 3609-3613.

Esteban,P.F., Yoon,H.Y., Becker,J., Dorsey,S.G., Caprari,P., Palko,M.E., Coppola,V., Saragovi,H.U., Randazzo,P.A., and Tessarollo,L. (2006). A kinase-deficient TrkC receptor isoform activates Arf6-Rac1 signaling through the scaffold protein tamalin. J Cell Biol *173*, 291-299.

Frank,S.R., Hatfield,J.C., and Casanova,J.E. (1998). Remodeling of the actin cytoskeleton is coordinately regulated by protein kinase C and the ADP-ribosylation factor nucleotide exchange factor ARNO. Mol. Biol. Cell *9*, 3133-3146.

Frigerio,G., Grimsey,N., Dale,M., Majoul,I., and Duden,R. (2007). Two human ARFGAPs associated with COP-I-coated vesicles. Traffic *8*, 1644-1655.

Fuss,B., Becker,T., Zinke,I., and Hoch,M. (2006). The cytohesin Steppke is essential for insulin signalling in Drosophila. Nature *444*, 945-948.

Geiger,C., Nagel,W., Boehm,T., van Kooyk,Y., Figdor,C.G., Kremmer,E., Hogg,N., Zeitlmann,L., Dierks,H., Weber,K.S.C., and Kolanus,W. (2000). Cytohesin-1 regulates beta-2 integrin-mediated adhesion through both ARF-GEF function and interaction with LFA-1. EMBO J *19*, 2525-2536.

Gillingham,A.K. and Munro,S. (2007). The Small G Proteins of the Arf Family and Their Regulators. Ann Rev Cell Dev Biol *23*, 579-611.

Goldberg,J. (1999). Structural and functional analysis of the ARF1-ARFGAP complex reveals a role for coatomer in GTP hydrolysis. Cell *96*, 893-902.

Goldberg,J. (2000). Decoding of sorting signals by coatomer through a GTPase switch in the COPI coat complex. Cell *100*, 671-679.

Goldfinger,L.E., Han,J., Kiosses,W.B., Howe,A.K., and Ginsberg,M.H. (2003). Spatial restriction of alpha 4 integrin phosphorylation regulates lamellipodial stability and alpha 4 beta 1-dependent cell migration. J Cell Biol *162*, 731-741.

Ha,V.L., Bharti,S., Inoue,H., Vass,W.C., Campa,F., Nie,Z.Z., de Gramont,A., Ward,Y., and Randazzo,P.A. (2008a). ASAP3 is a focal adhesion-associated Arf GAP that functions in cell migration and invasion. J. Biol. Chem. *283*, 14915-14926.

Ha,V.L., Luo,R.B., Nie,Z.Z., and Randazzo,P.A. (2008b). Contribution of AZAP-Type Arf GAPs to Cancer Cell Migration and Invasion. Advances in Cancer Research, *101*, 1-16

Habermann,B. (2004). The BAR-domain family of proteins: a case of bending and binding? The membrane bending and GTPase-binding functions of proteins from the BAR-domain family. EMBO Reports *5*, 250-255.

Hafner,M., Schmitz,A., Grune,I., Srivatsan,S.G., Paul,B., Kolanus,W., Quast,T., Kremmer,E., Bauer,I., and Famulok,M. (2006). Inhibition of cytohesins by SecinH3 leads to hepatic insulin resistance. Nature *444*, 941-944.

Hernandez-Deviez,D.J., Roth,M.G., Casanova,J.E., and Wilson,J.M. (2004). ARNO and ARF6 regulate axonal elongation and branching through downstream activation of phosphatidylinositol 4-phosphate 5-kinase alpha. Mol. Biol. Cell *15*, 111-120.

Hsu,V.W. (2011). Role of ARFGAP1 in COPI vesicle biogenesis. Cellular Logistics *1,55-56*.

Hsu,V.W., Lee,S.Y., and Yang,J.S. (2009). The evolving understanding of COPI vesicle formation. Nature Rev Mol Cell Biol *10*, 360-364.

Hurtado-Lorenzo,A., Skinner,M., El Annan,J., Futai,M., Sun-Wada,G.H., Bourgoin,S., Casanova,J., Wildeman,A., Bechoua,S., Ausiello,D.A., Brown,D., and Marshansky,V. (2006). V-ATPase interacts with ARNO and Arf6 in early endosomes and regulates the protein degradative pathway. Nature Cell Biol *8*, 124-1U8.

Ismail,S.A., Vetter,I.R., Sot,B., and Wittinghofer,A. (2010). The Structure of an Arf-ArfGAP Complex Reveals a Ca(2+) Regulatory Mechanism. Cell *141*, 812-821.

Jian,X.Y., Brown,P., Schuck,P., Gruschus,J.M., Balbo,A., Hinshaw,J.E., and Randazzo,P.A. (2009). Autoinhibition of Arf GTPase-activating Protein Activity by the BAR Domain in ASAP1. J. Biol. Chem. *284*, 1652-1663.

Kahn,R.A. (2011). GAPs:Terminator versus effector functions and the role(s) of ArfGAP1 in vesicle biogenesis. Cellular Logistics *1, 49-51*.

Kahn,R.A., Bruford,E., Inoue,H., Logsdon,J.M., Nie,Z., Premont,R.T., Randazzo,P.A., Satake,M., Theibert,A.B., Zapp,M.L., and Cassel,D. (2008). Consensus nomenclature for the human ArfGAP domain containing proteins. J. Cell Biol. *182*, 1039-1044.

Kahn,R.A., Cherfils,J., Elias,M., Lovering,R.C., Munro,S., and Schurmann,A. (2006). Nomenclature for the human Arf family of GTP-binding proteins: ARF, ARL, and SAR proteins. J Cell Biol *172*, 645-650.

Kahn,R.A. (2009). Toward a model for Arf GTPases as regulators of traffic at the Golgi. FEBS Lett *583*, 3872-3879.

Kam,J.L., Miura,K., Jackson,T.R., Gruschus,J., Roller,P., Stauffer,S., Clark,J., Aneja,R., and Randazzo,P.A. (2000). Phosphoinositide-dependent activation of the ADP-ribosylation factor GTPase-activating protein ASAP1 - Evidence for the pleckstrin homology domain functioning as an allosteric site. J. Biol. Chem. *275*, 9653-9663.

Kliouchnikov,L., Bigay,J., Mesmin,B., Parnis,A., Rawet,M., Goldfeder,N., Antonny,B., and Cassel,D. (2009). Discrete Determinants in ArfGAP2/3 Conferring Golgi Localization and Regulation by the COPI Coat. Mol. Biol. Cell *20*, 859-869.

Kruljac-Letunic,A., Moelleken,J., Kallin,A., Wieland,F., and Blaukat,A. (2003). The tyrosine kinase Pyk2 regulates Arf1 activity by phosphorylation and inhibition of the Arf-GTPase-activating protein ASAP1. J. Biol. Chem. *278*, 29560-29570.

Lee,S.Y., Yang,J.S., Hong,W.J., Premont,R.T., and Hsu,V.W. (2005). ARFGAP1 plays a central role in coupling COPI cargo sorting with vesicle formation. J Cell Biol *168*, 281-290.

Luo,R., Ahvazi,B., Amariei,D., Shroder,D., Burrola,B., Losert,W., and Randazzo,P.A. (2007). Kinetic analysis of GTP hydrolysis catalysed by the Arf1-GTP-ASAP1 complex. Biochem J *402*, 439-447.

Luo,R., Ha,V.L., Hayashi,R., and Randazzo,P.A. (2009). Arf GAP2 is positively regulated by coatomer and cargo. Cellular Signalling *21*, 1169-1179.

Luo,R., Jenkins,L.M.M., Randazzo,P.A., and Gruschus,J. (2008). Dynamic interaction between Arf GAP and PH domains of ASAP1 in the regulation of GAP activity. Cellular Signalling *20*, 1968-1977.

Luo,R.B. and Randazzo,P.A. (2008). Kinetic analysis of Arf GAP1 indicates a regulatory role for coatomer. J. Biol. Chem. *283*, 21965-21977.

Mazelova,J., stuto-Gribble,L., Inoue,H., Tam,B.M., Schonteich,E., Prekeris,R., Moritz,O.L., Randazzo,P.A., and Deretic,D. (2009). Ciliary targeting motif VxPx directs assembly of a trafficking module through Arf4. EMBO J *28*, 183-192.

McMahon,H.T. and Gallop,J.L. (2005). Membrane curvature and mechanisms of dynamic cell membrane remodelling. Nature *438*, 590-596.

Merkulova,M., Bakulina,A., Thaker,Y.R., Gruber,G., and Marshansky,V. (2010). Specific motifs of the V-ATPase a2-subunit isoform interact with catalytic and regulatory domains of ARNO. Biochim Biophys ACTA *1797*, 1398-1409.

Merkulova,M., Hurtado-Lorenzo,A., Hosokawa,H., Zhuang,Z.J., Brown,D., Ausiello,D.A., and Marshansky,V. (2011). Aldolase directly interacts with ARNO and modulates cell morphology and acidic vesicle distribution. Am J Physiol *300*, C1442-C1455.

Nagel,W., Zeitlmann,L., Schilcher,P., Geiger,C., Kolanus,J., and Kolanus,W. (1998). Phosphoinositide 3-OH kinase activates the beta(2) integrin adhesion pathway and induces membrane recruitment of cytohesin-1. J. Biol. Chem. *273*, 14853-14861.

Nie,Z., Boehm,M., Boja,E., Vass,W., Bonifacino,J., Fales,H., and Randazzo PA (2003). Specific Regulation of the Adaptor Protein Complex AP-3 by the Arf GAP AGAP1. Dev Cell *5*, 513-521.

Nie,Z., Hirsch,D.S., Luo,R., Jian,X., Stauffer,S., Cremesti,A., Andrade,J., Lebowitz,J., Marino,M., Ahvazi,B., Hinshaw,J.E., and Randazzo,P.A. (2006). A BAR domain in the N Terminus of the Arf GAP ASAP1 affects membrane structure and trafficking of Epidermal Growth Factor Receptor. Curr Biol *16*, 130-139.

Nie,Z.Z., Fei,J., Premont,R.T., and Randazzo,P.A. (2005). The Arf GAPs AGAP1 and AGAP2 distinguish between the adaptor protein complexes AP-1 and AP-3. J Cell Sci *118*, 3555-3566.

Nie,Z.Z. and Randazzo,P.A. (2006). Arf GAPs and membrane traffic. J Cell Sci *119*, 1203-1211.

Nie,Z.Z., Stanley,K.T., Stauffer,S., Jacques,K.M., Hirsch,D.S., Takei,J., and Randazzo,P.A. (2002). AGAP1, an endosome-associated, phosphoinositide-dependent ADP-ribosylation factor GTPase-activating protein that affects actin cytoskeleton. J. Biol. Chem. *277*, 48965-48975.

Onodera,Y., Hashimoto,S., Hashimoto,A., Morishige,M., Yamada,A., Ogawa,E., Adachi,M., SakuraiT., Manabe,T., Wada,H., Matsuura,N., and Sabe,H. (2005). Expression of AMAP1, an ArfGAP, provides novel targets to inhibit breast cancer invasive activities. EMBO J *24*, 963-973.

Randazzo,P.A. (1997). Resolution of two ADP-ribosylation factor 1 GTPase-activating proteins from rat liver. Biochem J *324*, 413-419.

Randazzo,P.A., Andrade,J., Miura,K., Brown,M.T., Long,Y.Q., Stauffer,S., Roller,P., and Cooper,J.A. (2000). The Arf GTPase-activating protein ASAP1 regulates the actin cytoskeleton. Proc Nat'l Acad Sci USA*97*, 4011-4016.

Randazzo,P.A. and Hirsch,D.S. (2004). Arf GAPs: multifunctional proteins that regulate membrane traffic and actin remodelling. Cellular Signalling *16*, 401-413.

Randazzo,P.A. and Kahn,R.A. (1994). GTP hydrolysis by ADP-ribosylation factor (Arf) is dependent on both an Arf GAP and acid phospholipids. J. Biol. Chem. *269*, 10758-10763.

Renault,L., Guibert,B., and Cherfils,J. (2003). Structural snapshots of the mechanism and inhibition of a guanine nucleotide exchange factor. Nature *426*, 525-530.

Sabe,H., Onodera,Y., Mazaki,Y., and Hashimoto,S. (2006). ArfGAP family proteins in cell adhesion, migration and tumor invasion. Curr Opin Cell Biol *18*, 558-564.

Sakurai,A., Gavard,J., nnas-Linhares,Y., Basile,J.R., Amornphimoltham,P., Palmby,T.R., Yagi,H., Zhang,F., Randazzo,P.A., Li,X.R., Weigert,R., and Gutkind,J.S. (2010). Semaphorin 3E Initiates Antiangiogenic Signaling through Plexin D1 by Regulating Arf6 and R-Ras. Mol. Cell. Biol. *30*, 3086-3098.

Sakurai,A., Jian,X., Lee,C.J., Manavski,Y., Chavakis,E., Donaldson,J., Randazzo,P.A., and Gutkind,J.S. (2011). Phosphatidylinositol-4-phosphate 5-Kinase and GEP100/Brag2 Protein Mediate Antiangiogenic Signaling by Semaphorin 3E-Plexin-D1 through Arf6 Protein. J. Biol. Chem. *286*, 34335-34345.

Santy,L.C. and Casanova,J.E. (2001). Activation of ARF6 by ARNO stimulates epithelial cell migration through downstream activation of both Rac1 and phospholipase D. J Cell Biol *154*, 599-610.

Scheffzek,K., Ahmadian,M.R., and Wittinghofer,A. (1998). GTPase-activating proteins: helping hands to complement an active site. Trends in Biochemical Sciences *23*, 257-262.

Shiba,Y., Luo,R.B., Hinshaw,J.E., Szul,T., Hayashi,R., Sztul,E., Nagashima,K., Baxa,U., and Randazzo,P.A. (2011). ArfGAP1 promotes COPI vesicle formation by facilitating coatomer polymerization . Cellular Logistics *1, 139-154.*.

Spang,A., Shiba,Y., and Randazzo,P.A. (2010). Arf GAPs: Gatekeepers of vesicle generation. FEBS Lett *584*, 2646-2651.

Terui,T, Kahn,RA., and Randazzo,P.A. (1994). Effects of acid phospholipids on nucleotide exchange properties of ADP-ribosylation factor-1 - Evidence for specific interaction with phosphatidylinositol 4,5-bisphosphate. J. Biol. Chem. *269*, 28130-28135.

Ward,H.H., Brown-Glaberman,U., Wang,J., Morita,Y., Alper,S.L., Bedrick,E.J., Gattone,V.H., Deretic,D., and Wandinger-Ness,A. (2011). A conserved signal and GTPase complex are required for the ciliary transport of polycystin-1. Mol. Biol. Cell *22*, 3289-3305.

Weimer,C., Beck,R., Eckert,P., Reckmann,I., Moelleken,J., Brugger,B., and Wieland,F. (2008). Differential roles of ArfGAP1, ArfGAP2, and ArfGAP3 in COPI trafficking. J Cell Biol *183*, 725-735.

Ye,K.Q. and Snyder,S.H. (2004). PIKE GTPase: a novel mediator of phosphoinositide signaling. J Cell Sci *117*, 155-161.

Zhu,G.Y., Chen,J., Liu,J., Brunzelle,J.S., Huang,B., Wakeham,N., Terzyan,S., Li,X.M., Rao,Z., Li,G.P., and Zhang,X.J.C. (2007). Structure of the APPL1 BAR-PH domain and characterization of its interaction with Rab5. EMBO J *26*, 3484-3493.

10

Metal Ion Homeostasis Mediated by NRAMP Transporters in Plant Cells – Focused on Increased Resistance to Iron and Cadmium Ion

Toshio Sano[1], Toshihiro Yoshihara[1], Koichi Handa[2], Masa H. Sato[3], Toshiyuki Nagata[1] and Seiichiro Hasezawa[2,4]
[1]*Faculty of Bioscience and Applied Chemistry, Hosei University*
[2]*Graduate School of Frontier Sciences, The University of Tokyo*
[3]*Graduate School of Life and Environmental Sciences, Kyoto Prefectural University*
[4]*Advanced Measurement and Analysis, Japan Science and Technology Agency (JST)*
Japan

1. Introduction

Plants have developed several adaptive systems to control the cellular concentrations of essential metals in which the ion transporters play significant roles. At the cell surface, transporters localized on plasmamembrane controlled metal ion uptake and release whereas inside of the plant cells, those localized on endomembrane sequestered and remobilized metal ions in organella, such as vacuoles and plastids (Pilon et al. 2009, Puig and Peñarrubia, 2009).

Iron is one of several essential nutrients but a problematic one for living organisms (Conte and Walker 2011). At the cellular level, iron is used as a cofactor in enzymatic activities based on the reversible reaction between Fe^{2+} (ferrous) and Fe^{3+} (ferric) ions (Hell and Stephan 2003). In plants, it is essential for chlorophyll synthesis and hence iron deficiency results in chlorosis and pale-yellow or white leaves (Wiedenhoeft 2006). Usually, iron is chelated to organic matter in insoluble forms in soils that causes iron deficiency whereas in anaerobic and acidic conditions, iron toxicity occurs because of the increase of iron solubility (Ricachenevsky et al. 2010). The basis of iron toxicity was usually discussed to be oxidative stress by generation of reactive-hydroxyl radicals (Neyens and Baeyens 2003). Iron homeostasis in plant cells is partly achieved through the control of iron transport across membranous structures, and several families of putative iron transporters, including ZIP (ZRT, IRT-like proteins) and NRAMP (natural resistance associated macrophage protein), have been described (Guerinot 2000, Curie and Briat 2003, Kim and Guerinot 2007). Among the ZIP transporters, the *Arabidopsis AtIRT1* was the first iron transport molecule identified in plants (Eide et al. 1996) and was shown to be the major transporter mediating iron uptake into roots (Vert et al. 2002). Recently, IRT2, a close homolog of IRT1 in *Arabidopsis,* was suggested to compartmentalize iron into vesicles to prevent toxicity from excess free iron in the cytosol (Vert et al. 2009). On the other hand, among the ubiquitous NRAMP family of metal transporters, it was the mouse *Nramp1* that was first cloned as the gene responsible for resistance to mycobacterial infection (Nevo and Nelson 2006). In *Arabidopsis*, six NRAMP transporters, AtNRAMP1-6, have been identified and categorized by phylogenic analysis into two subfamilies: AtNRAMP1 and 6 forming the first group and AtNRAMP2 through 5 comprising the second group (Mäser et al. 2001). Of these, *AtNRAMP1*, *3*, *4* and *6* have been shown to encode functional plant metal transporters (Krämer et al. 2007, Cailliatte et al. 2009). *AtNRAMP1* can complement the *fet3fet4* yeast mutant that is defective in both low- and high-affinity iron transporters, while overexpression of *AtNRAMP1* in *Arabidopsis* increases plant resistance to toxic iron concentrations (Curie et al. 2000). AtNRAMP3 and AtNRAMP4 mediate the remobilization of iron from the vacuolar store and are essential for seed germination under low iron conditions (Thomine et al. 2003, Lanquar et al. 2005).

In addition to iron transport activities, these transporters can mediate the transport of a wide range of metal cations because of their similar chemical characteristics (Hall and Williams 2003, Krämer et al. 2007). AtNRAMP1 can functionally complement a manganese-uptake defective mutant and confer cadmium sensitivity to yeast (Thomine et al. 2000). This transporter was recently demonstrated to act as a physiological manganese transporter in *Arabidopsis* (Cailliatte et al. 2010). Similarly, TcNRAMP3 and TcNRAMP4 from the metal hyperaccumulator, *Thlaspi caerulescens*, can transport various metal cations, including Fe^{2+}, Mn^{2+}, Cd^{2+}, Ni^{2+} and Zn^{2+} when expressed in yeast, and MbNRAMP1 from apple trees of *Malus baccata* was found to mediate Mn^{2+} uptake in addition to Fe^{2+} (Oomen et al. 2008, Xiao et al. 2008, Wei et al. 2009). Recently, rice Nrat1 that belongs to the NRAMP family has been reported to transport trivalent aluminum ion, but not other divalent ions such as manganese, iron and cadmium (Xia et al. 2010).

In the present chapter, we demonstrate that tobacco NtNRAMP1 is a plasma membrane transporter, and that overexpression of this protein in tobacco BY-2 cells increases the resistance of the cells to both iron and cadmium ions. We propose that NtNRAMP1 moderates metal ion-uptake and prevents toxicity resulting from excess iron or cadmium application.

2. Results

2.1 Excess iron application induces cell death and arrests cell cycle progression

To examine the effect of excess iron application to plant cells, we monitored the growth of tobacco BY-2 cells in medium containing high amounts of iron. The cells took up about 50 to 90 μg iron g $cells^{-1}$ 6 h after transfer to a medium containing 1.0, 2.0 or 5.0 mM $FeSO_4$ and lacking other divalent cations (Mg^{2+}, Ca^{2+}, Mn^{2+}, Zn^{2+} and Co^{2+}), but only about 5 μg iron g $cells^{-1}$ in standard medium with 0.1 mM $FeSO_4$ (Fig. 1A).

Under these conditions, about 50 % or more than 80 % of the cells died 24 h after transfer to medium containing 1.0 and 2.0 or 5.0 mM $FeSO_4$, respectively, whereas only a few percent of the cells died in the standard medium (Fig. 1B).

As cell death is known to relate to the arrest of cell cycle progression (Kadota et al. 2004, Sano et al. 2006), we monitored the latter by flow cytometric and mitotic index (MI) analyses upon excess iron application.

In the control condition containing 0.1 mM $FeSO_4$, the cell cycle progressed from the S to G_2 phase 2 h after aphidicolin release, and entered mitosis at 8 h and the G_1 phase at 10 h (Fig.

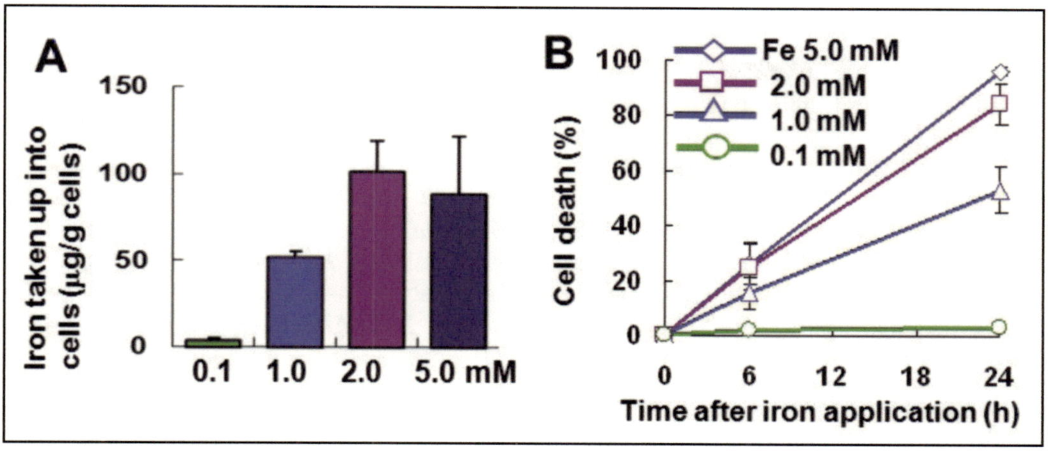

Fig. 1. Effect of excess iron application on tobacco BY-2 cell growth. (A) Amount of iron taken up into cells. After transfer of 7-day-old cells to medium containing 0.1, 1.0, 2.0 or 5.0 mM $FeSO_4$, the amount of iron taken up into the cells in 6 h was measured. Values shown are those after subtraction of measurements taken just after transfer to the medium as background. Data show the means ± SE of three independent experiments. (B) In the culture conditions in (A), the population undergoing cell death was measured. Data represent means ± SE of three independent experiments with more than 400 cells in each experiment.

2A, B). When the cells were additionally treated with 1.0 mM $FeSO_4$ after aphidicolin release, with the cell cycle restarting from S phase, cell cycle progression was delayed and the percentage of cells entering mitosis decreased (Fig. 2A). Further flow cytometric analysis demonstrated the cell cycle arrest of these cells in the S to G_2 phase (Fig. 2B).

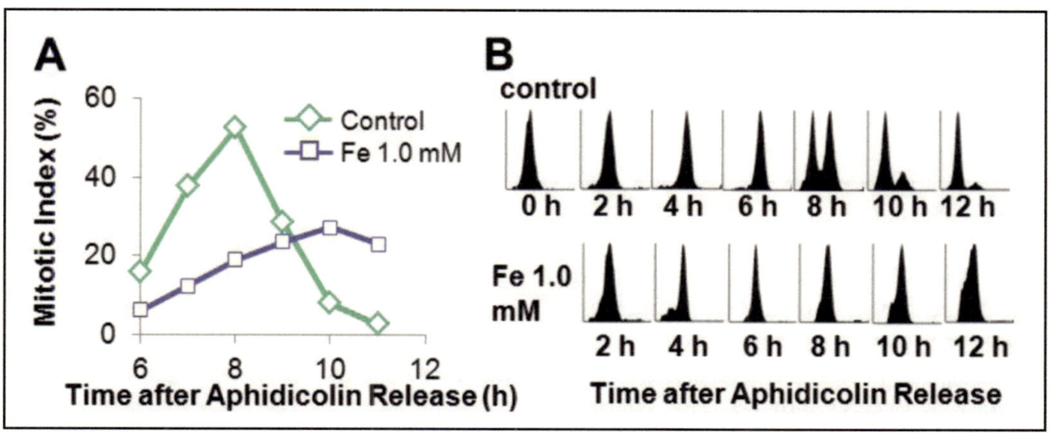

Fig. 2. Effects of excess iron application on cell cycle progression. Cell cycle progression of cells in control conditions (open diamonds) and those cultured with 1.0 mM $FeSO_4$ after aphidicolon treatment (open squares). Cell cycle progression was monitored by the mitotic index (A) and by flow cytometry (B). The data show representatives of three independent experiments.

2.2 Overexpression of *NtNRAMP1* decreases sensitivity to excess iron application

To investigate the molecular mechanisms of iron uptake and cell death of tobacco BY-2 cells, we identified and characterized several tobacco iron transporter genes. As the ZIP and NRAMP family proteins are known as iron/metal transporters in plants, we identified two tobacco cDNA clones that encoded proteins with high sequence similarity to ZIP or NRAMP and named them *NtZIP1* and *NtNRAMP1*, respectively. The amino acid sequence of NtZIP1 was 61 % identical to the MtZIP3 of *Medicago truncatula* (Lopez-Millan et al. 2004) and 53 % to AtZIP5 of *Arabidopsis thaliana*, whereas NtNRAMP1 was 71 % identical to *Arabidopsis* AtNRAMP1 and AtNRAMP6. Gene expression analysis revealed that 1.0 mM $FeSO_4$ application increased the relative transcript levels of *NtNRAMP1* but decreased those of *NtZIP1* (Fig. 3A, B).

As the increased level of *NtNRAMP1* gene expression upon iron application implied an involvement of this transporter under these culture conditions, we prepared transgenic tobacco BY-2 cell lines that overexpressed *NtNRAMP1* by placing the gene under control of the cauliflower mosaic virus 35S promoter. In one (NR1) of the four transgenic lines obtained, *NtNRAMP1* transcript levels were about 2-fold those of the non-transformed BY-2 cells whereas the *NtZIP1* transcript levels were reduced (Fig. 3A, B). Similar increases in *NtNRAMP1* transcript levels were also observed in the other three transgenic lines obtained (data not shown).

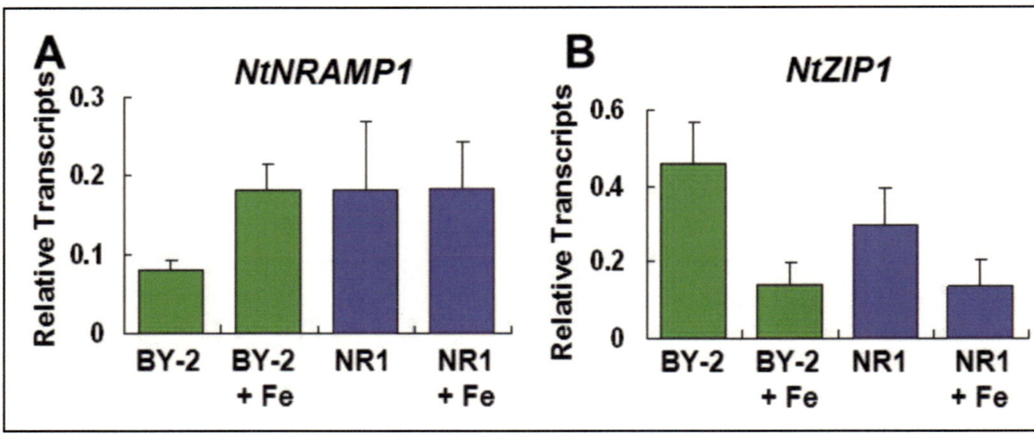

Fig. 3. Gene expression of the *NtNRAMP1* and *NtZIP1* iron transporters. *NtNRAMP1* (A) and *NtZIP1* (B) gene expression in non-transformed BY-2 cells cultured in control conditions for 24 h (BY-2) or with 1.0 mM $FeSO_4$ for 24 h (BY-2 + Fe), and in *NtNRAMP1* overexpressing cells cultured in control conditions (NR1) or with 1.0 mM $FeSO_4$ for 24 h (NR1 + Fe). Gene expression was monitored by real-time quantitative PCR and the data show relative transcripts normalized with GAPdH gene expression. The data show the means ± SE of three independent experiments.

When the iron uptake activities of NtNRAMP1 and NtZIP1 were measured in yeast cells, the amount of iron accumulated in the yeast cells expressing NtNRAMP1 or NtZIP1 was about 1.5 times high compared to control cells expressing LacZ. The amout was comparable to those expressing *Arabidopsis* AtNRAMP1 or AtNRAMP3 whereas that expressing an effulux pump AtHMA4 (Verret et al. 2004, Mills et al. 2005) was comparable to that expressing LacZ (Fig. 4). Therefore, both NtNRAMP1 and NtZIP1 could have the iron uptake activity comparable to *Arabidopsis* AtNRAMP1 and AtNRAMP3.

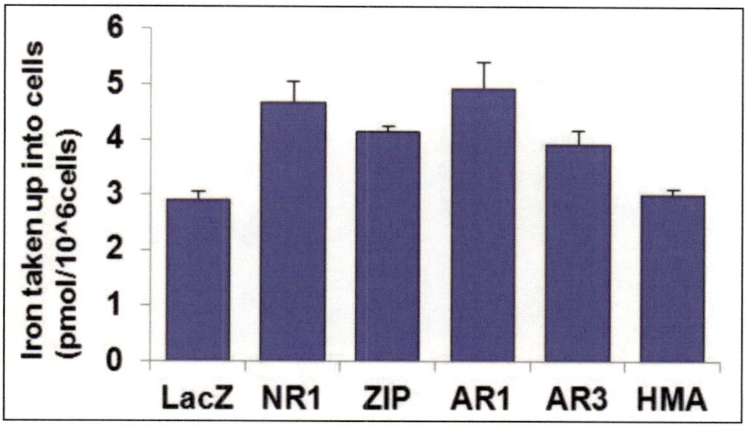

Fig. 4. Iron uptake activity of metal transporters in yeast cells. Iron accumulation in yeast cells transformed with the vector containing LacZ, NtNRAMP1 (NR1), NtZIP1 (ZIP), AtNRAMP1 (AR1), AtNRAMP3 (AR3) and AtHMA4 (HMA) was measured by atomic absorption spectrograph after incubation for 18 h in a medium with 0.2 mM $FeCl_3$. The data show the means ± SE of four independent experiments.

In the NtNRAMP1-overexpressing tobacco line (NR1), cell cycle progression was similar to that of the non-transformed BY-2 cells under control conditions in which 0.1 mM $FeSO_4$ was included. When 0.3 mM $FeSO_4$ was applied at the S phase, cell cycle progression of the non-transformed cells was delayed and the value of the peak MI reduced (Fig. 5A). In contrast, in the NR1 cells, the tendency of cell cycle progression was comparable to that in the control condition even though these cells took up as much as amounts of iron compared to the non-transformed cells (Fig. 5A, B). Furthermore, the proportion of NR1 cells undergoing cell death was reduced in comparison with the non-transformed cells following 1.0 mM $FeSO_4$ application (Fig. 5C), even though the amount of iron taken up by the NR1 cells was comparable to that in the non-transformed cells under these conditions (Fig. 5D).

To investigate the role of NtNRAMP1 on the suppression of cell cycle arrest and cell death upon excess iron application, we examined the subcellular localization of NtNRAMP1 by transient expression of NtNRAMP1-GFP fusion proteins. The GFP fluorescence was localized primarily on the plasma membrane, and confirmed by the plasma membrane marker, SYP132 (Enami et al. 2009, Fig. 6A-C). In contrast, cells transiently expressing GFP only showed cytoplasmic-localized fluorescence (Fig. 6D).

As the above results suggested that NtNRAMP1 is a plasma membrane transporter, we examined the effect of *NtNRAMP1* overexpression on iron uptake. The total amount of iron taken up into cells 24 h after 1.0 mM $FeSO_4$ application was comparable in the NR1 and non-transformed cells (Fig. 7A). However, when calculated on the basis of the rate of iron uptake, the non-transformed BY-2 cells had about 3-fold higher rates than the NR1 cells in the initial 1 hour after iron application (Fig. 7B). In subsequent periods, uptake rates were comparable in the two cell lines (Fig. 7B).

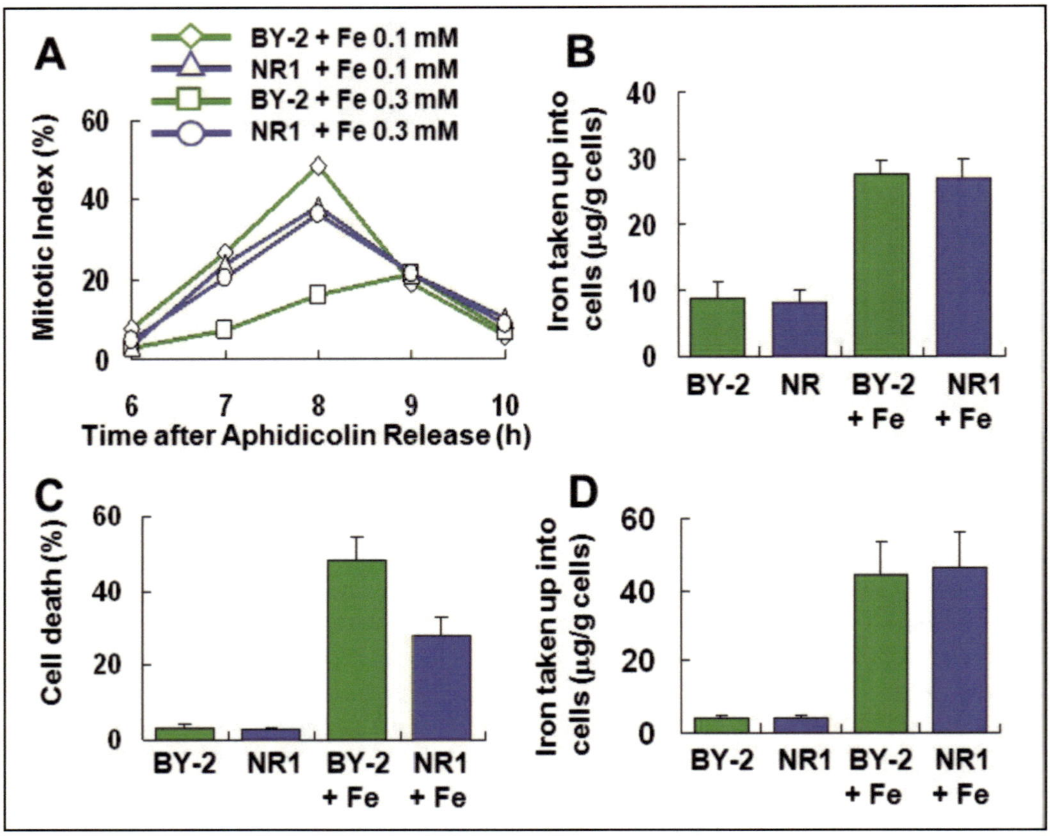

Fig. 5. Cell cycle progression and the population of NtNRAMP1 overexpressing cells undergoing cell death. (A) Cell cycle progression of non-transformed BY-2 cells (BY-2 + Fe 0.1 mM) and NtNRAMP1 overexpressing cells (NR1 + Fe 0.1 mM) cultured in control conditions or with 0.3 mM $FeSO_4$, respectively (BY-2 + Fe 0.3 mM, NR1 + Fe 0.3 mM). The data show a representative sample of three independent experiments. (B) Amount of iron taken up into cells cultured for 6 h in the culture conditions shown in (A). (C) Population of cells undergoing cell death in non-transformed BY-2 cells (BY-2) and NtNRAMP1 overexpressing cells cultured in control conditions (NR1) for 24 h, or those cultured with 1.0 mM $FeSO_4$ for 24 h, respectively (BY-2 + Fe, NR1 + Fe). (D) Amount of iron taken up into cells cultured for 24 h in culture conditions in (C). In (B), (C) and (D), the data show the means ± SE of three independent experiments.

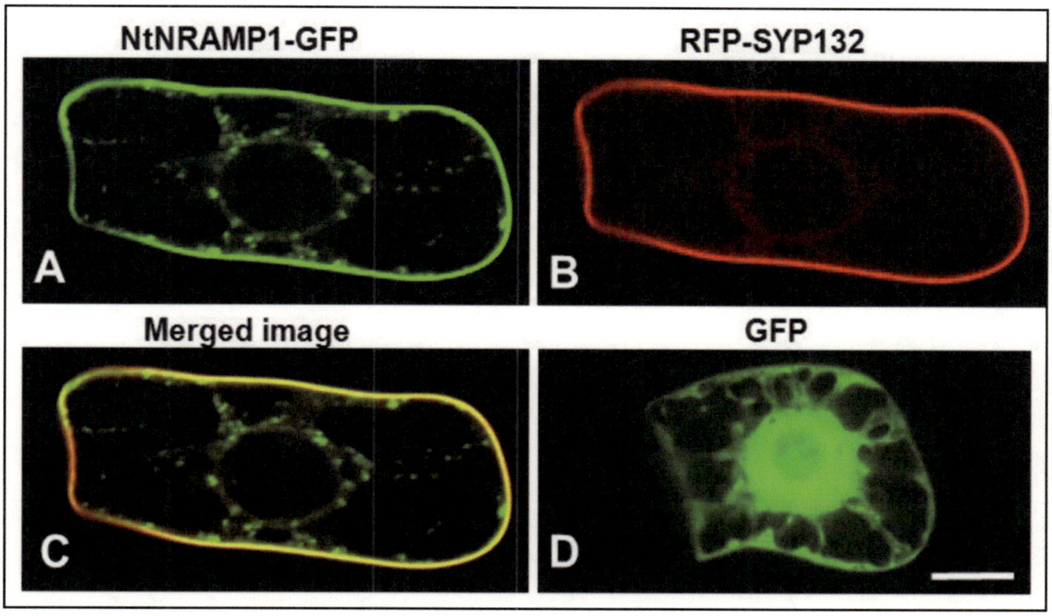

Fig. 6. Subcellular localization of NtNRAMP1. (A) NtNRAMP1 localization was monitored in a tobacco BY-2 cell transiently expressing NtNRAMP1-GFP. (B) Plasma membrane localization of the syntaxin, SYP132, monitored in cells transiently expressing tagRFP-SYP132. (C) Merged image of images (A) and (B). (D) GFP fluorescence in a tobacco BY-2 cell transiently expressing GFP. Scale bar represents 20 μm.

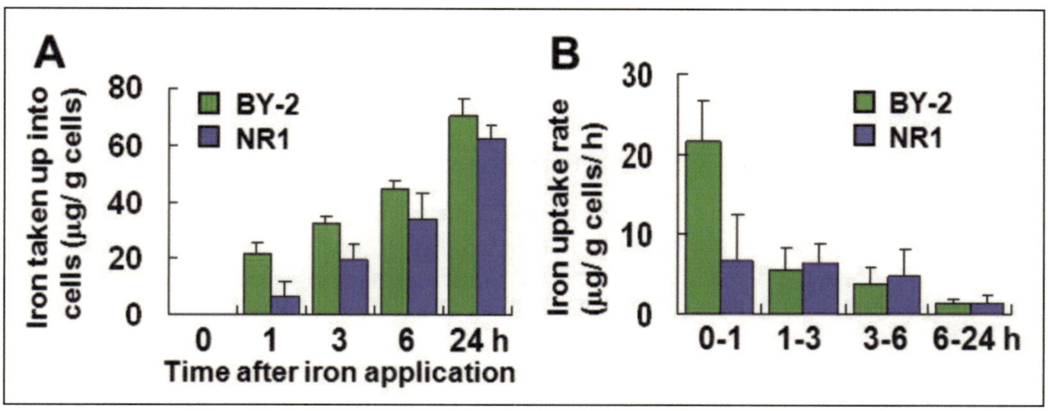

Fig. 7. Effect of NtNRAMP1 overexpression on iron uptake. (A) Changes in the amount of iron taken up into control BY-2 cells (BY-2) and NtNRAMP1 overexpressing cells (NR1). (B) Changes in iron uptake rate calculated from the data in (A). Data show the means ± SE of four independent experiments.

To further characterize NtNRAMP1, we examined the effects of cadmium on cell growth since NRAMP transporters are known to transport a variety of metal ions (Nevo and Nelson 2006, Krämer et al. 2007). In control medium without cadmium, both non-transformed and

NR1 cells proliferated about 70 times per week (Fig. 8A). When 1.0 or 10 μM $CdSO_4$ was added to the medium, the growth rate of controls cells decreased to 40 or 20 times per week, respectively (Fig. 8A). In contrast, the NR1 cell growth rates in 1.0 μM $CdSO_4$ were comparable to those without cadmium treatment, and were still about 55 times per week in 10 μM $CdSO_4$ (Fig 8A). After 10 μM $CdSO_4$ application, the amount of cadmium taken up into the control BY-2 cells increased in 24 h but decreased thereafter (Fig. 8B). In the NR1 cells, the amount of cadmium was smaller than that in the control BY-2 cells in 24 h (Fig. 8B).

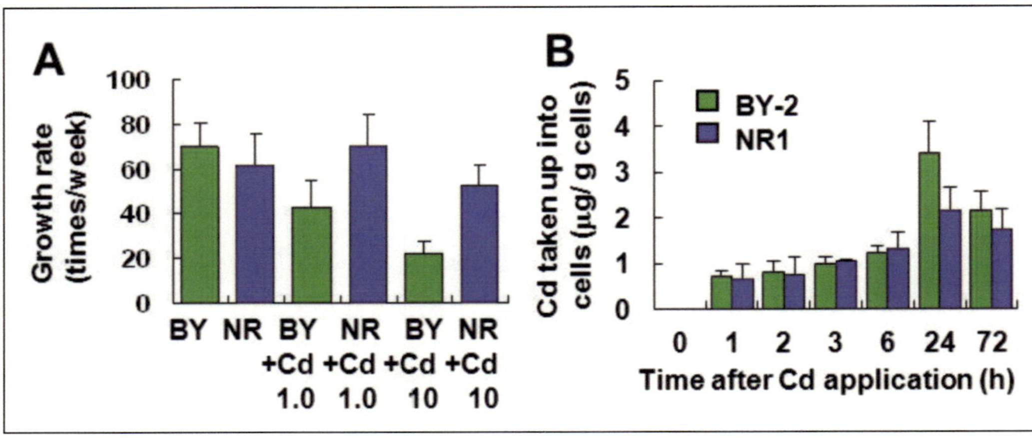

Fig. 8. Effect of NtNRAMP1 overexpression on cell growth following cadmium application. (A) Growth rate of control BY-2 cells and NtNRAMP1 overexpressing cells during culture for 7 days in control conditions (BY, NR), or with 1.0 μM (BY + Cd 1.0, NR + Cd 1.0) or 10 μM $CdSO_4$ (BY + Cd 10, NR + Cd 10). (B) Changes in the amount of cadmium taken up into cells cultured with 10 μM $CdSO_4$. Data show the means ± SE of four independent experiments.

3. Discussion

3.1 Role of the NtNRAMP1 transporter following excess iron application

The NRAMP family transporters function as general metal ion transporters (Nevo and Nelson 2006), and we have shown in this study that NtNRAMP1 overexpression in tobacco BY-2 cells suppressed cell cycle arrest and cell death upon excess iron application (Fig. 5). Plasma membrane localization of NtNRAMP1 (Fig. 6) and the decreased rate of iron uptake in the NtNRAMP1 overexpressing cells (Fig. 7B) implies the role of NtNRAMP1 as a modulator of iron uptake or an iron exporter.

Concerning the latter hypothes, the metal ion efflux activity of the NRAMP family members are somewhat uncommon. The TcNRAMP3 transporter was found to exclude Ni when expressed in yeast but transported iron and cadmium into cells in yeast and plants (Wei et al. 2009). Our iron uptake experiments in yeast implies the iron uptake of NtNRAMP1 rather than the export since the amout of iron uptake was higher than that in the cells expressing the effulux pump of AtHMA4 (Fig. 4).

In this context, the increaed iron resistance upon NtNRAMP1 overexpression might be explaned by the role of AtNRAMP1 as a physiological manganese (Mn) transporter

(Cailliatte et al. 2010). Although the AtNRAMP1 was capable of transporting both iron and Mn in yeast cells (Curie et al. 2000, Thomine et al. 2000), Cailliatte et al. (2010) discussed the competence of iron uptake by Mn uptake increased the resistance to iron toxicity. Similar competence of iron uptake might be occured in the NtNRAMP1 overexpressing cells (Fig. 9). In this model, the supposed metal transporters other than NtNRAMP1 that actively mediate iron uptake are remained to be determined.

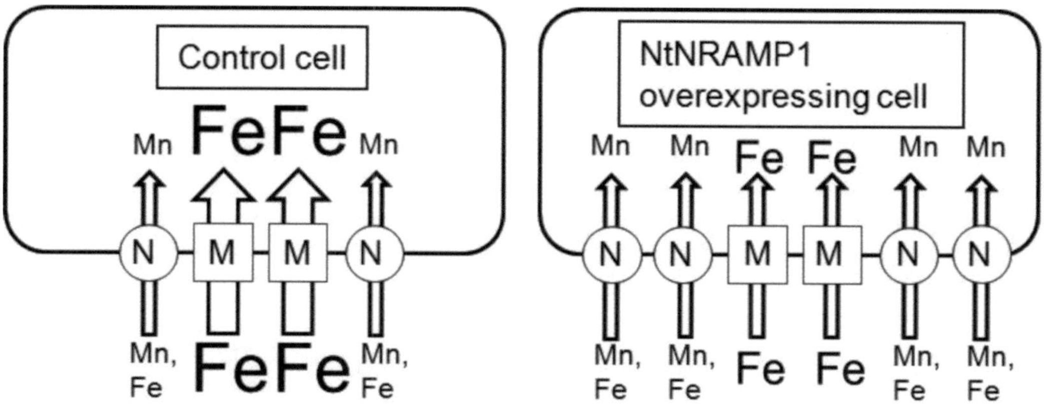

Fig. 9. A model for the increased resistance to iron in NtNRAMP1 overexpressing cells. In the control cells (left), excess iron application increases the rate of iron uptake by metal transporters (M) with high iron uptake activity other than NtNRAMP1 (N). In contrast in the NtNRAMP1 overexpressing cells (right), iron uptake is competed by manganese uptake through the increased number of the NtNRAMP1 proteins.

In graminaceous plants, the enhanced tolerance upon excess iron application was achieved by overproduction of a metal chelator, nicotianamine (Lee et al. 2009). The chelated iron was discussed to be an inactive form for reactive-hydroxyl radical generation as well as to be easily transported from roots to aerial organs (Curie et al. 2009). The increased translocation of iron to rice seeds was expected to provide iron-fortified plants and improve human health (Lee et al. 2009, Wirth et al. 2009, Zheng et al. 2010). Recently, transporters involved in iron translocation was identified in which iron-nocotianamine complex was transported to the rice shoots and phytosiderophore for iron acquisition was secreted to the soil (Ishimaru et al. 2010, Nozoye et al. 2011). In dicot plants, loss of nicotianamine synthase genes did not to fully supply iron to flowers and seeds (Klatte et al. 2009) whereas overaccumulation of nicotianamine did not affect iron translocation (Cassin et al. 2009). The role of metal chelator in dicot plants on iron translocation and resistance against iron application has still been controversial. The combination of the iron uptake moduration and the enhanced iron translocation could enhance the iron fortification and torelance in dicot plants.

3.2 Increased resistance of NtNRAMP1 overexpressing cells to cadmium

In plants, cadmium has various effects, such as the inhibition of photosynthesis, respiration and metabolism, and may finally lead to plant growth inhibition (Deckert 2005). NRAMP family members can potentially transport toxic heavy metals, including cadmium, and further characterization of the NtNRAMP1 overexpressing cells in this study revealed their enhanced resistance to cadmium application (Fig. 8A). Changes in plant cadmium sensitivity as a consequence of NRAMP transporter activity have also been reported in which overexpression of AtNRAMP3 or AtNRAMP6 resulted in cadmium hypersensitivity of *Arabidopsis* growth (Thomine et al. 2000, Cailliatte et al. 2009). These proteins were considered to remobilize cadmium to cytoplasm from a detoxifying compartment such as a vacuole and an endomembrane compartment (Thomine et al. 2003, Cailliatte et al. 2009). The plasmamembrane localization of NtNRAMP1 may explain the increased resistance of NtNRAMP1 overexpressing cells to cadmium by the moderation of Cd^{2+} uptake similar to iron uptake.

The basis of cadmium toxicity is not completely understood, but it appears to affect cellular metabolism through its high affinity for sulfydryl compounds that therefore leads to the misfolding of enzymes, while its chemical similarity to other divalent cations reduces the activity of enzymes similar to the divalent trace metals described above (DalCorso et al. 2008, Verbruggen et al. 2009). In addition, cadmium is thought to be related to ROS generation and subsequent oxidative stress, although primarily through reduced antioxidative capacities rather than a direct effect on ROS generation (Schützendübel and Polle 2002, Deckert 2005, Heyno et al. 2008). In tobacco BY-2 cells, H_2O_2 production and subsequent cell death was reported upon application of 3 or 5 mM $CdCl_2$ (Olmos et al. 2003, Garnier et al. 2006). Upon application of 50 μM $CdSO_4$, cell cycle phase-specific death was also observed in these cells (Kuthanova et al. 2008). In our observation, as cell death was not clearly observed 24 h after application of 10 μM $CdSO_4$, effect of cadmium application was monitored by measurement of cell growth (Fig. 8A). Although we can not exclude the possible effects of cadmium on ROS generation, the reduced cellular growth rates shown in this study may have resulted from the reduced enzymatic activities of metabolic pathways since the amounts of cadmium accumulated in the NtNRAMP1 overexpressing cells decreased (Fig. 8B).

4. Conclusion

Upon excess metal ion application, plant cells modulated transporter activities. The activation of the plasma membrane transporters upon excess metal application might be inconsistent to keep the cellular metal homeostasis. Our findings suggest that activation of transporters with low affinity to metal ions could be involved in avoiding metal ion toxicity.

5. Material and methods

5.1 Plant materials and culture conditions

A tobacco BY-2 cell line (*Nicotiana tabacum* L. cv. Bright Yellow 2) was maintained by weekly subculture in a modified Linsmaier and Skoog medium supplemented with 2,4-D (LSD medium), in which KH_2PO_4 and thiamine HCl were increased to 370 and 1 mg l^{-1}, respectively. To this basal medium, sucrose and 2,4-D were supplemented to 3 % and 0.2 mg l^{-1}, respectively, and the pH was adjusted to 5.8 before autoclaving (Nagata et al. 1992). The cell suspension was cultured on a rotary shaker at 130 rpm and 27°C in the dark.

Cell synchrony was established by treatment with 5 μg l^{-1} aphidicolin (Sigma Chemical Co., St. Louis, MO, USA) essentially as described by Kumagai-Sano et al. (2006). After 24 h of aphidicolin treatment, the cell culture was washed with 1 L of LSD medium on a glass filter and then incubated further in this medium. The cell culture was divided into two to four portions, and several different $FeSO_4$ concentrations were applied as described in Results before cell cycle and cell death analyses were conducted as described below.

5.2 Cell cycle and cell death analyses

The mitotic index (MI) was determined by fluorescence microscopy after the nuclei were stained with 1 μM of SYTOX (Molecular Probes Inc., Eugene, OR, USA). For flow cytometry, cells were fixed with 100 % ethanol, then rehydrated in Galbraith's buffer (45 mM $MgCl_2$, 30 mM Na-Citrate, 20 mM MOPS and 1 g l^{-1}l Triton X-100, pH 7.0, Galbraith et al. 1983), and finally treated with 20 μg l^{-1} RNase A (Sigma) and 10 μg l^{-1} propidium iodide (Sigma) for 1 h at room temperature. Cytometric analysis was performed on 5×10^3 cells with a laser scanning cytometer (LSC101, Olympus, Tokyo, Japan) as described by Sano et al. (2006). Cell death was determined after staining the cells with 0.05 % Evans Blue (Sigma) as described in Kadota et al. (2004).

5.3 Quantification of iron and cadmium concentrations

Intracellular iron and cadmium concentrations were measured by atomic absorption spectrograph. Cells were sedimented by centrifugation to determine their packed cell volumes, and were then washed with 3 % sucrose on a glass filter before being resuspended in distilled water. For iron or cadmium extraction, cells were disrupted by a bead cell disrupter (MS-100, Tomy Seiko Co. Tokyo, Japan) and the iron or cadmium concentrations determined by atomic absorption spectrograph (AA-6800, Shimadzu Co., Kyoto, Japan).

5.4 Molecular cloning of tobacco iron transporter genes

Tobacco total RNA was isolated with the E.Z.N.A. Plant RNA Kit (Omega Bio-tek, Inc. Doraville, GA, USA), and cDNA synthesized using M-MLV reverse transcriptase (Promega, Heidelberg, Germany) with oligo-dT primers. Tobacco BY-2 *NRAMP* cDNA fragments were amplified with degenerate primers of 5′-CCNCAYAAYCTNTTYCTNCAYTSNGC-3′ and 5′-TGNCCNGCRTANGTNCCNGTDATNGT-3′ designed from homologous regions of known plant NRAMP proteins. *NtZIP1* cDNA was obtained based on the sequence information with high homology to *AtZIP* gene families deposited in the tobacco BY-2 EST database (TAB, Transcriptome Analysis of BY-2, http://mrg.psc.riken.jp/strc/). Amplification of the 5′ and 3′ cDNA ends was performed by RACE (SMART RACE cDNA Amplification kit, Clontech, Palo Alto, CA, USA), and the amplified fragments then subcloned into the pCR2.1 vector (Invitrogen Corp., Carlsbad, CA, USA).

5.5 Gene expression analysis by quantitative RT-PCR

Real-time quantitative PCR was performed in a Smart Cycler II System (Takara Bio Inc., Shiga, Japan) using the SYBR Green Real time PCR Master Mix (Toyobo Co., LTD., Osaka, Japan). *NtNRAMP1* and *NtZIP1* fragments from nucleotides 635 to 833 and 313 to 487 were amplified with primers 5′-TCTTCAAGGGATTCCCAGGA-3′ (NRAMP1 FW) and 5′-TGTTATCCCACGGCATGCAAC-3′ (NRAMP1 RV) or 5′- TCGCCATGTTTG AAAGAGAATCC-3′ (ZIP1 FW) and 5′- CCAGACTGAGCCACCAATCCA-3′ (ZIP1 RW), respectively. As internal standards of the cDNA amounts, GAPdH fragments were amplified with primers 5′-CCGGACAAGGCTGCTGCTAC-3′ (GAP FW) and 5′-GACCCTCCACAATGCCAAACC-3′ (GAP RW), designed on the basis of the tobacco GAPdH (cytosolic glyceraldehyde-3-phosphate dehydrogenase) gene (Accession number: M14419, Dambrauskas et al., 2003) and the relative transcript values then calculated.

5.6 Transformation of tobacco BY-2 cells

The coding region of *NtNRAMP1* was amplified by PCR using gene specific primers of 5′-CACCATGGCGGCGAACTCGTCCCC-3′ and 5′-ATTAGTGGTCCTCTGCTGAGGCAA-3′, then cloned into the pENTR/D-TOPO vector (Invitrogen) and finally introduced into the pGWB502 binary vector (Nakagawa et al. 2007) by the Gateway cloning system using LR clonase (Invitrogen). The pGWB502 vector gave the cauliflower 35S promoter sequence to the PCR products. *Agrobacterium*-mediated transformation of the tobacco BY-2 cells was performed as described by Mayo et al. (2006). Transformants were selected with 50 mg l^{-1} hygromycin.

Transient gene expression was carried out by particle bombardment. A cell suspension of 2 d-old BY-2 cells was filtrated onto filter paper, and the cells bombarded with gold particles (1.0 μm) coated with the appropriate vector constructs using a particle delivery system (PDS-1000/He, Bio-Rad, Hercules, CA, USA) according to the manufacturer's recommendations. Filtrated BY-2 cells were placed at a distance of 6 cm under the stopping screen and were bombarded in a vacuum of 28 inches Hg at a helium pressure of 1100 psi. Following bombardment, the cells were diluted in LSD medium and kept in the dark at 27 °C for 6 to 12 h before observation. The GFP fluorescence was detected on the inverted platform of a fluorescence microscope equipped with a spinning disc confocal laser scanning system (CSU-X1, Yokogawa, Tokyo, Japan) and a cooled CCD camera (Cool-SNAP HQ, PhotoMetrics, Huntington Beach, Canada).

5.7 Yeast experiments

Yeast cells INVSc1 (Invtrogen) were transformed by pYES2.1/V5-His-TOPO vectors (Invitrogen) containing an entire ORF region of the respective metal transporter cDNAs according to standard procedures (Invitrogen). The transformants were selected on synthesic complete medium omitted uracil (SC-uracil) containing 2 % glucose, 0.67 % yeast nitrogen base (without amino acids, Difco), amino acids omitting uracil (-Ura DO Supplement, Clontech Laboratories Inc.), 0.5 % ammmoniumu sulfate and 2 % agar. The transporter proteins were induced by application of 2 % galactose instead of glucose in the SC-uracil medium. For iron uptake measurements, yeast cells precultured in the SC-uracil medium were diluted to OD_{600} of 0.3 and cultured in the medium supplied with 2 % galactose and 0.2 mM $FeCl_3$. After 18 h incubation, OD_{600} were measured and the yeast culture was washed with deionized water twice. For iron extraction, yeast cells were digested with 2N HCl and the iron concentrations were determined by atomic absorption spectrograph (AA-660, Shimadzu Co., Kyoto, Japan).

6. Acknowledgment

We are grateful to Dr. T. Nakagawa (Shimane University) for the kind gift of the pGWB502 binary vector.

The nucleotide sequences reported in this paper have been submitted to GenBank as accession numbers AB505625 for NtNRAMP1 and AB505626 for NtZIP1.

This work was financially supported in part by a Grant-in-Aid for Scientific Research on Priority Areas to S.H. (No. 23012009), a Grant-in-Aid for Scientific Research on Innovative Areas to S.H. (No. 22114505) from the Japanese Ministry of Education, Science, Culture, Sports and Technology, an Advanced Measurement and Analysis grant from the Japan Science and Technology Agency (JST) to S.H. and Wada Kunkokai Foundation, Japan to T.S.

7. References

Cailliatte, R., Lapeyre, B., Briat, J.F., Mari, S. & Curie, C. (2009) The NRAMP6 metal transporter contributes to cadmium toxicity. *Biochem. J.* 422:217-228.

Cailliatte, R., Schikora, A., Briat, J.F., Mari, S. & Curie, C. (2010) High-affinity manganese uptake by the metal transporter NRAMP1 is essential for *Arabidopsis* growth in low manganese conditions. *Plant Cell* 22: 904-917.

Cassin, G., Mari, S., Curie, C., Briat, J.F. & Czernic, P. (2009) Increased sensitivity to iron deficiency in Arabidopsis thaliana overaccumulating nicotianamine. *J. Exp. Bot.* 60: 1249-1259.

Conte, S.S. & Walker, E.L. (2011) Transporters contributing to iron trafficking in plants. *Mol. Plant* 4: 464-476.

Curie, C., Alonso, J.M., Le, J.M., Ecker, R. & Briat, J.F. (2000) Involvement of NRAMP1 from *Arabidopsis thaliana* in iron transport. *Biochem. J.* 347: 749–755.

Curie, C. & Briat, J.F. (2003) Iron transport and signaling in plants. *Annu. Rev. Plant Biol.* 54:183-206.

Curie, C., Cassin, G., Couch, D., Divol, F., Higuchi, K., Le Jean, M., Misson, J., Schikora, A., Czernic, P. & Mari, S. (2009) Metal movement within the plant: contribution of nicotianamine and yellow stripe 1-like transporters. *Ann. Bot.* 103:1-11.

DalCorso, G., Farinati, S., Maistri, S. & Furini, A. (2008) How plants cope with cadmium: staking all on metabolism and gene expression. *J. Integr. Plant Biol.* 50:1268-1280.

Dambrauskas, G., Aves, S.J., Bryant, J.A., Francis, D. & Rogers, H.J. (2003) Genes encoding two essential DNA replication activation proteins, Cdc6 and Mcm3, exhibit very different patterns of expression in the tobacco BY-2 cell cycle. *J. Exp. Bot.* 54: 699-706.

Deckert, J. (2005) Cadmium toxicity in plants: is there any analogy to its carcinogenic effect in mammalian cells? *Biometals.* 18:475-481.

Eide, D., Broderius, M., Fett, J. & Guerinot, M.L. (1996) A novel iron regulated metal transporter from plants identified by functional expression in yeast. *Proc. Natl. Acad. Sci. USA.* 93: 5624–5628.

Enami, K., Ichikawa, M., Uemura, T., Kutsuna, N., Hasezawa, S., Nakagawa, T., Nakano, A. & Sato, M.H. (2009) Differential Expression Control and Polarized Distribution of Plasma Membrane-Resident SYP1 SNAREs in *Arabidopsis thaliana. Plant Cell Physiol.* 50: 280-289.

Galbraith, D.W., Harkins, K.R., Maddox, J.M., Ayres, N.M., Sharma, D.P. & Firoozabady, E. (1983) Rapid flow cytometric analysis of the cell-cycle in intact plant-tissues. *Science* 220: 1049-1051.

Garnier, L., Simon-Plas, F., Thuleau, P., Agnel, J.P., Blein, J.P., Ranjeva, R. & Montillet J.L. (2006) Cadmium affects tobacco cells by a series of three waves of reactive oxygen species that contribute to cytotoxicity. *Plant Cell Environ.* 29:1956-1969.

Guerinot, M.L. (2000) The ZIP family of metal transporters. *Biochim. Biophys. Acta.* 1465:190-198.

Hall, J.L. & Williams, L.E. (2003) Transition metal transporters in plants. *J. Exp. Bot.* 54: 2601-2613.

Hell, R. & Stephan, U.W. (2003) Iron uptake, trafficking and homeostasis in plants. *Planta* 216: 541-551.

Heyno, E., Klose, C. & Krieger-Liszkay, A. (2008) Origin of cadmium-induced reactive oxygen species production: mitochondrial electron transfer versus plasma membrane NADPH oxidase. *New Phytol.* 179:687-699.

Ishimaru, Y., Masuda, H., Bashir, K., Inoue, H., Tsukamoto, T., Takahashi, M., Nakanishi, H., Aoki, N., Hirose, T., Ohsugi, R. & Nishizawa, N.K. (2010) Rice metal-nicotianamine transporter, OsYSL2, is required for the long-distance transport of iron and manganese. *Plant J.* 62: 379-390.

Kadota, Y., Watanabe, T., Fujii, S., Higashi, K., Sano, T., Nagata, T. Hasezawa, S. & Kuchitsu, K. (2004) Crosstalk between elicitor-induced cell death and cell cycle regulation in tobacco BY-2 cells. *Plant J.* 40: 131-142.

Kim, S.A. & Guerinot, M.L. (2007) Mining iron: iron uptake and transport in plants. *FEBS Lett.* 581:2273-2280.

Klatte, M., Schuler, M., Wirtz, M., Fink-Straube, C., Hell, R. & Bauer, P. (2009) The analysis of Arabidopsis nicotianamine synthase mutants reveals functions for nicotianamine in seed iron loading and iron deficiency responses. *Plant Physiol.* 150: 257-271.

Krämer, U., Talke, I.N. & Hanikenne, M. (2007) Transition metal transport. *FEBS Lett.* 581:2263-2272.

Kumagai-Sano, F., Hayashi, T., Sano, T. & Hasezawa, S. (2006) Cell cycle synchronization of tobacco BY-2 cells. *Nat. Protoc.* 1:2621-2627.

Kuthanova, A., Fischer, L., Nick, P. & Opatrny, Z. (2008) Cell cycle phase-specific death response of tobacco BY-2 cell line to cadmium treatment. *Plant Cell Environ.* 31:1634-1643.

Lanquar, V., Lelièvre, F., Bolte, S., Hamès, C., Alcon, C., Neumann, D. Vansuyt, G., Curie, C., Schröder, A., Krämer, U., Barbier-Brygoo, H. & Thomine, S. (2005) Mobilization of vacuolar iron by AtNRAMP3 and AtNRAMP4 is essential for seed germination on low iron. *EMBO J.* 24: 4041–4051.

Lee, S., Jeon, U.S., Lee, S.J., Kim, Y.K., Persson, D.P., Husted, S., Schjørring, J.K., Kakei, Y., Masuda, H., Nishizawa, N.K. & An, G. (2009) Iron fortification of rice seeds through activation of the nicotianamine synthase gene. *Proc. Natl. Acad. Sci. USA.* 106:22014-22019.

Lopez-Millan, A.F., Ellis, D.R. & Grusak, M.A. (2004) Identification and characterization of several new members of the ZIP family of metal ion transporters in *Medicago truncatula. Plant Mol.Biol.* 54: 583-596.

Mäser, P., Thomine, S., Schroeder, J.I., Ward, J.M,, Hirschi, K., Sze, H. Talke, I.N., Amtmann A., Maathuis, F.J., Sanders, D., Harper, J.F., Tchieu, J., Gribskov, M., Persans, M.W., Salt, D.E., Kim, S.A. & Guerinot, M.L. (2001) Phylogenetic relationships within cation transporter families of Arabidopsis. *Plant Physiol.* 126:1646-1667.

Mayo, K.J., Gonzales, B.J. & Mason, H.S. (2006) Genetic transformation of tobacco NT1 cells with *Agrobacterium tumefaciens*. *Nat. Protoc.* 1:1105–1111.

Mills, R.F., Francini, A., Ferreira da Rocha, P.S., Baccarini, P.J., Aylett, M., Krijger, G.C. & Williams, L.E. (2005) The plant P1B-type ATPase AtHMA4 transports Zn and Cd and plays a role in detoxification of transition metals supplied at elevated levels. *FEBS Lett.* 579:783-791.

Nagata, T., Nemoto, Y. & Haswzawa, S. (1992) Tobacco BY-2 cell line as the "HeLa" cell in the cell biology of higher plants. *Int. Rev. Cytol.* 132:1-30.

Nakagawa, T., Suzuki, T., Murata, S., Nakamura, S., Hino, T., Maeo, K., Tabata, R., Kawai, T., Tanaka, K., Niwa, Y., Watanabe, Y., Nakamura, K., Kimura, T. & Ishiguro, S. (2007) Improved Gateway binary vectors: high-performance vectors for creation of fusion constructs in transgenic analysis of plants. *Biosci. Biotechnol. Biochem.* 71: 2095-2100.

Nevo, Y. & Nelson, N. (2006) The NRAMP family of metal-ion transporters. *Biochim. Biophys. Acta.* 1763:609-620.

Neyens, E. & Baeyens, J. (2003) A review of classic Fenton's peroxidation as an advanced oxidation technique. *J. Hazard. Mater.* 98: 33-50.

Nozoye, T., Nagasaka, S., Kobayashi, T., Takahashi, M., Sato, Y., Uozumi, N., Nakanishi, H. & Nishizawa, N.K. (2011) Phytosiderophore efflux transporters are crucial for iron acquisition in graminaceous plants. *J. Biol. Chem.* 286: 5446-5454.

Olmos, E., Martínez-Solano, J.R., Piqueras, A. & Hellín, E. (2003) Early steps in the oxidative burst induced by cadmium in cultured tobacco cells (BY-2 line). *J. Exp. Bot.* 54:291-301.

Oomen, R.J., Wu, J., Lelièvre, F., Blanchet, S., Richaud, P., Barbier-Brygoo, H., Aarts, M.G. & Thomine, S. (2008) Functional characterization of NRAMP3 and NRAMP4 from the metal hyperaccumulator *Thlaspi caerulescens*. *New Phytol.* 181: 637-650.

Pilon, M., Cohu, C.M., Ravet, K., Abdel-Ghany, S.E. & Gaymard, F. (2009) Essential transition metal homeostasis in plants. *Curr. Opin. Plant Biol.* 12:347-357.

Puig, S. & Peñarrubia, L. (2009) Placing metal micronutrients in context: transport and distribution in plants. *Curr. Opin. Plant Biol.* 12: 299-306.

Ricachenevsky, F.K., Sperotto, R.A., Menguer, P.K. & Fett, J.P. (2010) Identification of Fe-excess-induced genes in rice shoots reveals a WRKY transcription factor responsive to Fe, drought and senescence. *Mol. Biol. Rep.* 37: 3735-3745.

Sano, T., Higaki, T., Handa, K., Kadota, Y., Kuchitsu, K., Hasezawa, S. Hoffmann, A., Endter, J., Zimmermann, U., Hedrich, R. & Roitsch, T. (2006) Calcium ions are involved in the delay of plant cell cycle progression by abiotic stresses. *FEBS Lett.* 580: 597-602.

Schützendübel, A. & Polle, A. (2002) Plant responses to abiotic stresses: heavy metal-induced oxidative stress and protection by mycorrhization. *J. Exp. Bot.* 53:1351-1365.

Thomine, S., Wang, R., Ward, J.M., Crawford, N.M. & Schroeder, J.I. (2000) Cadmium and iron transport by members of a plant metal transporter family in *Arabidopsis* with homology to Nramp genes. *Proc. Natl. Acad. Sci. USA.* 97: 4991-4996.

Thomine, S., Lelièvre, F., Debarbieux, E., Schroeder, J.I. & Barbier-Brygoo, H. (2003) AtNRAMP3, a multispecific vacuolar metal transporter involved in plant responses to iron deficiency. *Plant J.* 34:685-695.

Verbruggen, N., Hermans, C. & Schat, H. (2009) Mechanisms to cope with arsenic or cadmium excess in plants. *Curr. Opin. Plant Biol.* 12:364-372.

Verret, F., Gravot, A., Auroy, P., Leonhardt, N., David, P., Nussaume, L., Vavasseur, A. & Richaud, P. (2004) Overexpression of AtHMA4 enhances root-to-shoot translocation of zinc and cadmium and plant metal tolerance. *FEBS Lett.* 576:306-312.

Vert, G., Grotz, N., Dédaldéchamp, F., Gaymard, F., Guerinot, M.L., Briat, J.F. & Curie, C. (2002) IRT1, an *Arabidopsis* transporter essential for iron uptake from the soil and for plant growth. *Plant Cell* 14:1223-1233.

Vert, G., Barberon, M., Zelazny, E., Séguéla, M., Briat, J.F. & Curie, C. (2009) *Arabidopsis* IRT2 cooperates with the high-affinity iron uptake system to maintain iron homeostasis in root epidermal cells. *Planta* 229:1171-1179.

Wei, W., Chai, T., Zhang, Y., Han, L., Xu, J. & Guan, Z. (2009) The *Thlaspi caerulescens* NRAMP homologue TcNRAMP3 is capable of divalent cation transport. *Mol. Biotechnol.* 41:15-21.

Wiedenhoeft, A.C. (2006) Micronutrients, In: *Plant nutrition*, William G. Hopkins, pp. 26-35., Chelsea House Publisher, New York.

Wirth, J., Poletti, S., Aeschlimann, B., Yakandawala, N., Drosse, B., Osorio, S., Tohge, T., Fernie, A.R., Günther, D., Gruissem, W. & Sautter, C. (2009) Rice endosperm iron biofortification by targeted and synergistic action of nicotianamine synthase and ferritin. *Plant Biotechnol. J.* 7: 631-644.

Xia, J., Yamaji, N., Kasai, T. & Ma, J.F. (2010) Plasma membrane-localized transporter for aluminum in rice. *Proc. Natl. Acad. Sci. USA.* 107: 18381-18385.

Xiao, H., Yin, L., Xu, X., Li, T. & Han, Z. (2008) The iron-regulated transporter, MbNRAMP1, isolated from *Malus baccata* is involved in Fe, Mn and Cd trafficking. *Ann. Bot.* 102:881-889.

Zheng, L., Cheng, Z., Ai, C., Jiang, X., Bei, X., Zheng, Y., Glahn, R.P., Welch, R.M., Miller, D.D., Lei, X.G. & Shou, H. (2010) Nicotianamine, a novel enhancer of rice iron bioavailability to humans. *PLoS One.* 5: e10190.

11

Analysis of SNARE-Mediated Exocytosis Using a Cell Fusion Assay

Chuan Hu, Nazarul Hasan and Krista Riggs
Department of Biochemistry and Molecular Biology,
University of Louisville School of Medicine, Louisville, KY
USA

1. Introduction

Exocytosis is the fusion of transport vesicles with the plasma membrane. By exocytosis, eukaryotic cells secrete soluble proteins and endogenous chemicals to the extracellular space, and deliver new membrane proteins and lipids to the plasma membrane. A large body of work has demonstrated that the interactions of SNARE (soluble N-ethylmaleimide-sensitive factor attachment protein receptor) proteins on vesicles (v-SNAREs) and on target membranes (t-SNAREs) catalyze intracellular vesicle fusion events, including exocytosis (Bonifacino and Glick, 2004; Jahn et al., 2003; Rothman, 1994) (Fig. 1). The vesicle-associated membrane proteins (VAMPs), *i.e.*, VAMPs 1, 2, 3, 4, 5, 7 and 8, are v-SNAREs that reside in various post-Golgi vesicular compartments, and have been implicated in exocytosis. In this chapter, we review recent progress of using a novel cell fusion assay to analyze the specificity and membrane fusion activities of VAMPs (Hasan et al., 2010).

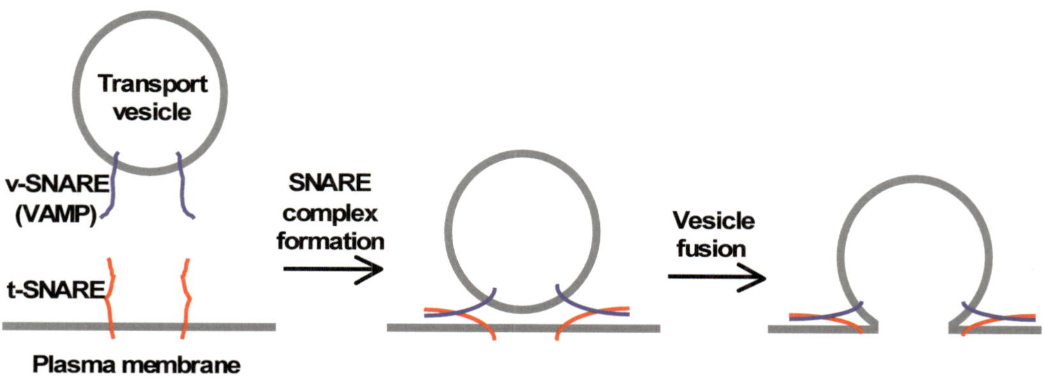

Fig. 1. Interactions of VAMPs and plasma membrane t-SNAREs drive exocytosis.

2. SNAREs – Core machinery of vesicle fusion

SNAREs are cytoplasmic oriented type I membrane proteins. SNAREs share one homologous domain, the 'SNARE motif,' which contains eight heptad repeats ready for coiled-coil formation. The SNARE proteins that mediate the fusion of synaptic vesicles with the presynaptic plasma membrane are well studied (Sollner et al., 1993b). In synapses, the v-

SNARE VAMP2 resides in synaptic vesicles, whereas t-SNAREs syntaxin1 and synaptosomal-associated protein of 25 kD (SNAP-25) are located in the plasma membrane. Syntaxin1 and SNAP-25 constitute an acceptor complex for VAMP2 (Fasshauer and Margittai, 2004). The cytoplasmic domains of VAMP2, syntaxin1 and SNAP-25 form an extremely stable complex that is resistant to sodium dodecyl sulfate (SDS) (Hayashi et al., 1994) and heat stable up to ~90°C (Yang et al., 1999), indicating that SNARE complex formation is thermodynamically favorable. One α-helix from VAMP2, one α-helix from syntaxin1 and two α-helices from SNAP-25 intertwine to form a four-helix bundle (Sutton et al., 1998). Assembly of SNARE complexes is initiated at the N-termini and proceeds to the transmembrane domains at the C-termini in a zipper-like fashion (Stein et al., 2009). When v- and t-SNARE proteins are incorporated into liposomes, they spontaneously drive liposome fusion (McNew et al., 2000b; Weber et al., 1998), demonstrating that SNAREs form the minimal machinery for membrane fusion. Using a cell fusion assay, we showed that v- and t-SNARE proteins ectopically expressed on the cell surface spontaneously drive cell-cell fusion (Hu et al., 2003; Hu et al., 2007), providing further proof that SNAREs form the core machinery for intracellular membrane fusion. After membrane fusion, the adapter protein SNAP (soluble NSF attachment protein) and the ATPase NSF (N-ethylmaleimide-sensitive factor) dissociate v-/t-SNARE complexes at the expense of ATP (Mayer et al., 1996; Sollner et al., 1993a) to free SNAREs for the next round of fusion.

Genomic analysis indicates that there are 36 SNAREs in humans (Bock et al., 2001). Individual members of the SNARE family localize to distinct subcellular organelles (Chen and Scheller, 2001), suggesting that each SNARE has a selective role in vesicle trafficking. Using yeast SNARE proteins as models, a series of experiments showed that to a remarkable degree the specificity of intracellular membrane fusion can be predicted from the pattern of liposome fusion mediated by isolated v- and t-SNARE proteins (McNew et al., 2000a; Parlati et al., 2002). However, membrane fusion by SNAREs in mammalian cells is more promiscuous (Brandhorst et al., 2006; Shen et al., 2007). Here we show that with the exception of VAMP5, VAMPs are essentially redundant in mediating membrane fusion with plasma membrane t-SNAREs (Hasan et al., 2010).

3. Roles of VAMPs in exocytosis

VAMPs have been implicated in vesicle fusion with the plasma membrane, the *trans*-Golgi network (TGN) and endosomes. VAMP1 (synaptobrevin 1) and VAMP2 (synaptobrevin 2) mediate regulated exocytosis in neurons and endocrine cells (Hanson et al., 1997; Kesavan et al., 2007; Morgenthaler et al., 2003). In addition, VAMP2 is involved in the exocytosis of the water channel aquaporin 2 (Procino et al., 2008) and α5β1 integrin (Hasan and Hu, 2010), as well as insulin-stimulated translocation of the glucose transporter GLUT4 (Randhawa et al., 2000). Enriched in recycling endosomes and endosome-derived vesicles (Galli et al., 1994; McMahon et al., 1993), VAMP3 (cellubrevin) mediates the recycling of transferrin receptors to the cell surface (Galli et al., 1994), integrin trafficking (Luftman et al., 2009; Proux-Gillardeaux et al., 2005; Skalski and Coppolino, 2005), and the secretion of α-granules in platelets (Feng et al., 2002; Polgar et al., 2002). Present primarily in the TGN, VAMP4 participates in the transport between the TGN and endosomes (Mallard et al., 2002; Steegmaier et al., 1999), as well as in homotypic fusion of early endosomes (Brandhorst et al., 2006). Expressed in muscle cells, VAMP5 (myobrevin) is associated with the plasma membrane and intracellular vesicles (Zeng et al., 1998). In addition to vesicular transport from endosomes to lysosomes (Advani et al., 1999), the tetanus neurotoxin-insensitive VAMP (VAMP7) is involved in apical exocytosis in polarized epithelial cells (Galli et al., 1998; Pocard et al., 2007). Associated with early endosomes (Advani et al., 1998; Wong et al., 1998), VAMP8 (endobrevin) is required in regulated exocytosis in pancreatic acinar cells (Wang et al., 2004).

VAMPs have high sequence homology in their SNARE motifs (Fig. 2). All VAMPs possess a conserved arginine residue at the center of SNARE motifs (Fig. 2), and have been classified as R-SNAREs based on crystal structures (Fasshauer et al., 1998). The N-terminal 51 residues of VAMP4 contain a dominant signal for targeting to the TGN, while the SNARE motif of VAMP5 is responsible for its targeting to the plasma membrane (Zeng et al., 2003). In VAMP7, the N-terminal 'longin domain' regulates subcellular targeting (Pryor et al., 2008).

VAMPs 3, 4, 7 and 8 have broad tissue distribution (Advani et al., 1998; McMahon et al., 1993). Originally identified in nervous tissues, VAMPs 1 and 2 are also detected in skeletal muscle, fat and other tissues (Jagadish et al., 1996; Martin et al., 1998; Procino et al., 2008;

```
VAMP1    ------------------------------------------------------------
VAMP2    ------------------------------------------------------------
VAMP3    ------------------------------------------------------------
VAMP4    ------------------------------------------------------------
VAMP5    ------------------------------------------------------------
VAMP7    MAILFAVVARGTTILAKHAWCGGNFLEVTEQILAKIPSENNKLTYSHGNYLFHYICQDRI 60
VAMP8    ------------------------------------------------------------

VAMP1    ------------------------------MSAPAQPPAEGTEGTAPGGGPPGPPPNMTS 30
VAMP2    ------------------------------MSATAATAPP--AAPAGEGGPPAPPPNLTS 28
VAMP3    ------------------------------MST-------------------GPTAATGS 11
VAMP4    -------------MPPKFKRHLNDDDVTGSVKSERRNLLEDDSDEEEDFFLRGPSGPRFG 47
VAMP5    ---------------------------------------------------------MAG 3
VAMP7    VYLCITDDDFERSRAFNFLNEIKKRFQTTYGSRAQTALPYAMNSEFSSVLAAQLKHHSEN 120
VAMP8    -----------------------------------------------------MEEASEG 7

             <------------------- SNARE motifs ------------------->
VAMP1    --NRRLQQTQAQVEEVVDIIRVNVDKVLERDQKLSELDDRADALQAGASQFESSAAKLKR 88
VAMP2    --NRRLQQTQAQVDEVVDIMRVNVDKVLERDQKLSELDDRADALQAGASQFETSAAKLKR 86
VAMP3    --NRRLQQTQNQVDEVVDIMRVNVDKVLERDQKLSELDDRADALQAGASQFETSAAKLKR 69
VAMP4    PRNDKIKHVQNQVDEVIDVMQENITKVIERGERLDELQDKSESLSDNATAFSNRSKQLRR 107
VAMP5    ---IELERCQQQANEVTEIMRNNFGKVLERGVKLAELQQRSDQLLDMSSTFNKTTQNLAQ 60
VAMP7    KGLDKVMETQAQVDELKGIMVRNIDLVAQRGERLELLIDKTENLVDSSVTFKTTSRNLAR 180
VAMP8    GGNDRVRNLQSEVEGVKNIMTQNVERILARGENLEHLRNKTEDLEATSEHFKTTSQKVAR 67
             .: . * :.: :  ::  *.  :  *. .*  * :::: *   :  *.. : :: :

VAMP1    KYWWKNCKMMIMLGAICAIIVVVIVRRD---------------------------- 116
VAMP2    KYWWKNLKMMIILGVICAIILIIIIVYFST-------------------------- 116
VAMP3    KYWWKNCKMWAIGITVLVIFIIIIIVWVVSS------------------------- 100
VAMP4    QMWWRGCKIKAIMALVAAILLLVIIILIVMKYRT---------------------- 141
VAMP5    KKCWENIRYRICVGLVVVGVLLIIILIVLLVVFLPQSSDSSSAPRTQDAGIASGPGN 116
VAMP7    AMCMKNLKLTIIIIIVSIVFIYIIVSPLCGGFTWPSCVKK---------------- 220
VAMP8    KFWWKNVKMIVLICVIVFIIILFIVLFATGAFS----------------------- 100
             .. :       :   .: .*:
```

Fig. 2. Sequence alignment of human VAMP proteins. The conserved arginine residues in the center of SNARE motifs are labeled red.

Randhawa et al., 2000; Veale et al., 2010). Therefore, multiple VAMPs are co-expressed in mammalian cells. However, it is not clear if the seven VAMPs have differential membrane fusion activities. Using a cell fusion assay, we compare the membrane fusion activities of VAMPs.

3. Cell fusion assays

3.1 Expression of flipped SNAREs at the cell surface

By fusing the pre-prolactin signal sequence, which specifies translocation across the endoplasmic reticulum, to N-termini of the neuronal SNAREs VAMP2, syntaxin1 and SNAP-25, the cell fusion assay was originally developed in Dr. James Rothman's lab (Hu et al., 2003). The engineered SNAREs are called 'flipped' SNAREs because the orientation of their SNARE motifs against cellular membranes is flipped. A Myc tag is inserted between the signal sequence and N-termini of SNAREs to detect flipped SNARE proteins (Fig. 3A). When COS-7 cells are transfected with flipped SNARE constructs, flipped SNARE proteins are expressed at the cell surface (Fig. 3B). To express plasma membrane t-SNAREs, flipped syntaxins 1 or 4 are cotransfected with flipped SNAP-25. SNAP-25 does not contain a transmembrane domain. After expression, flipped SNAP-25 proteins are anchored to the cell surface by assembling with flipped syntaxin proteins.

Flow cytometry is used to measure the expression levels of SNARE proteins at the cell surface (Figs. 3C and D). When flipped VAMPs 1, 3, 4, 5, 7 and 8, and syntaxins 1 and 4 plasmids are transfected at the same concentration, cell surface expression of VAMPs 5 and 8 is higher than VAMPs 1, 3, 4 and 7, and cell surface expression of syntaxin4 is higher than syntaxin1 (Hasan et al., 2010). To express the v- and t-SNAREs at the same level, we optimized the concentration of each flipped SNARE plasmid used in transfection (concentrations see Section 4.2), so that VAMPs 1, 3, 4, 5, 7 and 8, and syntaxins 1 and 4 are expressed at the same level at the cell surface, respectively (Fig. 3D).

3.2 Microscopic cell fusion assay

Multiple readout systems have been developed to detect fusion of the cells that express flipped v-SNAREs (v-cells) and the cells that express flipped t-SNARE proteins (t-cells) (Hasan et al., 2010; Hu et al., 2003; Hu et al., 2007). In the microscopic assay shown in Fig. 4A, flipped v-SNARE constructs are cotransfected with a plasmid that encodes the green fluorescent protein EGFP. In t-cells, the flipped t-SNARE constructs are cotransfected with a plasmid that encodes the red fluorescent protein DsRed2. Fusion of the v- and t-cells results in fused cells that contain both EGFP and DsRed2. In the merged channel, the cytoplasm of the fused cells is yellow (arrows, Fig. 4A).

3.3 Enzymatic cell fusion assay

The microscopic cell fusion assay becomes less efficient when used to analyze multiple v-/t-SNARE combinations quantitatively. To develop a quantitative cell fusion assay, we take advantage of the strong transcriptional activation by binding of the tetracycline-controlled transactivator (tTA) to the tetracycline-response element (TRE) (Gossen and Bujard, 1992). Two plasmids in CLONTECH's Tet-Off gene expression system are used (Hasan et al., 2010). The first plasmid pTet-Off encodes the transcriptional activator tTA, and the second

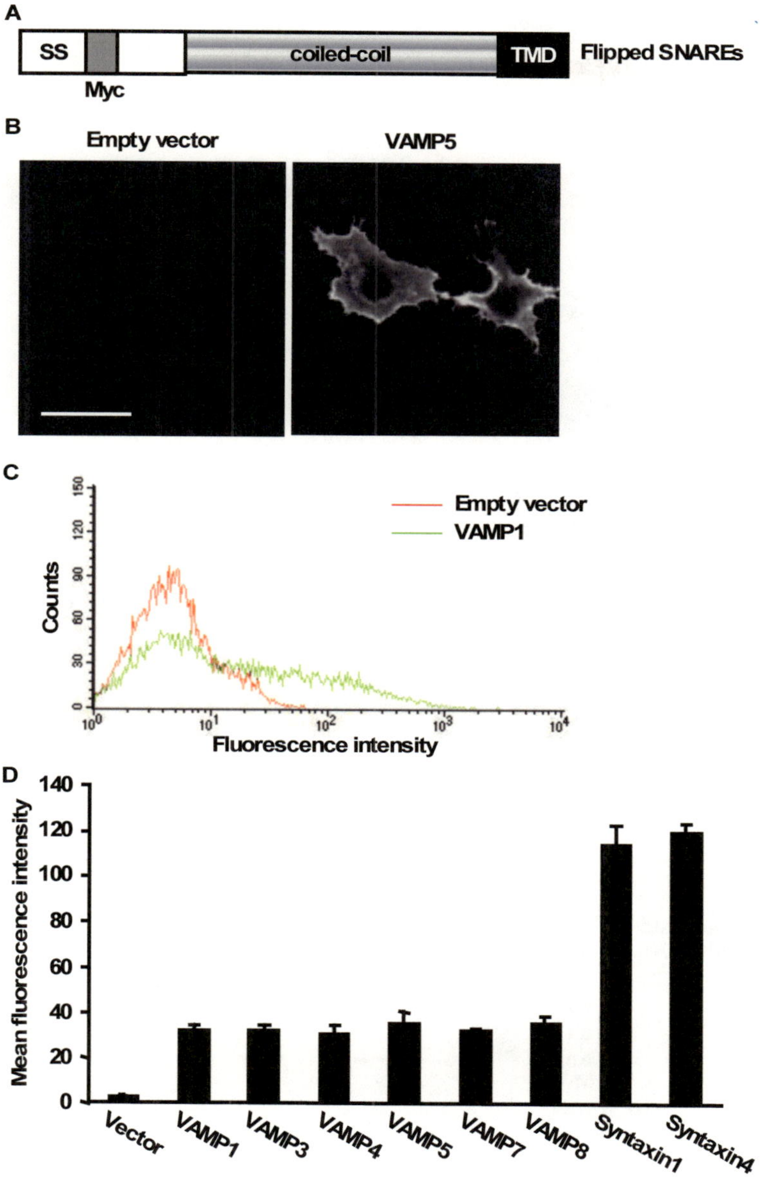

Fig. 3. Expression of flipped SNAREs at the cell surface. (A) Domain structure of flipped SNAREs. (B) Twenty-four hours after transfection with empty vector pcDNA3.1(+) or flipped VAMP5 plasmid, unpermeabilized COS-7 cells are stained with an anti-Myc antibody. Representative confocal images are shown. Scale bar, 50 μm. (C and D) Twenty-four hours after transfection with empty vector or flipped SNARE plasmids, unpermeabilized cells are stained with the anti-Myc antibody and analyzed by flow cytometry. (C) Representative FACS profiles of the cells transfected with empty vector or flipped VAMP1. (D) To express VAMPs and syntaxins at the same level at the cell surface, flipped SNARE plasmids are transfected at titrated concentrations. The mean fluorescence intensity of staining is obtained using CellQuest Pro software.

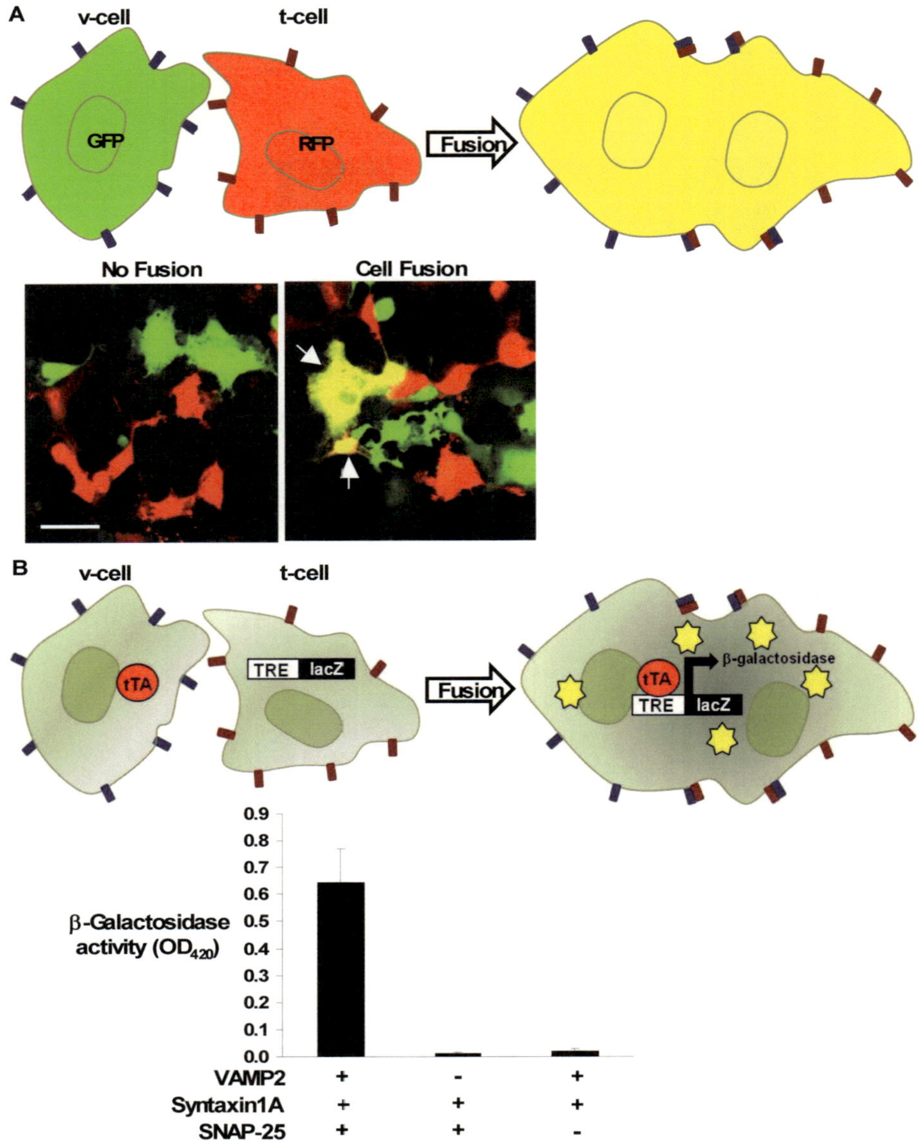

Fig. 4. Cell fusion assays. (A) Microscopic cell fusion assay. Cells that express flipped v-SNAREs (v-cells) are labeled by the green fluorescent protein EGFP, whereas cells that express flipped t-SNAREs (t-cells) are labeled by the red fluorescent protein DsRed2. Fusion of v- and t-cells results in fused cells (arrows) whose cytoplasm is yellow under fluorescence microscope. Scale bar, 50 μm. (B) Enzymatic cell fusion assay. The tetracycline-controlled transactivator (tTA) is expressed in v-cells, and a reporter plasmid that encodes β-galactosidase under control of the tetracycline-response element (TRE-LacZ) is transfected into t-cells. Fusion of the v- and t-cells leads to the binding of tTA to TRE and the expression of β-galactosidase, which is measured using a colorimetric method by absorbance at 420 nm. Only baseline β-galactosidase activity is detected when either flipped VAMP2 or SNAP-25 is not expressed.

plasmid pBI-G encodes the *LacZ* gene under control of the tetracycline-response element (TRE-*LacZ*). In the absence of tTA, transcription of the *LacZ* gene in TRE-*LacZ* is silent. When tTA is present, it binds to the TRE and activates the transcription of *LacZ*, resulting in the expression of β-galactosidase. We hypothesize that if tTA is located in v-cells and TRE-*LacZ* is located in t-cells, β-galactosidase will not be expressed. Fusion of the v- and t-cells would result in the binding of tTA to TRE and transcriptional activation of *LacZ* (Fig. 4B).

The neuronal SNAREs are used to test feasibility of the assay. VAMP2 is coexpressed with tTA in v-cells, and syntaxin1 and SNAP-25 are coexpressed with TRE-*LacZ* in t-cells. When the v- and t-cells are combined, robust β-galactosidase expression is indeed detected (Fig. 4B). However, when either VAMP2 or SNAP-25 is not expressed, only baseline β-galactosidase activity is detected, indicating that cell fusion and expression of β-galactosidase rely on interactions of v- and t-SNAREs.

3.4 Experimental procedures

3.4.1 Cell culture and reagents

COS-7 cells were obtained from the American Type Culture Collection, and cultured in Dulbecco Modified Eagle's Medium (DMEM) supplemented with 4.5 g/l glucose and 10% fetal bovine serum (FBS). The anti-Myc monoclonal antibody 9E10, developed by Dr. Bishop, was obtained from the Developmental Studies Hybridoma Bank maintained by the University of Iowa. pEGFP-N3, pDsRed2-N1, pTet-Off and pBI-G were obtained from CLONTECH. Plasmid transfection is done with Lipofectamine according to the manufacturer's instructions (Invitrogen).

3.4.2 FACS analysis

Expression levels of SNAREs at the cell surface are measured using immunostaining and flow cytometry as we previously reported (Hasan and Hu, 2010). Briefly, COS-7 cells are seeded in 6-well plates. Twenty-four hours after transfection with the flipped SNARE, pTet-Off and pBI-G plasmids, cells are fixed with 1% paraformaldehyde in PBS++ (PBS supplemented with 0.1 g/l $CaCl_2$ and 0.1 g/l $MgCl_2$). After labeling with the anti-Myc monoclonal antibody 9E10 and FITC-conjugated secondary antibodies, the cells are scraped off the plates with a cell scraper. 15,000 cells are analyzed using a FACSCalibur flow cytometer (BD Biosciences). The mean fluorescence intensity of each sample is obtained using CellQuest Pro software.

3.4.3 Microscopic cell fusion assay

The day before transfection, 1.2×10^6 COS-7 cells are seeded in each 100-mm cell culture dish, and 5×10^4 COS-7 cells are seeded on sterile 12-mm glass coverslips contained in 24-well plates. For v-cells, 5 μg of flipped v-SNARE and pEGFP-N3 are cotransfected into the cells grown in each 100-mm culture dish. For t-cells, 0.25 μg each of flipped syntaxin, SNAP-25 and pDsRed2-N1 are cotransfected into the cells seeded in the 24-well plates. Twenty-four hours after transfection, the v-cells are detached from culture dishes with EDTA (Enzyme-free Cell Dissociation Buffer (Invitrogen)). Detached cells are counted with a hemocytometer and resuspended in HEPES-buffered DMEM supplemented with 10% FBS,

6.7 μg/ml tunicamycin and 0.67 mM DTT. v-Cells (1.2×10^5) are added to each coverslip already containing the t-cells. After 24 hours at 37°C in 5% CO_2, the cells are gently washed once with PBS++, then fixed with 4% paraformaldehyde. Confocal images are collected on an Olympus laser scanning confocal microscope. The 488 nm argon laser line is used to excite EGFP and the 543 nm HeNe laser line is used to excite DsRed2. To prevent cross-contamination between EGFP and DsRed2, each channel is imaged sequentially before merging the images.

3.4.4 Enzymatic cell fusion assay

The day before transfection, 1.2×10^6 COS-7 cells are seeded in each 100-mm cell culture dish, and 2×10^5 COS-7 cells are seeded in each well of 6-well plates. For v-cells, 5 μg each of flipped v-SNARE plasmid are cotransfected with 5 μg of pTet-Off into the cells in each 100-mm culture dish. Control cells are cotransfected with empty vector pcDNA3.1(+) and pTet-Off. For t-cells, 1 μg each of flipped syntaxins 1 or 4, SNAP-25 and pBI-G are cotransfected into the cells in each well of the 6-well plates. There are putative N-glycosylation motifs (Asn-X-Ser/Thr) in VAMPs. Our previous studies (Hu et al., 2003; Hu et al., 2007) showed that N-glycosylation disrupts the function of flipped SNAREs, and that treatment of tunicamycin, an antibiotic that inhibits N-glycosylation (Elbein, 1984), restores the fusion activities of flipped SNAREs. To prevent N-glycosylation of VAMPs 1, 4, 5, 7 and 8, v-cells expressing these VAMP proteins and control cells are incubated in cell culture medium containing 10 μg/ml of tunicamycin during transfection. Since we have mutated the putative N-glycosylation sites in flipped VAMP2 (Hu et al., 2003) and VAMP3 proteins (Hu et al., 2007), v-cells that express VAMPs 2 or 3 are not treated with tunicamycin. Since flipped syntaxins 1 or 4, SNAP-25 proteins are not N-glycosylated (Hu et al., 2003; Hu et al., 2007), the t-cells are not treated with tunicamycin.

Twenty-four hours after transfection, the v-cells are detached from the culture dishes with Enzyme-free Cell Dissociation Buffer, and added (4.8×10^5 cells) to each well already containing the t-cells. After 6, 12 or 24 hours at 37°C in 5% CO_2, the expression of β-galactosidase is measured using the β-Galactosidase Enzyme Assay System with Reporter Lysis Buffer according to the manufacturer's instructions (Promega). The cells are washed twice with PBS, and then lysed in Reporter Lysis Buffer. Cell lysates are mixed with equal volume of Assay 2× Buffer. As a blank control, Reporter Lysis Buffer is mixed with Assay 2× Buffer. After 90 minutes, the colorimetric reaction is stopped by adding 1 M sodium carbonate, and absorbance at 420 nm is measured using a HITACHI 100-40 spectrophotometer.

4. Membrane fusion by VAMPs and plasma membrane t-SNAREs

4.1 Fusogenic interactions of VAMPs and plasma membrane t-SNAREs

Using the enzymatic fusion assay (Fig. 4B), we examine the fusogenic pairings between the seven VAMPs and two plasma membrane t-SNARE complexes, syntaxin1/SNAP-25 (Fasshauer and Margittai, 2004) and syntaxin4/SNAP-25 (Reed et al., 1999). Robust β-galactosidase expression is detected when VAMPs 1, 2, 3, 4, 7 or 8 are combined with syntaxin1/SNAP-25 (Fig. 5A) or syntaxin4/SNAP-25 (Fig. 5B), indicating that these VAMPs mediate membrane fusion with plasma membrane t-SNAREs. With syntaxin1/SNAP-25, the six VAMPs drive fusion to a similar degree. With syntaxin4/SNAP-25, VAMP8 fuses less efficiently than VAMPs 1, 2, 3 and 4 (31% lower fusion activity and $P = 0.046$ vs. VAMP1, Fig. 5B). In contrast, when VAMP5 is combined with the t-SNAREs, only baseline β-galactosidase activity is detected (Figs. 5A and B), suggesting that VAMP5 does not drive membrane fusion with the t-SNAREs. The stronger fusion activities of syntaxin4/SNAP-25 than syntaxin1/SNAP-25 (compare Figs. 5A and B) are likely caused by higher cell surface expression of syntaxin4/SNAP-25 than syntaxin1/SNAP-25 and higher fusion activity of syntaxin4 than syntaxin1 (Hasan et al., 2010). Together, the data shown in Fig. 5 indicate that VAMPs 1, 2, 3, 4, 7 and 8, but not VAMP5, drive membrane fusion when partnering with plasma membrane t-SNAREs.

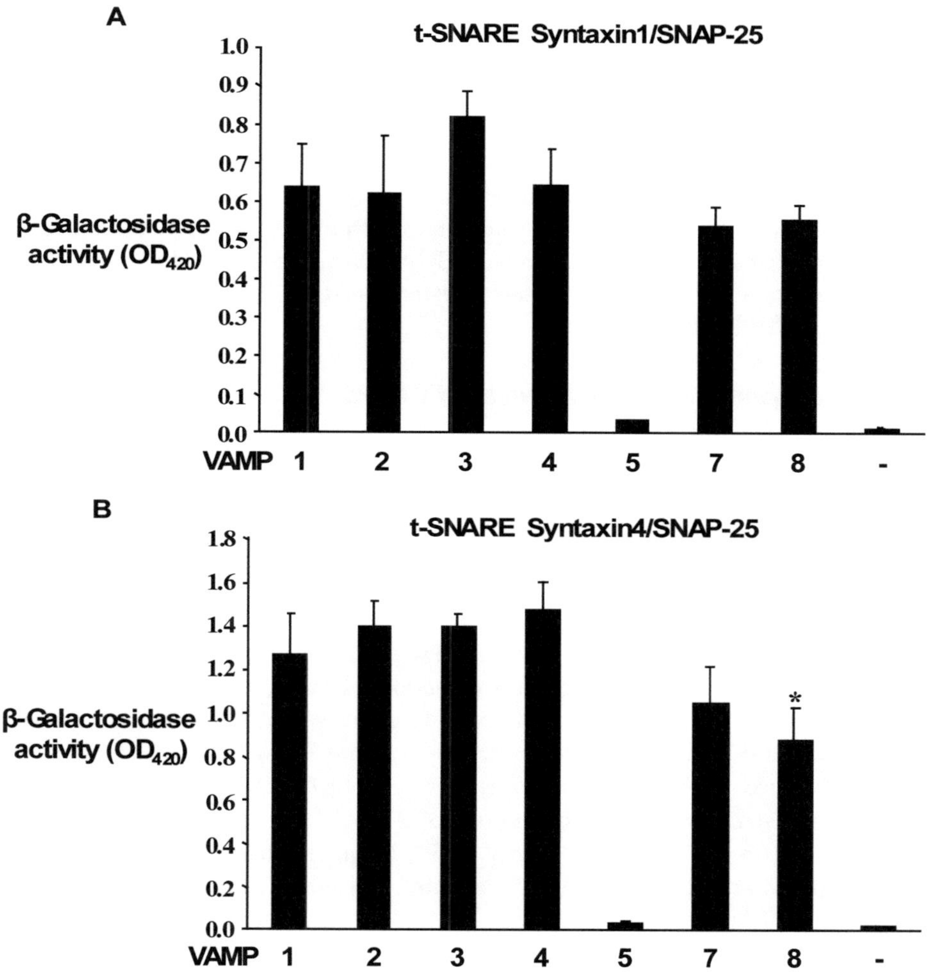

Fig. 5. Cell fusion by VAMPs and plasma membrane t-SNAREs. Twenty-four hours after combining v-cells that express VAMPs 1, 2, 3, 4, 5, 7 or 8 and t-cells that express (A) syntaxin1/SNAP-25 or (B) syntaxin4/SNAP-25, cell fusion is quantified using the enzymatic cell fusion assay. Control cells (-VAMP) are transfected with the empty vector. The flipped SNARE plasmids are transfected at the same concentration. Error bars represent standard deviation of three independent experiments. * $P < 0.05$ vs. VAMP1.

VAMPs 1, 2, 3, 7 and 8 are known to drive vesicle fusion with the plasma membrane. The above data provide additional evidence that vesicles that carry either one of these five VAMPs are capable of fusing with the plasma membrane. Since these VAMPs are functionally redundant, they may compensate each other in loss of function studies (Borisovska et al., 2005).

To rule out the possibility that the baseline β-galactosidase activity detected in the VAMP5 combinations (Fig. 5) is caused by residual membrane fusion activity of VAMP5, we perform the microscopic cell fusion assay, which analyzes individual cell fusion events and can detect rare fusion events. In multiple experiments, no cell fusion is observed in the microscopic assay when VAMP5 is combined with syntaxin1/SNAP-25 or syntaxin4/SNAP-25, whereas VAMP4 drives fusion efficiently with both t-SNARE complexes (Hasan et al., 2010) (data not shown). Together, the results using enzymatic and microscopic cell fusion assays suggest that VAMP5 is unable to mediate membrane fusion with the plasma membrane t-SNAREs.

VAMP5 is expressed in muscle cells, in which it is mainly associated with the plasma membrane as well as intracellular vesicles (Zeng et al., 1998; Zeng et al., 2003). It is still possible that VAMP5 forms fusogenic interactions with other plasma membrane t-SNAREs to mediate exocytosis in muscle cells. Indeed, SNAP-23, a ubiquitously expressed SNAP-25 homolog, is expressed in skeletal muscle (Bostrom et al., 2010). The role of VAMP5 in exocytosis needs further investigation.

4.2 Differential membrane fusion activities of VAMPs

Do VAMPs have differential membrane fusion capacities? An ideal experimental system to answer this question will require equal expression of VAMP proteins and a quantitative membrane fusion assay. Using anti-Myc staining and flow cytometry as measurement (Figs. 3C and D), we express VAMP proteins at the same level at the cell surface. Furthermore, using the quantitative enzymatic cell fusion assay, we compare their membrane fusion activities.

When COS-7 cells are transfected with the flipped SNARE plasmids at the same concentration, cell surface expression of VAMPs 5 and 8 is more than 2 fold higher than VAMPs 1, 3, 4 and 7, and cell surface expression of syntaxin4 is 1.8 fold higher than syntaxin1 (Hasan et al., 2010) (data not shown). To express the v- and t-SNAREs at the same level, we titrate and optimize the concentration of each flipped SNARE plasmid used in transfection. Flipped SNARE plasmids are transfected at the following concentrations (per 10 cm^2 growth area, *i.e.*, per well in 6-well plates): VAMP1, 0.2 μg; VAMP3, 0.5 μg; VAMP4, 0.5 μg; VAMP5, 0.05 μg; VAMP7, 1.0 μg; VAMP8, 0.1 μg; syntaxin1, 0.5 μg; syntaxin4, 0.05 μg. tTA, TRE-LacZ and flipped SNAP-25 are cotransfected at 1 μg per 10 cm^2 growth area. Under such conditions, VAMPs 1, 3, 4, 5, 7 and 8 are expressed at same level at the cell surface, while syntaxins 1 and 4 are expressed at the same level (Fig. 3D). Because the flipped VAMP2 protein does not contain a Myc tag (Hu et al., 2003), we are not able to compare its expression level with the other VAMPs.

After expressing VAMPs 1, 3, 4, 5, 7 and 8 at the same level, we compare their membrane fusion activities using the enzymatic cell fusion assay. With syntaxin1/SNAP-25, VAMPs 1, 3, and 8 have comparable and the highest fusion activities, whereas VAMPs 4 and 7 have 50% and 30% lower fusion activities, respectively (Fig. 6A). With syntaxin4/SNAP-25,

VAMPs 1 and 3 have comparable and the highest fusion activities, whereas VAMPs 4, 7 and 8 have 36%, 26% and 54% lower fusion activities, respectively (Fig. 6B). As expected, only baseline β-galactosidase activity is detected when VAMP5 is paired with the t-SNAREs. These data indicate that VAMPs have differential membrane fusion activities with plasma membrane t-SNAREs. However, when expressed at higher levels, VAMP4 drives membrane fusion as efficiently as VAMPs 1 and 3 (Fig. 5). Since the differences of fusion activities among the fusogenic VAMPs are within a factor of 2, these results imply that with the exception of VAMP5, VAMPs are essentially redundant in mediating membrane fusion with plasma membrane t-SNAREs.

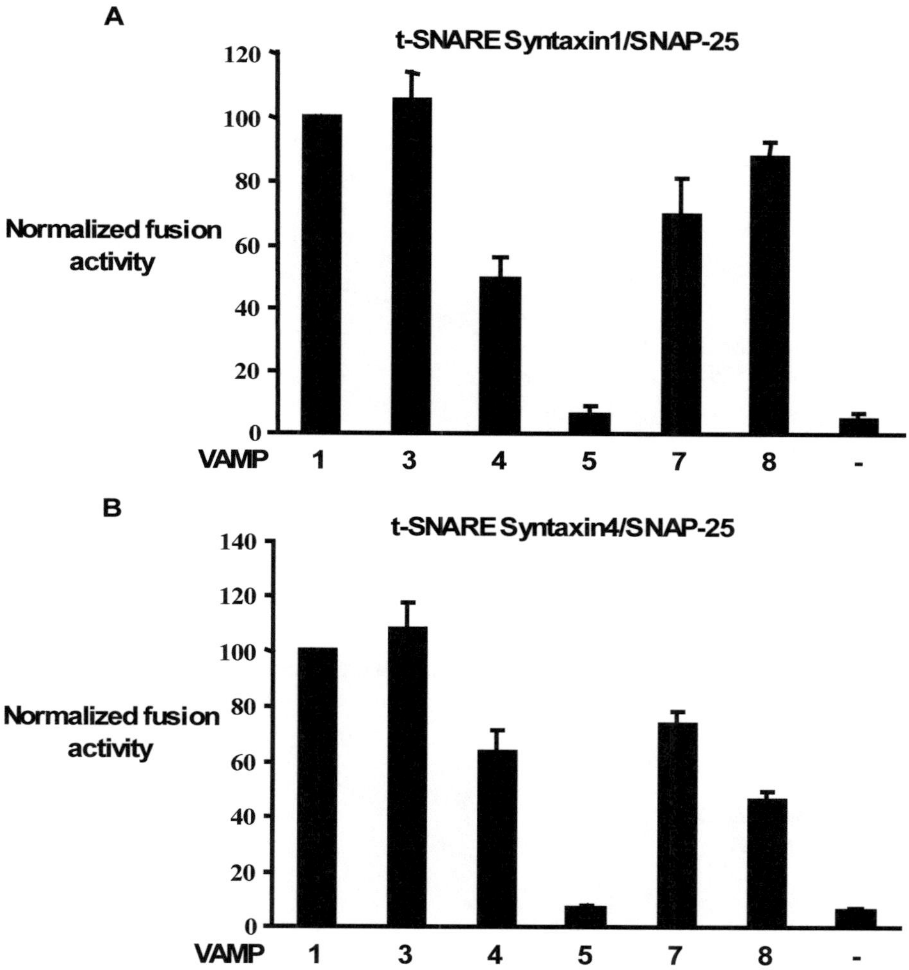

Fig. 6. Comparison of fusion activities of VAMPs. With optimized transfection conditions, VAMPs 1, 3, 4, 5, 7 and 8 are expressed at the same level at the cell surface, while syntaxin1/SNAP-25 (A) and syntaxin4/SNAP-25 (B) are expressed at the same level. Cell fusion is quantified using the enzymatic fusion assay. The fusion activities (OD_{420}) of control cells (-VAMP) and the v-cells expressing VAMPs 3, 4, 5, 7 or 8 are normalized to the fusion activity of the v-cells expressing VAMP1. Error bars represent standard deviation of four independent experiments.

4.3 Cooperativity of VAMP proteins in the cell fusion reaction

The number of SNARE complexes that cooperate to mediate vesicle fusion is under active investigation. To investigate the cooperativity of VAMPs in membrane fusion, we determine the dependence of cell fusion activity on cell surface expression level of VAMP1. We choose VAMP1 in this experiment because it has high membrane fusion activity (Fig. 6). In t-cells, the cell surface expression levels of syntaxin1 and SNAP-25 are kept constant. v-Cells are transfected with increasing concentrations of the flipped VAMP1 plasmid. At each concentration, we measure the cell surface expression level of VAMP1 proteins using flow cytometry, and determine cell fusion activity of VAMP1 with syntaxin1/SNAP-25 using the enzymatic fusion assay. Cell fusion activity is then plotted as a function of the mean fluorescence intensity of VAMP1 staining (Fig. 7A). The correlation is best fit with a polynomial regression. The hyperbolic instead of sigmoidal correlation (Fig. 7A) suggests that there is no cooperativity of VAMP1 proteins in driving cell fusion.

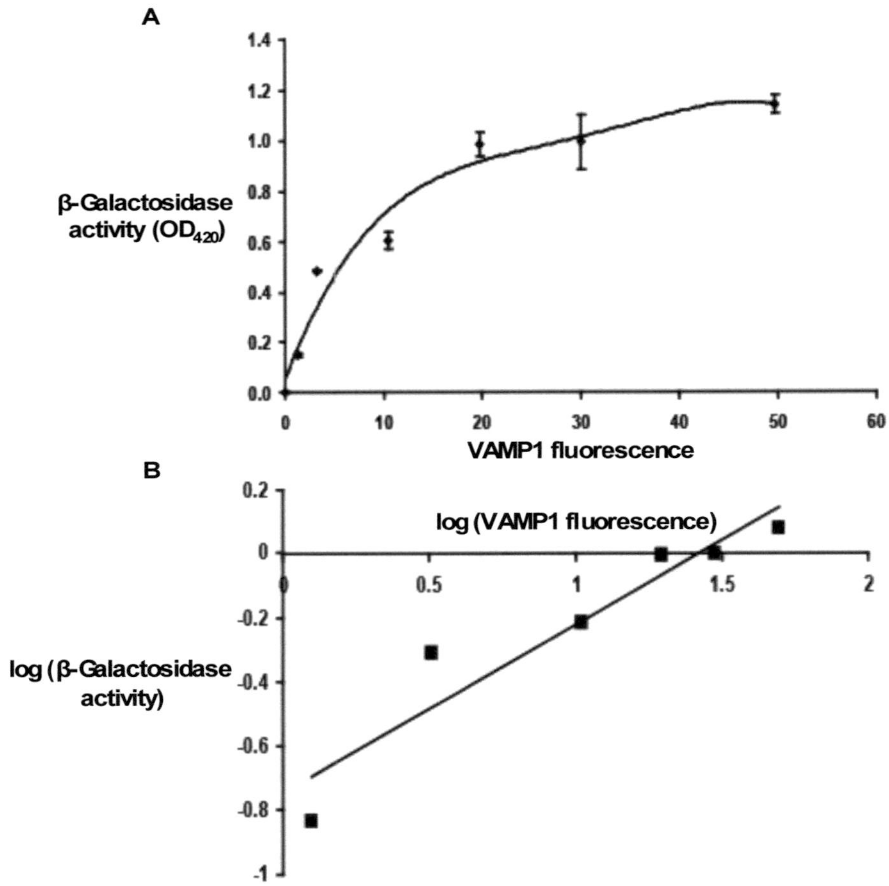

Fig. 7. Dependence of cell fusion activity on cell surface density of VAMP1. (A) v-cells expressing increasing amount of VAMP1 at the cell surface are combined with t-cells expressing syntaxin1/SNAP-25. Cell fusion activities are quantified and correlated with the mean fluorescence intensity of VAMP1 staining. (B) Log-log plot of cell fusion activity vs. mean fluorescence intensity of VAMP1 staining.

The log-log plot has been used to determine the cooperativity of viral fusion proteins in membrane fusion (Danieli et al., 1996). Using log-log plot, we further analyze the cooperativity of VAMP1 proteins. If two VAMP1 (V) proteins take part in the cell fusion reaction, *i.e.*, V + V → Fusion, the rate of fusion (F) = k $[V]^2$. Therefore, log (F) = log (k) + 2 log [V], and the slope of the resulting log-log plot will be 2. A log-log analysis of the dependence of cell fusion activity on VAMP1 cell surface density is shown in Fig. 7B. Linear regression is performed to model the log-log correlation (Fig. 7B), and the resulting slope is 0.52. These results further suggest that the cell fusion reaction does not involve concerted action of VAMP1 proteins.

Previous studies have estimated that 1 to 11 SNARE complexes are needed for membrane fusion (Domanska et al., 2009; Hua and Scheller, 2001; Karatekin et al., 2010; van den Bogaart et al., 2010). In the cell fusion reaction, we do not observe cooperativity of VAMP1 proteins, suggesting that concerted action of multiple SNARE complexes is not required to fuse cellular membranes. However, to achieve fast exocytosis in intact cells, concerted action of multiple SNARE complexes is clearly needed (Mohrmann et al., 2010). Such cooperativity of SNARE complexes may be organized by the binding of regulatory proteins such as synaptotagmins and Munc18 (Chicka et al., 2008; Shen et al., 2007).

5. Conclusion

To examine v-/t-SNARE interactions quantitatively, we developed an enzymatic cell fusion assay that utilizes activated expression of β-galactosidase and spectrometric measurement. Using this assay, we show that VAMPs 1, 2, 3, 4, 7 and 8 mediate membrane fusion efficiently with plasma membrane t-SNAREs syntaxin1/SNAP-25 and syntaxin4/SNAP-25, whereas VAMP5 does not drive fusion with the t-SNAREs. By expressing VAMPs 1, 3, 4, 7 and 8 at the same level, we further compare their membrane fusion activities. VAMPs 1 and 3 exhibit comparable and the highest fusion activities, whereas VAMPs 4, 7 and 8 have 30 - 50% lower fusion activities. Collectively, these data indicate that VAMPs have differential membrane fusion activities, and imply that with the exception of VAMP5, VAMPs are essentially redundant in mediating membrane fusion with plasma membrane t-SNAREs. Furthermore, no cooperativity of VAMP1 proteins is observed in the cell fusion reaction, suggesting that concerted action of multiple SNARE complexes is not required to fuse cellular membranes.

6. Acknowledgement

We thank Adrienne Bushau and David Humphrey for critically reading the manuscript. This work was supported by startup funds from the University of Louisville and CA135123 from the National Institutes of Health (to C.H.).

7. References

Advani, R.J., Bae, H.R., Bock, J.B., Chao, D.S., Doung, Y.C., Prekeris, R., Yoo, J.S., and Scheller, R.H. (1998). Seven novel mammalian SNARE proteins localize to distinct membrane compartments. *Journal of Biological Chemistry*, Vol.273, pp. 10317-10324.

Advani, R.J., Yang, B., Prekeris, R., Lee, K.C., Klumperman, J., and Scheller, R.H. (1999). VAMP-7 mediates vesicular transport from endosomes to lysosomes. *Journal of Cell Biology*, Vol.146, pp. 765-776.

Bock, J.B., Matern, H.T., Peden, A.A., and Scheller, R.H. (2001). A genomic perspective on membrane compartment organization. *Nature*, Vol.409, pp. 839-841.

Bonifacino, J.S., and Glick, B.S. (2004). The mechanisms of vesicle budding and fusion. *Cell*, Vol.116, pp. 153-166.

Borisovska, M., Zhao, Y., Tsytsyura, Y., Glyvuk, N., Takamori, S., Matti, U., Rettig, J., Sudhof, T., and Bruns, D. (2005). v-SNAREs control exocytosis of vesicles from priming to fusion. *EMBO Journal*, Vol.24, pp. 2114-2126.

Bostrom, P., Andersson, L., Vind, B., Haversen, L., Rutberg, M., Wickstrom, Y., Larsson, E., Jansson, P.A., Svensson, M.K., Branemark, R., Ling, C., Beck-Nielsen, H., Boren, J., Hojlund, K., and Olofsson, S.O. (2010). The SNARE protein SNAP23 and the SNARE-interacting protein Munc18c in human skeletal muscle are implicated in insulin resistance/type 2 diabetes. *Diabetes*, Vol.59, pp. 1870-1878.

Brandhorst, D., Zwilling, D., Rizzoli, S.O., Lippert, U., Lang, T., and Jahn, R. (2006). Homotypic fusion of early endosomes: SNAREs do not determine fusion specificity. *Proceedings of the National Academy of Sciences of the United States of America*, Vol.103, pp. 2701-2706.

Chen, Y.A., and Scheller, R.H. (2001). SNARE-mediated membrane fusion. *Nature Reviews Molecular Cell Biology*, Vol.2, pp. 98-106.

Chicka, M.C., Hui, E., Liu, H., and Chapman, E.R. (2008). Synaptotagmin arrests the SNARE complex before triggering fast, efficient membrane fusion in response to Ca2+. *Nature Structural and Molecular Biology*, Vol.15, pp. 827-835.

Danieli, T., Pelletier, S.L., Henis, Y.I., and White, J.M. (1996). Membrane fusion mediated by the influenza virus hemagglutinin requires the concerted action of at least three hemagglutinin trimers. *Journal of Cell Biology*, Vol.133, pp. 559-569.

Domanska, M.K., Kiessling, V., Stein, A., Fasshauer, D., and Tamm, L.K. (2009). Single vesicle millisecond fusion kinetics reveals number of SNARE complexes optimal for fast SNARE-mediated membrane fusion. *Journal of Biological Chemistry*, Vol.284, pp. 32158-32166.

Elbein, A.D. (1984). Inhibitors of the biosynthesis and processing of N-linked oligosaccharides. *CRC Critical Reviews in Biochemistry*, Vol.16, pp. 21-49.

Fasshauer, D., and Margittai, M. (2004). A transient N-terminal interaction of SNAP-25 and syntaxin nucleates SNARE assembly. *Journal of Biological Chemistry*, Vol.279, pp. 7613-7621.

Fasshauer, D., Sutton, R.B., Brunger, A.T., and Jahn, R. (1998). Conserved structural features of the synaptic fusion complex: SNARE proteins reclassified as Q- and R-SNAREs. *Proceedings of the National Academy of Sciences of the United States of America*, Vol.95, pp. 15781-15786.

Feng, D., Crane, K., Rozenvayn, N., Dvorak, A.M., and Flaumenhaft, R. (2002). Subcellular distribution of 3 functional platelet SNARE proteins: human cellubrevin, SNAP-23, and syntaxin 2. *Blood*, Vol.99, pp. 4006-4014.

Galli, T., Chilcote, T., Mundigl, O., Binz, T., Niemann, H., and De Camilli, P. (1994). Tetanus toxin-mediated cleavage of cellubrevin impairs exocytosis of transferrin receptor-containing vesicles in CHO cells. *Journal of Cell Biology*, Vol.125, pp. 1015-1024.

Galli, T., Zahraoui, A., Vaidyanathan, V.V., Raposo, G., Tian, J.M., Karin, M., Niemann, H., and Louvard, D. (1998). A novel tetanus neurotoxin-insensitive vesicle-associated membrane protein in SNARE complexes of the apical plasma membrane of epithelial cells. *Molecular Biology of the Cell*, Vol.9, pp. 1437-1448.

Gossen, M., and Bujard, H. (1992). Tight control of gene expression in mammalian cells by tetracycline-responsive promoters. *Proceedings of the National Academy of Sciences of the United States of America*, Vol.89, pp. 5547-5551.

Hanson, P.I., Heuser, J.E., and Jahn, R. (1997). Neurotransmitter release - four years of SNARE complexes. *Current Opinion in Neurobiology*, Vol.7, pp. 310-315.

Hasan, N., Corbin, D., and Hu, C. (2010). Fusogenic Pairings of Vesicle-Associated Membrane Proteins (VAMPs) and Plasma Membrane t-SNAREs - VAMP5 as the Exception. *PLoS One*, Vol.5, pp. e14238.

Hasan, N., and Hu, C. (2010). Vesicle-associated membrane protein 2 mediates trafficking of alpha5beta1 integrin to the plasma membrane. *Experimental Cell Research*, Vol.316, pp. 12-23.

Hayashi, T., McMahon, H., Yamasaki, S., Binz, T., Hata, Y., Sudhof, T.C., and Niemann, H. (1994). Synaptic vesicle membrane fusion complex: action of clostridial neurotoxins on assembly. *EMBO Journal*, Vol.13, pp. 5051-5061.

Hu, C., Ahmed, M., Melia, T.J., Sollner, T.H., Mayer, T., and Rothman, J.E. (2003). Fusion of cells by flipped SNAREs. *Science*, Vol.300, pp. 1745-1749.

Hu, C., Hardee, D., and Minnear, F. (2007). Membrane fusion by VAMP3 and plasma membrane t-SNAREs. *Experimental Cell Research*, Vol.313, pp. 3198-3209.

Hua, Y., and Scheller, R.H. (2001). Three SNARE complexes cooperate to mediate membrane fusion. *Proceedings of the National Academy of Sciences of the United States of America*, Vol.98, pp. 8065-8070.

Jagadish, M.N., Fernandez, C.S., Hewish, D.R., Macaulay, S.L., Gough, K.H., Grusovin, J., Verkuylen, A., Cosgrove, L., Alafaci, A., Frenkel, M.J., and Ward, C.W. (1996). Insulin-responsive tissues contain the core complex protein SNAP-25 (synaptosomal-associated protein 25) A and B isoforms in addition to syntaxin 4 and synaptobrevins 1 and 2. *Biochemical Journal*, Vol.317, pp. 945-954.

Jahn, R., Lang, T., and Sudhof, T.C. (2003). Membrane fusion. *Cell*, Vol.112, pp. 519-533.

Karatekin, E., Di Giovanni, J., Iborra, C., Coleman, J., O'Shaughnessy, B., Seagar, M., and Rothman, J.E. (2010). A fast, single-vesicle fusion assay mimics physiological SNARE requirements. *Proceedings of the National Academy of Sciences of the United States of America*, Vol.107, pp. 3517-3521.

Kesavan, J., Borisovska, M., and Bruns, D. (2007). v-SNARE actions during Ca(2+)-triggered exocytosis. *Cell*, Vol.131, pp. 351-363.

Luftman, K., Hasan, N., Day, P., Hardee, D., and Hu, C. (2009). Silencing of VAMP3 inhibits cell migration and integrin-mediated adhesion. *Biochemical and Biophysical Research Communications*, Vol.380, pp. 65-70.

Mallard, F., Tang, B.L., Galli, T., Tenza, D., Saint-Pol, A., Yue, X., Antony, C., Hong, W., Goud, B., and Johannes, L. (2002). Early/recycling endosomes-to-TGN transport involves two SNARE complexes and a Rab6 isoform. *Journal of Cell Biology*, Vol.156, pp. 653-664.

Martin, L.B., Shewan, A., Millar, C.A., Gould, G.W., and James, D.E. (1998). Vesicle-associated membrane protein 2 plays a specific role in the insulin-dependent trafficking of the facilitative glucose transporter GLUT4 in 3T3-L1 adipocytes. *Journal of Biological Chemistry*, Vol.273, pp. 1444-1452.

Mayer, A., Wickner, W., and Haas, A. (1996). Sec18p (NSF)-driven release of Sec17p (alpha-SNAP) can precede docking and fusion of yeast vacuoles. *Cell*, Vol.85, pp. 83-94.

McMahon, H.T., Ushkaryov, Y.A., Edelmann, L., Link, E., Binz, T., Niemann, H., Jahn, R., and Sudhof, T.C. (1993). Cellubrevin is a ubiquitous tetanus-toxin substrate homologous to a putative synaptic vesicle fusion protein. *Nature*, Vol.364, pp. 346-349.

McNew, J.A., Parlati, F., Fukuda, R., Johnston, R.J., Paz, K., Paumet, F., Sollner, T.H., and Rothman, J.E. (2000a). Compartmental specificity of cellular membrane fusion encoded in SNARE proteins. *Nature*, Vol.407, pp. 153-159.

McNew, J.A., Weber, T., Parlati, F., Johnston, R.J., Melia, T.J., Sollner, T.H., and Rothman, J.E. (2000b). Close is not enough: SNARE-dependent membrane fusion requires an active mechanism that transduces force to membrane anchors. *Journal of Cell Biology*, Vol.150, pp. 105-117.

Mohrmann, R., de Wit, H., Verhage, M., Neher, E., and Sorensen, J.B. (2010). Fast Vesicle Fusion in Living Cells Requires at Least Three SNARE Complexes. *Science*, Vol.330, pp. 502-505.

Morgenthaler, F.D., Knott, G.W., Floyd Sarria, J.C., Wang, X., Staple, J.K., Catsicas, S., and Hirling, H. (2003). Morphological and molecular heterogeneity in release sites of single neurons. *European Journal of Neuroscience*, Vol.17, pp. 1365-1374.

Parlati, F., Varlamov, O., Paz, K., McNew, J.A., Hurtado, D., Sollner, T.H., and Rothman, J.E. (2002). Distinct SNARE complexes mediating membrane fusion in Golgi transport based on combinatorial specificity. *Proceedings of the National Academy of Sciences of the United States of America*, Vol.99, pp. 5424-5429.

Pocard, T., Le Bivic, A., Galli, T., and Zurzolo, C. (2007). Distinct v-SNAREs regulate direct and indirect apical delivery in polarized epithelial cells. *Journal of Cell Science*, Vol.120, pp. 3309-3320.

Polgar, J., Chung, S.H., and Reed, G.L. (2002). Vesicle-associated membrane protein 3 (VAMP-3) and VAMP-8 are present in human platelets and are required for granule secretion. *Blood*, Vol.100, pp. 1081-1083.

Procino, G., Barbieri, C., Tamma, G., De Benedictis, L., Pessin, J.E., Svelto, M., and Valenti, G. (2008). AQP2 exocytosis in the renal collecting duct -- involvement of SNARE isoforms and the regulatory role of Munc18b. *Journal of Cell Science*, Vol.121, pp. 2097-2106.

Proux-Gillardeaux, V., Gavard, J., Irinopoulou, T., Mege, R.M., and Galli, T. (2005). Tetanus neurotoxin-mediated cleavage of cellubrevin impairs epithelial cell migration and integrin-dependent cell adhesion. *Proceedings of the National Academy of Sciences of the United States of America*, Vol.102, pp. 6362-6367.

Pryor, P.R., Jackson, L., Gray, S.R., Edeling, M.A., Thompson, A., Sanderson, C.M., Evans, P.R., Owen, D.J., and Luzio, J.P. (2008). Molecular basis for the sorting of the SNARE VAMP7 into endocytic clathrin-coated vesicles by the ArfGAP Hrb. *Cell*, Vol.134, pp. 817-827.

Randhawa, V.K., Bilan, P.J., Khayat, Z.A., Daneman, N., Liu, Z., Ramlal, T., Volchuk, A., Peng, X.R., Coppola, T., Regazzi, R., Trimble, W.S., and Klip, A. (2000). VAMP2, but not VAMP3/cellubrevin, mediates insulin-dependent incorporation of GLUT4 into the plasma membrane of L6 myoblasts. *Molecular Biology of the Cell*, Vol.11, pp. 2403-2417.

Reed, G.L., Houng, A.K., and Fitzgerald, M.L. (1999). Human platelets contain SNARE proteins and a Sec1p homologue that interacts with syntaxin 4 and is phosphorylated after thrombin activation: implications for platelet secretion. *Blood*, Vol.93, pp. 2617-2626.

Rothman, J.E. (1994). Mechanisms of intracellular protein transport. *Nature*, Vol.372, pp. 55-63.

Shen, J., Tareste, D.C., Paumet, F., Rothman, J.E., and Melia, T.J. (2007). Selective activation of cognate SNAREpins by Sec1/Munc18 proteins. *Cell*, Vol.128, pp. 183-195.

Skalski, M., and Coppolino, M.G. (2005). SNARE-mediated trafficking of alpha5beta1 integrin is required for spreading in CHO cells. *Biochemical and Biophysical Research Communications*, Vol.335, pp. 1199-1210.

Sollner, T., Bennett, M.K., Whiteheart, S.W., Scheller, R.H., and Rothman, J.E. (1993a). A protein assembly-disassembly pathway in vitro that may correspond to sequential steps of synaptic vesicle docking, activation, and fusion. *Cell*, Vol.75, pp. 409-418.

Sollner, T., Whiteheart, S.W., Brunner, M., Erdjument-Bromage, H., Geromanos, S., Tempst, P., and Rothman, J.E. (1993b). SNAP receptors implicated in vesicle targeting and fusion. *Nature*, Vol.362, pp. 318-324.

Steegmaier, M., Klumperman, J., Foletti, D.L., Yoo, J.S., and Scheller, R.H. (1999). Vesicle-associated membrane protein 4 is implicated in trans-Golgi network vesicle trafficking. *Molecular Biology of the Cell*, Vol.10, pp. 1957-1972.

Stein, A., Weber, G., Wahl, M.C., and Jahn, R. (2009). Helical extension of the neuronal SNARE complex into the membrane. *Nature*, Vol.460, pp. 525-528.

Sutton, R.B., Fasshauer, D., Jahn, R., and Brunger, A.T. (1998). Crystal structure of a SNARE complex involved in synaptic exocytosis at 2.4 A resolution. *Nature*, Vol.395, pp. 347-353.

van den Bogaart, G., Holt, M.G., Bunt, G., Riedel, D., Wouters, F.S., and Jahn, R. (2010). One SNARE complex is sufficient for membrane fusion. *Nature Structural and Molecular Biology*, Vol.17, pp. 358-364.

Veale, K.J., Offenhauser, C., Whittaker, S.P., Estrella, R.P., and Murray, R.Z. (2010). Recycling endosome membrane incorporation into the leading edge regulates lamellipodia formation and macrophage migration. *Traffic*, Vol.11, pp. 1370-1379.

Wang, C.C., Ng, C.P., Lu, L., Atlashkin, V., Zhang, W., Seet, L.F., and Hong, W. (2004). A role of VAMP8/endobrevin in regulated exocytosis of pancreatic acinar cells. *Developmental Cell*, Vol.7, pp. 359-371.

Weber, T., Zemelman, B.V., McNew, J.A., Westermann, B., Gmachl, M., Parlati, F., Sollner, T.H., and Rothman, J.E. (1998). SNAREpins: minimal machinery for membrane fusion. *Cell*, Vol.92, pp. 759-772.

Wong, S.H., Zhang, T., Xu, Y., Subramaniam, V.N., Griffiths, G., and Hong, W. (1998). Endobrevin, a novel synaptobrevin/VAMP-like protein preferentially associated with the early endosome. *Molecular Biology of the Cell*, Vol.9, pp. 1549-1563.

Yang, B., Gonzalez, L., Jr., Prekeris, R., Steegmaier, M., Advani, R.J., and Scheller, R.H. (1999). SNARE interactions are not selective. Implications for membrane fusion specificity. *Journal of Biological Chemistry*, Vol.274, pp. 5649-5653.

Zeng, Q., Subramaniam, V.N., Wong, S.H., Tang, B.L., Parton, R.G., Rea, S., James, D.E., and Hong, W. (1998). A novel synaptobrevin/VAMP homologous protein (VAMP5) is increased during in vitro myogenesis and present in the plasma membrane. *Molecular Biology of the Cell*, Vol.9, pp. 2423-2437.

Zeng, Q., Tran, T.T., Tan, H.X., and Hong, W. (2003). The cytoplasmic domain of Vamp4 and Vamp5 is responsible for their correct subcellular targeting: the N-terminal extenSion of VAMP4 contains a dominant autonomous targeting signal for the trans-Golgi network. *Journal of Biological Chemistry*, Vol.278, pp. 23046-23054.